AF389382

Microfluidics for Biosensing

Microfluidics for Biosensing

Editors

Shilun Feng
Mohsen Asadnia
Ming Li

MDPI • Basel • Beijing • Wuhan • Barcelona • Belgrade • Manchester • Tokyo • Cluj • Tianjin

Editors
Shilun Feng
Chinese Academy of Sciences
China

Mohsen Asadnia
Macquarie University
Australia

Ming Li
Macquarie University,
Australia

Editorial Office
MDPI
St. Alban-Anlage 66
4052 Basel, Switzerland

This is a reprint of articles from the Special Issue published online in the open access journal *Biosensors* (ISSN 2079-6374) (available at: https://www.mdpi.com/journal/biosensors/special_issues/microfluidics_biosensing).

For citation purposes, cite each article independently as indicated on the article page online and as indicated below:

LastName, A.A.; LastName, B.B.; LastName, C.C. Article Title. *Journal Name* **Year**, *Volume Number*, Page Range.

ISBN 978-3-0365-2630-0 (Hbk)
ISBN 978-3-0365-2631-7 (PDF)

Contents

About the Editors

Shilun Feng focuses on different Point-of-care testing (POCT) research avenues for food, environmental water, and biomedical sensing. He is currently an associate professor in State Key Laboratory of Transducer Technology, Shanghai Institute of Microsystem and Information Technology, Chinese Academy of Sciences, Shanghai. He has just finished his Research Fellow journey for POCT microfluidics projects on environmental water in the School of Electrical and Electronic Engineering, Nanyang Technological University, Singapore. He completed his Ph.D. in Biomedical POCT microfluidics with Dr. David Inglis, who specialised in POCT microfluidic sampling probe and POCT on-chip cell concentrators, in the School of Engineering, Macquarie University, Australia. His research interests include biomedical microfluidics; microfabrication; Point-of-care (POC) sampling, manipulation, and testing with the developments of biodevice and instrumentation systems.

Mohsen Asadnia is an associate professor and ARC DECRA fellow and group leader at the School of Engineering, Macquarie University. He received his Ph.D. in Mechanical Engineering from Nanyang Technological University (NTU)-Singapore. He undertook his postdoctoral training at Massachusetts Institute of Technology (SMART center) and the University of Western Australia (UWA). He has been a group leader at Macquarie University since 2016. His research interests are ion selective membranes, chemical sensors, MEMS/NEMS bio-inspired sensory systems, biomimetic devices, artificial hair cells of the vertebrate inner ear and microfluidic devices.

Ming Li is a Senior Lecturer in the School of Engineering at Macquarie University. She completed her Ph.D. in Mechanical Engineering at the University of Wollongong in July 2013. Before joining MQ, she was a postdoctoral research fellow at the Department of Electrical and Computer Engineering, University of Houston, USA (2013–2014); and at the Department of Bioengineering and Department of Electrical Engineering, University of California, Los Angeles, USA (2014–2017). Dr Li has consistently been at the forefront of advanced biotechnologies in the field of microfluidics and plasmonic, especially for cell and molecular manipulation, detection and analysis. This is evidenced by her 34 peer-reviewed articles published in high-impact journals, which include Cell, Nature Protocols, Small, Analytical Chemistry, Lab on a Chip and Nanoscale with over 1400 citations and an h-index of 21 (5 January 2021, Google Scholar).

Preface to "Microfluidics for Biosensing"

We are pleased to introduce this Special Issue covering biosensor development and biosensing in a wide range of fields using microfluidic lab-on-a-chip platforms. This Special Issue will report the latest innovative microfluidic devices and technologies for biosensing applications.

Rapid, accurate, real-time, on-site and multiplexed detection and characterization have been the requirements in current biosensor technology development, especially during the COVID-19 outbreak. The development of reliable and stable microfluidic biosensors, point-of-care biosensing platforms, and advanced detection methods have attracted increasing attention from both academia and industry.

Microfluidics offers excellent platforms for biosensor development and biosensing. The platforms are useful for sample preparation, liquid handling, and cell/particle manipulation. Multiple functions can be designed and achieved in microfluidic chips along with different on-chip and off-chip detection and processing modules. This has been extensively adopted in both academic and industrial applications for a wide range of applications in healthcare, biochemistry, life science, food, water quality, etc.

We therefore invited contributions to this Special Issue from different fields about biosensing and microfluidics. The list of potential topics is broad, but we are particularly interested in studies on the design, development, and applications of microfluidics-based technologies in point-of-care biosensing and testing.

We are pleased that the collected high-quality publications with significant novelties that can potentially have high impacts on the abovementioned fields. The work are thoughtfully organized and of great interest to the diverse microfluidics and biosensing community.

Shilun Feng, Mohsen Asadnia, Ming Li
Editors

Article

A 3D-Printed Microfluidic Device for qPCR Detection of Macrolide-Resistant Mutations of *Mycoplasma pneumoniae*

Anyan Wang [1,2,†], Zhenhua Wu [2,†], Yuhang Huang [2,3], Hongbo Zhou [2], Lei Wu [2,*], Chunping Jia [2,*], Qiang Chen [1,*] and Jianlong Zhao [2]

[1] College of Metrology and Measurement Engineering, China Jiliang University, Hangzhou 310018, China; p1902085248@cjlu.edu.cn

[2] State Key Laboratory of Transducer Technology, Shanghai Institute of Microsystem and Information Technology, Chinese Academy of Sciences, Shanghai 200050, China; wuzhx@mail.sim.ac.cn (Z.W.); 1000479741@smail.shnu.edu.cn (Y.H.); zhouhb@mail.sim.ac.cn (H.Z.); jlzhao@mail.sim.ac.cn (J.Z.)

[3] College of Life Sciences, Shanghai Normal University, Shanghai 200233, China

[*] Correspondence: wulei@mail.sim.ac.cn (L.W.); jiachp@mail.sim.ac.cn (C.J.); chenqiang_cjlu@cjlu.edu.cn (Q.C.)

[†] These authors contributed equally to this work.

Abstract: Mycoplasma pneumonia (MP) is a common respiratory infection generally treated with macrolides, but resistance mutations against macrolides are often detected in *Mycoplasma pneumonie* in China. Rapid and accurate identification of *Mycoplasma pneumonie* and its mutant type is necessary for precise medication. This paper presents a 3D-printed microfluidic device to achieve this. By 3D printing, the stereoscopic structures such as microvalves, reservoirs, drainage tubes, and connectors were fabricated in one step. The device integrated commercial polymerase chain reaction (PCR) tubes as PCR chambers. The detection was a sample-to-answer procedure. First, the sample, a PCR mix, and mineral oil were respectively added to the reservoirs on the device. Next, the device automatically mixed the sample with the PCR mix and evenly dispensed the mixed solution and mineral oil into the PCR chambers, which were preloaded with the specified primers and probes. Subsequently, quantitative real-time PCR (qPCR) was carried out with the homemade instrument. Within 80 min, *Mycoplasma pneumonie* and its mutation type in the clinical samples were determined, which was verified by DNA sequencing. The easy-to-make and easy-to-use device provides a rapid and integrated detection approach for pathogens and antibiotic resistance mutations, which is urgently needed on the infection scene and in hospital emergency departments.

Keywords: *Mycoplasma pneumoniae*; macrolides; resistance mutations; microfluidic; 3D-printed; sample-to-answer; qPCR

Citation: Wang, A.; Wu, Z.; Huang, Y.; Zhou, H.; Wu, L.; Jia, C.; Chen, Q.; Zhao, J. A 3D-Printed Microfluidic Device for qPCR Detection of Macrolide-Resistant Mutations of *Mycoplasma pneumoniae*. *Biosensors* **2021**, *11*, 427. https://doi.org/10.3390/bios11110427

Received: 15 September 2021
Accepted: 26 October 2021
Published: 29 October 2021

Publisher's Note: MDPI stays neutral with regard to jurisdictional claims in published maps and institutional affiliations.

1. Introduction

Respiratory infection is the most common type of infection in the clinic and the risk of illness is consistent year-round. The pathogens causing pneumonia are often *Mycoplasma pneumonie* [1], especially in children and adolescents [2,3]. Mycoplasma pneumonia (MP) has a longer disease duration and more severe symptoms of the lung compared with other pneumonia. Macrolide is the preferred drug to treat MP [4], but macrolide resistance is becoming a widespread problem because of antibiotic abuses. According to the recent reports, the detection rate of the drug-resistant gene of *Mycoplasma pneumonie* is over 90% in China [5,6]. The primary mechanism leading to macrolide resistance to *Mycoplasma pneumonie* is the point mutation at sites 2063 and 2064 in domain V of 23S ribosomal RNA, especially the mutation of A2063G [7]. The difference of mutant types and sites can cause different macrolides resistance [8]: both A2063G and A2064G mutant strains are highly resistant to erythromycin and azithromycin; the A2064G mutant strain is more resistant to thromycin than A2063G; the A2063T mutant strain is also highly resistant to erythromycin, but has a low resistance to azithromycin and thyromycin. Therefore, it is pivotal to rapidly detect *Mycoplasma pneumonie* and its drug-resistant sites, which can help to diagnose respiratory infection and provide precise guidance on treatment and the use of medication.

Traditionally, the golden standard of microbial pathogen identification is to culture human sputum, blood or other samples collected from the human body with specific media and afterwards judge the type of pathogen by its appearance and morphology with experience [9]. The method is time-consuming, and it is easy to cause false-negative results because of missed detections. The emerging microbial pathogen detection technology can overcome the shortcomings of the culture method, which is divided into three categories: molecular diagnostic technology [10], immunoassay [11] and biosensors technology [12,13]. The detecting targets of molecular diagnostic technology are nucleic acids, so it can directly reveal the genetic information of the pathogen. The detection can be implemented by hybridization [14], polymerase chain reaction (PCR) [10], isothermal amplification [15,16], CRISPR [17] or high-throughput sequencing [18]. Among these methods, PCR is the most widely used, especially quantitative real-time PCR (qPCR), because it provides fast, quantitative, specific and sensitive detection, which can precisely distinguish a single-base mutation in the gene sequence. However, qPCR has a cumbersome multi-step preparation before nucleic acid amplification. And there are three challenges during the operations, including personnel operating deviations, sample contaminations and the risk of operator infections. With the development of microfluidics technology, these problems can be solved by building an integrated automation platform.

Recently, several microfluidic approaches have been developed for the identification of pathogen and antibiotic resistant mutations, such as single-cell trap, culture and imaging [19], identifying and detecting the pathogen markers in droplets by DNAzyme-based sensors [20], aptamer-Ag10NPs detection with bright field imaging [21] and nucleic acid samples preparation and amplification [10,22–24]. The sample-to-answer systems are also available on the market, such as the BioFire FilmArray and Cepheid GeneXpert, which are based on rapid, specific and sensitive PCR assays [10]. In order to further increase efficiency and reduce costs, newer microfluidic devices were developed to perform the complete procedure of the nucleic acid assay, including cell lysis, DNA purification, gene amplification and amplicon detection [10]. One category consists of centrifugal microfluidic devices [22], the other includes devices with micro-pumps, micro-valves, and reaction chambers [23,25]. The accelerated development in the field needs rapid and cheap fabrication approaches for a proof of concept.

3D printing is a promising technology to fabricate microfluidic device because of its many advantages, including the ability to create stereoscopic architectures directly and rapidly, as well as the cheap, quick implementation of the design and the optimization. Currently, microfluidic 3D printing for biological assays and clinical tests is still in a start-up stag, the resolution of 3D printing, the biocompatibility of resin and multi-materials integration still need investigation [26,27].

In this study, a novel 3D-printed device used for qPCR detection of macrolide-resistant genes of *Mycoplasma pneumonie* was proposed. The device automatically mixed the sample with the reagent and then the mixture was evenly dispensed into multiple PCR chambers in which the specified primers and probes had been preloaded. Afterwards, the device was transferred to the homemade qPCR system. Two kinds of single-base mutations of *Mycoplasma pneumoni* were identified with high speed, specificity and convenience.

2. Materials and Methods

2.1. Device Design and Fabrication

The device diagram is shown in Figure 1A, which consists of 6 layers. The pneumatic layer and the fluid layer were fabricated by stereolithographic (SLA)3D printing (Union-Tech Lite800, UnionTech Inc., Shanghai, China). A 200-µm-thick PDMS membrane was sandwiched between the pneumatic layer and the fluid layer. Two pieces of twin adhesive tapes (ARseal™ 90880, Adhesives Research Inc., Glen Rock, PA, USA) were patterned using laser etching. During assembly, two sides of the tape were adhered respectively to the 3D-printed layer and the PDMS membrane, both of which were treated with plasma in advance. After the pneumatic layer, the PDMS membrane and the fluid layer were assembled; a pressure sensitive adhesive tape (PSA, 3M 9795R) was pressed tightly onto the bottom of the fluid layer to close the flow channels.

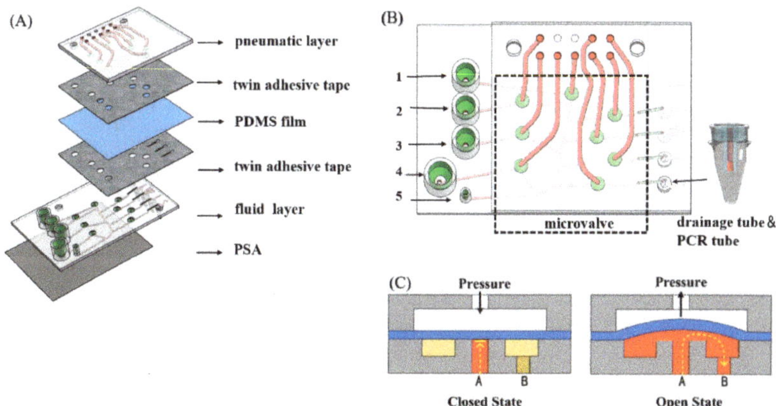

Figure 1. (**A**) Exploded diagram of the device. (**B**) Schematic illustration of the device. One to three are reagent reservoirs, 4 is a vent and 5 is a Luer taper connected to a syringe pump. Polypropylene PCR tubes can be fitted with the connectors on the device. (**C**) Cross section of a microvalve and its working principle.

The assembled device is shown in Figure 1B, with a length of 100 mm and a width of 75 mm. Five is a simple Luer taper, used to connect the flow channel to a syringe pump. One to three are sample reservoirs, which are respectively filled with mineral oil, the PCR mix, and a mixed solution containing sample. Four is a vent for balancing the internal and the external pressure of the device. There are eight microvalves in the device. Functions including mixing, dispensing and oil adding can be performed by manipulating the syringe pump and these microvalves. The cross-sectional structure of the microvalve is shown in Figure 1C. The upper pneumatic layer and the lower fluid layer are separated by a PDMS membrane, whose deformation direction determines the opening and closing of the microvalve. When a positive pressure is applied to the upper channel, the bend-down PDMS membrane blocks the fluid flow from the A channel. When a negative pressure is applied, the PDMS membrane bends up so the fluid can flow from A to B. On the right lower side of the device there are five drainage tubes and connectors which can fit PCR tubes tightly. Ventholes designed at the edge of the base of drainage tubes ensure that the solution can flow smoothly into PCR tubes.

2.2. System Setup and Workflow

In order to achieve the goal of sample-to-answer, a set of systems were developed for the automation of both the sample preparation and qPCR, which contained a mixing-dispensing module and a detection module. As shown in Figure 2A, the mixing-dispensing module was composed of a syringe pump, an air compressor, a vacuum pump and a control circuit. The syringe pump was connected to the fluid layer of the device to control the advance and retreat of the fluid, and the vacuum pump and the air compressor were connected to the pneumatic layer of the device to control the opening and closing of the microvalve. As shown in Figure 2C, the detection module was composed of a LED light source, a filter unit, a photomultiplier tube (PMT) detector and a temperature control unit. The spectral coverage of the light source was from the blue region to the green region. The beam from the light source was delivered with an optical fiber and then was expanded by a lens. Next, the expanded beam went through the filter cube and was delivered with an optical fiber to illuminate the PCR solutions in tubes. Afterwards, the excited fluorescent beam was delivered back into the filter tube and went through the dichroic mirror and the barrier filter of the tube and was finally detected by PMT. The PCR tubes on the device could be inserted into the metal slots on the copper base of the temperature control unit, where a Peltier module was used for precise temperature control (Figure S1). During each temperature cycle, the fluorescent signals of five PCR tubes were recorded sequentially with the PMT detector.

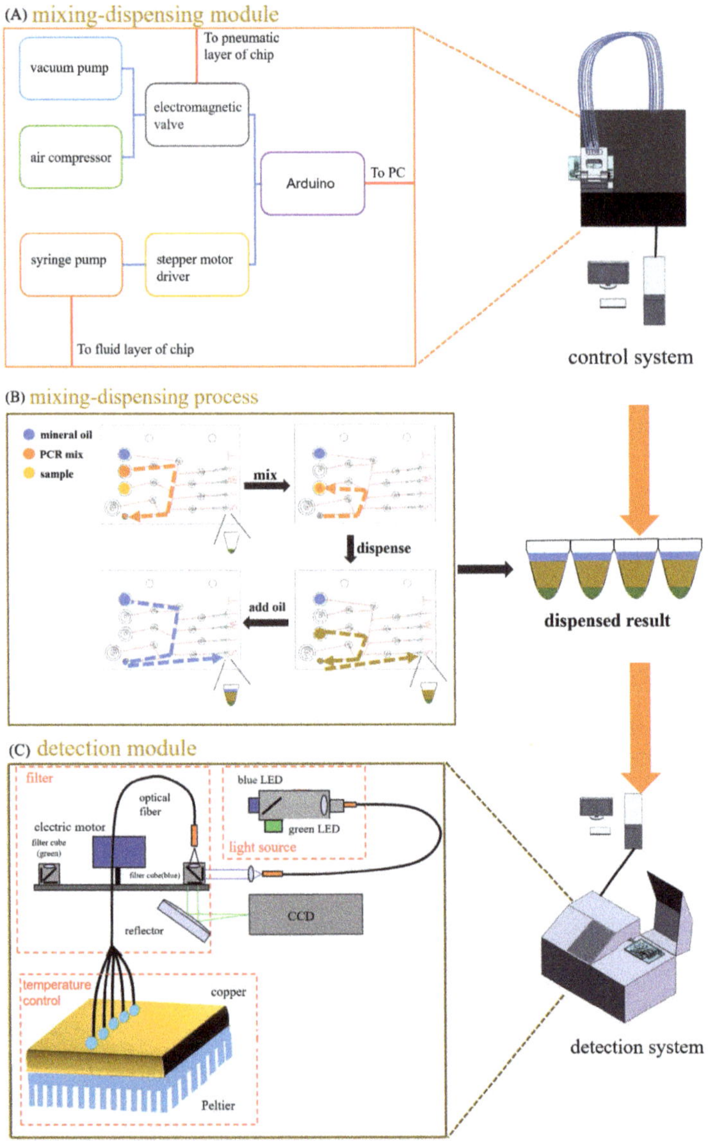

Figure 2. (**A**) Structure diagram of the mixing-dispensing module. (**B**) Schematic illustration of mixing and dispensing process. (**C**) Structure diagram of the detection module.

The workflow was divided into two steps. The first step was mixing the sample with PCR mix and then dispensing the mixtures into four tubes. The PCR mix was sucked into the tube connected with the syringe from reservoir 2 and then infused into reservoir 3 to mix with the sample. After the mixing was completed, a certain amount of the mixed solution was sucked by the syringe pump and dispensed to the 4 PCR tubes in sequence. The PCR tubes were preloaded with primers and fluorescent probes. In order to balance the air pressure inside and outside the device after each dispensing, the valve of the vent would be opened. The mineral oil was finally added to each PCR tube to prevent the aerosol contamination during PCR thermal cycling. At the second step, the device was transferred to the detection module, in which qPCR detection was performed. The curves of both the fluorescence intensity and temperature of the PCR tubes were displayed and recorded in real time with the homemade software.

2.3. Analyzing the Effects of Materials and Coatings on PCR

The photosensitive resins used in the device fabrication were Somos WaterClear Ultra 10122 and Somos WaterShed XC 11122. After 3D printing, some devices were treated with varnish to incease the transparancy. In order to analyze the effects of the resins and the coatings on the PCR, a piece of $2 \times 2 \times 3$ mm^3 was cut from the 3D-printed devices, which were made with different resins, coated or uncoated, and was immersed in 30 μL PCR reagent for 12 min. Also, a piece of 2×2 mm^2 PSA was immersed in 30 μL PCR reagent for 12 min. Next, the original and treated PCR reagents were put in LightCycler® 480 (Roche Diagnostics, Rotkreuz, Switzerland) to carry out qPCR analysis. In order to further investigate the effects of photosensitive resins on PCR, the materials were kept in the PCR reagents to carry out qPCR analysis. In all assays, the PCR regents mixed and dispensed by hand were used as the control.

2.4. qPCR Detection of Macrolide-Resistant Mutations of Mycoplasma pneumoniae

In the early stage of proof of concept, the gene sequences of *Mycoplasma pneumonie* with the mutation of A2063G in 23S rRNA was inserted into a pUC57 vector and cloned. The cloned plasmids were used as the substitute of the clinic samples. The plasmid sequence was described in the supplementary material. Finally, three clinical samples of throat swabs were analyzed, which were kindly provided by Prof. Min Li's group from Renji Hospital affiliated to Shanghai Jiao Tong University.

The primers and the probes were respectively designed according to the conserved sequence of the P1 gene, the mutation sequence of A2063G and A2064G in 23S rRNA. The TaqMan MGB probes were labeled with FAM. In the qPCR assay with the microfluidic device, 4 PCR tubes were preloaded with the different primers and probes and 1 tube, as the negative control, was preloaded with all PCR reagents except the sample. The final composition of the PCR solution after the reagents' preparation is depicted in Table 1.

Table 1. qPCR solution composition after the preparation with the device (besides PCR mix).

	Primer	Probe	Other	Sample Treated by the Device
Tube 1 (conserved sequence)	P1 gene	P1 gene	/	Plasmid/clinical sample
Tube 2 (A2063G mutation)	23S rRNA	A2063G	/	plasmid/clinical sample
Tube 3 (A2064G mutation)	23S rRNA	A2064G	/	plasmid/clinical sample
Tube 4 (positive control)	23S rRNA	A2063G	A2063G plasmid	plasmid/clinical sample
Tube 5 (negative control)	23S rRNA	A2063G	water	/

In the assay, the 103.5 μL of sample, 135 μL of the PCR mix and 300 μL of mineral oil were added into the different reagent reservoirs of the device. After the sample preparation was completed, the device was transferred to the detection module for qPCR. The thermocycling protocol included an initial denaturation at 95 °C for 2 min, followed by denaturation at 95 °C for 30 s, annealing at 55 °C for 40 s andan extension at 72 °C for 10 s, repeated for 45 cycles.

2.5. Sensitivity of Detection System

In order to test the detection sensitivity of the system, A2063G plasmid was used as a template for ten times dilution to obtain the reaction premixes containing the plasmids at different concentrations. In different premixes, the sample contents were 10,000 copies, 1000 copies, 100 copies and 30 copies, respectively. Then the PCR reaction solution was prepared with the above diluted samples and amplified in the homemade qPCR instrument.

3. Results

3.1. Characterization of Bonding Strength and Microvalve Performance

In order to integrate pneumatic microvalves in the device, it is necessary to assemble a pneumatic layer, a PDMS membrane and a fluid layer together. In the traditional manu-facturing method of PDMS chip [28], multi-layer PDMS can bond together after the plasma treatment, which can withstand the pressures up to 300 kPa without delamination [29]. However, the cured photosensitive resin cannot be easily bonded this way. Therefore, adhesive tapes were used to assemble the device. To avoid liquid leakage, the bonding strength between different layers was tested. In the fluid layer, PSA bore the pressure change caused by the movement of the syringe piston. In the pneumatic layer, the twin adhesive tape and PDMS bore the positive pressure applied when closing the microvalve.

Pressure test chips with only inlet but no outlet were fabricated (Figure S2). The inlet radius was 1 mm and the straight channel was 18 mm long and 0.8 mm wide. The total area of channel was 17.54 mm^2. Tapes were cut into 5 mm wide strips and bonded with the back of the chip to close the channel. The inlet was connected to the precision air pressure control system (MFCS^{TM}-EZ, Fluigent). Figure 3A–C shows that when the pressure reached the critical pressure, the measured pressure dropped off cliff-like, which means that the tapes separated from the test chip. All cases were tested three times, and the critical pressure was recorded for each (Figure 3D). As shown in Figure 3D, the minimum critical pressure of PSA is 4133mbar. The minimum force that the PSA can withstand was calculated to be 7.2N. According to the equation of the ideal gas state, pV = nRT, where p is the pressure, V is the volume, n is the amount of substance, R is the gas constant and T is the absolute temperature, the product of the pressure and the volume of a certain amount of gas is constant when the temperature is constant. The total volume of the syringe and the connecting pipe is 16.5 cm^3, ignoring the volume of the chip channel. When the mixing-dispensing module was working, the maximum distance that the syringe pump pushed the piston was 8mm and the volume was reduced by 1.23 cm^3, assuming that eight microvalves were closed at the same time. The pressure was increased by about 80 mbar. The channel area from the Luer taper to the eight microvalves is 137.6 mm^2, so the PSA bore force is 1.1N, much less than 7.2 N. This showed that the PSA could ensure the chip would not be layered or leak during the working process. In the experiment, no liquid leakage was observed.

The microvalve was a key structure to ensure that the samples can be mixed and dispensed correctly, so its performance was evaluated. According to the test results in Figure 3D, the minimum pressure that the PDMS can withstand was 723 mbar. Therefore, the calculated minimum force that the PDMS can withstand was 1.3 N. Because the area of the microvalve is 19.6 mm^2, the positive pressure applied to the microvalve cannot exceed 647 mbar. Since the pressure that the twin adhesive tape can withstand was much greater than the PDMS, we did not need to calculate. Based on the calculated results, 400 mbar, 500 mbar and 600 mbar were respectively chosen as the positive pressure to test the performance of the microvalve. As shown in Figure 3E, according to the difference of the positive pressure, the pressure that the microvalve can withstand was 272 mbar, 352 mbar and 446 mbar, respectively. Theoretically, the pressure that the microvalve can withstand should be equal to the applied positive pressure. However, the surface of the

3D-rinted products was very rough (Figure S3). This caused a decrease in the performance of the microvalve. The Quake valve based on the 3D printing also encountered the same problem, and the non-smooth wall surface would affect the closing performance of the Quake valve [30]. However, due to the advantages of the microvalve with this structure, the actual performance was about 70% of the theoretical value. Therefore, it is very worthwhile to sacrifice the closing performance to reduce the difficulty of microvalve construction. The positive pressure in the experiment was usually 500 mbar, which can withstand pressure of 352 mbar far greater than 80 mbar, so the microvalves can ensure the normal progress of the mixing and dispensing.

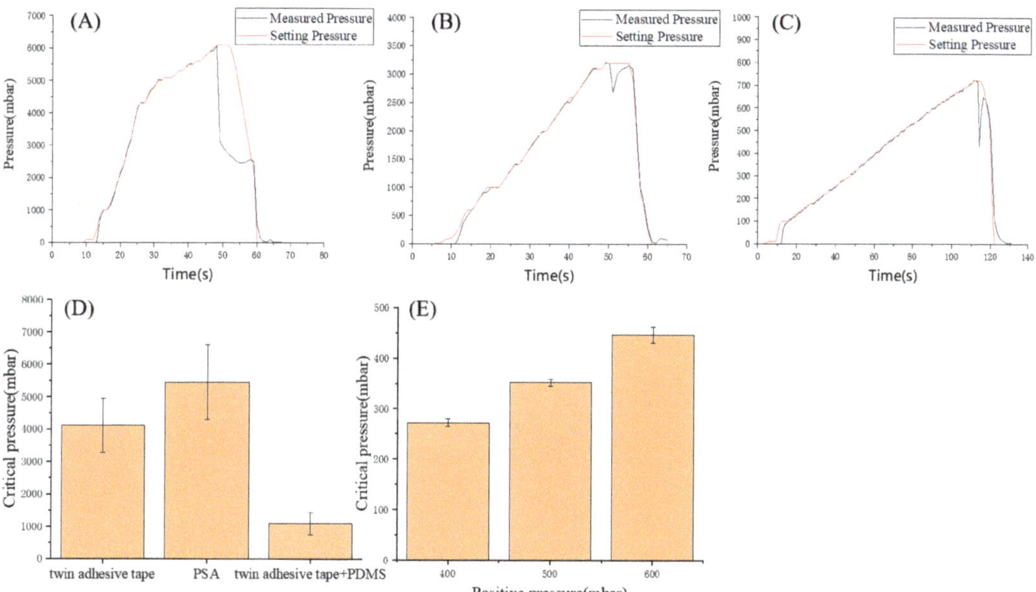

Figure 3. Typical bonding test curves of PSA (**A**), twin adhesive tape (**B**) and PDMS (**C**) with the chip. (**D**) The critical pressure that the different bonding can withstand. (**E**) The critical pressure that the microvalve can withstand at a given positive pressure. (n = 3).

3.2. Effects of Materials and Coatings on PCR

So far, researchers have developed many photosensitive resins used for 3D printing, but whether these resins can be used as the container for biochemical reactions has not been tested. In addition, in order to enhance the transparency of the photosensitive resin, the cured product will be sprayed with the varnish. The effect of the coating on the biochemical reactions is unknown. It has been found that photoinitiator in photosensitive resins can inhibit the PCR reaction [31], so ultraviolet light was irradiated before the experiment to remove the remaining photoinitiator. As shown in Figure 4A,B, the photosensitive resin and the PSA had a certain inhibitory effect on the PCR reaction. However, the Ct value of each curve was nearly same, indicating that the plasmid was hardly adsorbed by the channels of the device. From Figure 4C, it can be seen that the final fluorescence intensity decreased significantly compared with the control, which might be because the resin immered in the PCR agent affected the activity of the polymerase. The heating might make small molecules in the photosensitive resin escape and inactivate the enzyme. The photosensitive resin sprayed with varnish could slightly increase the final fluorescence intensity. It was spectuated that the varnish coating of the device made the internal surface of channels more smooth and reduced the enzyme adsorption and small molecule escape.

However, the improvement of the varnish coating was not obvious. In order to reduce the inhibitory effect of photosensitive resin on the PCR reaction, the reagent was prepared in the 3D-printed device and was then dispensed into the polypropylene PCR tubes to avoid contact between the reagent and the photosensitive resin during PCR amplification.

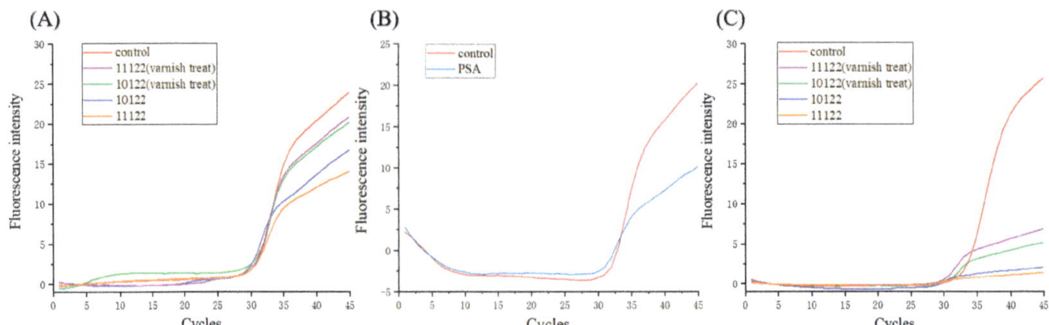

Figure 4. The typical qPCR curves after the PCR reagent contacted photosensitive resins (**A**) and PSA (**B**) for 12 min. (**C**) The qPCR curves of the PCR reagent in which photosensitive resins were immersed for PCR thermocycling.

3.3. Sample Preparation

To complete qPCR sensitively and correctly, it is crucial that the sample and the PCR mix are mixed well and dispensing evenly. The mixing performance was evaluated. First, the red dye and blue dye were put in the reagent reservoirs, respectively, and then the dye solutions were mixed automatically by pushing and pulling the piston of the syringe pump with the program. The mixing result is shown in Figure 5A. The mixed solutions produced with the device and by hand are shown in the left and in the right, respectively. It can be clearly seen that both mix effects are similar, so the mixing with the device can meet the requirements. The dispensing of the solution mainly depends on the synergy between the microvalve control and the syringe pump, so the operation steps and the parameters were optimized in the preliminary experiments and the automatic process program was determined. To evaluate the dispensing performance, two solutions of 110 µL each were added into the two reservoirs and the volumes of the solution dispensed into PCR tubes were analyzed. As shown in Figure 5B, the amount of dispensed solution had good uniformity. Figure 5C shows the photographs of the dispensing results before and after adding oil, where tube five was for the manual oil added, and the rest of four tubes were for the oil added with the device.

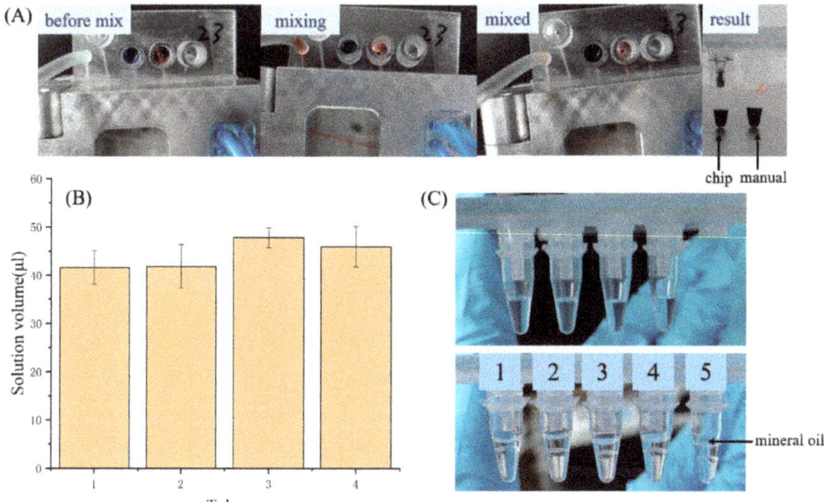

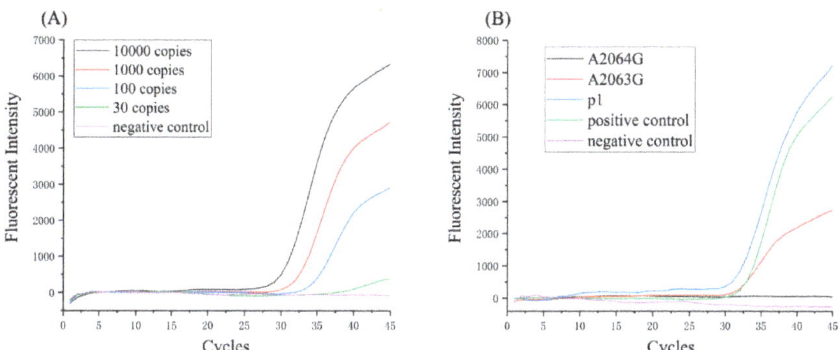

Figure 5. (**A**) Snapshots of dispensing process. (**B**) The reagent volume in PCR tubes after the preparation with the device. n = 3. (**C**) Photograph of the prepared reagents in the PCR tubes.

3.4. Macrolide-Resistant Mutations Detection

The sensitivity test result of the homemade qPCR instrument is shown in Figure 6A. The curve of 100 copies/reaction is an obvious S-shaped curve, which proves that the sensitivity of the detection system reaches 100 CFU/reaction.

Figure 6. (**A**) Detection sensitivity of the homemade detection system. (**B**) The typical qPCR curves of the plasmid sample.

The detection of macrolide-resistant mutations of *Mycoplasma pneumoniae* took 80 min in total, 12 min for preparation and 68 min for qPCR. Three tests were conducted and the results were all positive for A2063G (Figure S4). The typical qPCR curves are shown in Figure 6B. A2063G and the negative control showed negative curves. P1, A2063G resistance mutation and positive control showed obvious positive curves, whose Ct values were 30.8, 30.7 and 30.1, respectively. According to the detection criteria (see the supplementary material), the analyte was *Mycoplasma pneumoniae* and the type of mutation was A2063G, which were consistent with the A2063G plasmid sample.

3.5. Detection of the Clinical Samples

The detection based on the device demonstrated high sensitivity and accuracy in the plasmid assays. Subsequently, three clinical samples of throat swabs were analyzed with the device. The qPCR curves of one clinical sample are shown in Figure 7A. A2063G mutations of *Mycoplasma pneumonie* were determined. In order to verify the result, the amplified product was analyzed using DNA sequencing. The result also indicated the A2063G mutation (Figure 7B). Two more clinical samples were analyzed, both using the approach based on the device and DNA sequencing, and consistent results were obtained (Figure S5). This approach, based on a 3D-printed device, can accurately detect the macrolide-resistant mutation of *Mycoplasma pneumonie*.

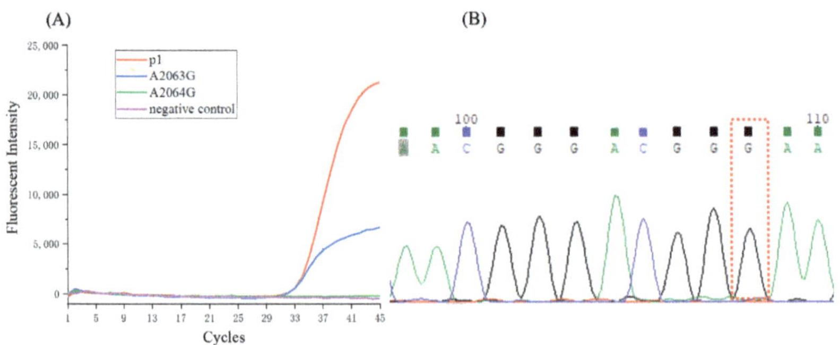

Figure 7. (**A**)The qPCR curves of the clinical sample. (**B**) The sequencing result of the clinical sample.

4. Discussion

Nowadays, *Mycoplasma pneumoniae* has a variety of macrolide-resistant mutations, and the mutations of A2063G and A2064G are the most common types. The detection of macrolide-resistant mutations using qPCR is very important for clinical treatment. However, traditional qPCR detection requires manual sample preparation and special PCR rooms. Also, it may cause the risks of sample contamination and operator infection. Therefore, it is a good choice to use an integrated microfluidic device instead of manual operations.

The 3D printed microfluidic device combined with the mixing-dispensing module and detection module is proposed to easily realize sample-to-answer detection. Using the modules, the device can mix reagents and dispense the mixture to different PCR tubes and achieve multi-channel qPCR detection. In this way, *Mycoplasma pneumoniae* and its mutation type can be identified.

Differing from the previous works, the PCR chambers were not constructed inside the device. Instead, the connectors at the end of channels were designed and fabricated. The commercially available PCR tubes could be fitted tightly with the connectors and performed as PCR chambers, which effectively reduced the inhibitory effect of the device materials on the PCR. Moreover, the mutually independent PCR chambers avoided the contamination problem, making the detection results more credible. However, limited by the size of the PCR tubes, the size of the device cannot be reduced, which is a problem to be addressed in the future.

The microvalve was an important part of the mixing-dispensing module. The proposed microvalve structure can be fabricated with one step using 3D printing. The PDMS membrane was sandwiched between the pneumatic layer and the fluid layer to construct the microvalve structure. The microvalve was a normally closed valve and the positive pressure applied to the pneumatic layer further ensured that the membrane fully covered the inlet of the microvalve and completely blocked the flow of liquid. Compared with the Quake valve, which is a normally open valve, less positive pressure is needed to close the valve.

In addition, with the cooperation of microvalves, only one syringe pump was needed to drive several fluids, which greatly reduced the number of syringe pumps used. At present, the dead volume of the valves cannot be eliminated and the fluid in the annular chamber of the valve cannot be discharged completely, which results in the loss of part of the solution during each dispensing. Due to the uncontrollable amount of lost solution, there are slight differences in the amount of dispensed solution, which in turn affects the uniformity of the dispensing result. More efforts should be made to improve the performance of the microvalve.

5. Conclusions

In summary, a sample-to-answer device and its periphery were developed to detect macrolide-resistant mutations of *Mycoplasma pneumoniae*, including A2063G and A2064G. Sample preparation, including mixing and dispensing, were achieved with the device on the mixed-dispensing module and, afterwards, the device was transferred to the fluorescence detection module to carry on qPCR. In the assays both of the plasmid and the clinical sample, the approach based on the device showed high sensitivity and accuracy.

In fabrication, the 3D printing technology provides conveniences such as rapid, easy-to-make and cheap production. Therefore, the design can be easily changed according to the requirements. The device has several similar structure units, so the detection of more channels is easy to achieve by increasing the numbers of structure units.

The device integrated different materials, taking advantages of their benefits. The photosensitive resin easily processes stereoscopic structures. The elastic PDMS membrane is used for the microvalve structure. The PCR tubes reduced the inhibitory effects of the material on the PCR.

Supplementary Materials: The following are available online at https://www.mdpi.com/article/10.3390/bios11110427/s1. Figure S1. (A)Temperature curve of the homemade instrument. The white line is the temperature condition of fluorescence detection. Whenever the temperature curve drops below the white line, the fluorescence intensity is measured once; Figure S2. (A) The 3D model of the test chip. scale bar is 10mm. Closed channel by PSA(B), twin adhesive tape(C) and PDMS&hollowed twin adhesive tape(D); Figure S3. Snapshot of microvalve. scale bar is 500µm; Figure S4. The qPCR curves of other two plasmids samples; Figure S5. The qPCR curves and sequencing results of other two clinical samples(No.129(A) and No.112(B)).

Author Contributions: Conceptualization, A.W., H.Z., L.W., C.J. and Q.C.; methodology, Z.W., L.W. and C.J.; validation, A.W., Z.W. and Y.H.; formal analysis, Z.W.; investigation, Y.H.; resources, J.Z.; data curation, A.W.; writing—original draft preparation, A.W.; writing—review and editing, L.W., C.J. and Q.C.; visualization, Q.C.; supervision, Q.C.; project administration, L.W. and C.J.; funding acquisition, J.Z. All authors have read and agreed to the published version of the manuscript.

Funding: This research was funded by the Program of Science and Technology Commission of Shanghai Municipality (No.17JC1401001).

Institutional Review Board Statement: The study was conducted according to the guidelines of the Declaration of Helsinki, and approved by Renji Hospital Ethics Committee, Shanghai Jiaotong University School of Medicine (protocol code: SK2020-051; date of approval: August 24th, 2020).

Informed Consent Statement: Informed consent was obtained from all subjects involved in the study.

Data Availability Statement: The data presented in this study are available upon request from the corresponding author.

References

1. Waites, K.B.; Talkington, D.F. Mycoplasma pneumoniae and its role as a human pathogen. *Clin. Microbiol. Rev.* **2004**, *17*, 697–728. [CrossRef]
2. Jain, S.; Self, W.H.; Wunderink, R.G.; Fakhran, S.; Balk, R.; Bramley, A.M.; Reed, C.; Grijalva, C.G.; Anderson, E.J.; Courtney, D.M. Community-acquired pneumonia requiring hospitalization among US adults. *N. Engl. J. Med.* **2015**, *373*, 415–427. [CrossRef] [PubMed]
3. Jain, S.; Williams, D.J.; Arnold, S.R.; Ampofo, K.; Bramley, A.M.; Reed, C.; Stockmann, C.; Anderson, E.J.; Grijalva, C.G.; Self, W.H. Community-acquired pneumonia requiring hospitalization among US children. *N. Engl. J. Med.* **2015**, *372*, 835–845. [CrossRef]

4. Pereyre, S.; Goret, J.; Bébéar, C. Mycoplasma pneumoniae: Current knowledge on macrolide resistance and treatment. *Front. Microbiol.* **2016**, *7*, 974. [CrossRef]

5. Zhao, F.; Liu, G.; Wu, J.; Cao, B.; Tao, X.; He, L.; Meng, F.; Zhu, L.; Lv, M.; Yin, Y. Surveillance of macrolide-resistant Mycoplasma pneumoniae in Beijing, China, from 2008 to 2012. *Antimicrob. Agents Chemother.* **2013**, *57*, 1521–1523. [CrossRef]

6. Liu, Y.; Ye, X.; Zhang, H.; Xu, X.; Li, W.; Zhu, D.; Wang, M. Antimicrobial susceptibility of Mycoplasma pneumoniae isolates and molecular analysis of macrolide-resistant strains from Shanghai, China. *Antimicrob. Agents Chemother.* **2009**, *53*, 2160–2162. [CrossRef] [PubMed]

7. Bébéar, C.; Pereyre, S. Mechanisms of drug resistance in Mycoplasma pneumoniae. *Curr. Drug Targets-Infect. Disord.* **2005**, *5*, 263–271. [CrossRef] [PubMed]

8. Xin, D.; Mi, Z.; Han, X.; Qin, L.; Li, J.; Wei, T.; Chen, X.; Ma, S.; Hou, A.; Li, G. Molecular mechanisms of macrolide resistance in clinical isolates of Mycoplasma pneumoniae from China. *Antimicrob. Agents Chemother.* **2009**, *53*, 2158–2159. [CrossRef]

9. Daxboeck, F.; Krause, R.; Wenisch, C. Laboratory diagnosis of Mycoplasma pneumoniae infection. *Clin. Microbiol. Infect.* **2003**, *9*, 263–273. [CrossRef]

10. Shin, D.J.; Andini, N.; Hsieh, K.; Yang, S.; Wang, T.H. Emerging Analytical Techniques for Rapid Pathogen Identification and Susceptibility Testing. *Annu. Rev. Anal. Chem.* **2019**, *12*, 41–67. [CrossRef] [PubMed]

11. Ohst, C.; Saschenbrecker, S.; Stiba, K.; Steinhagen, K.; Probst, C.; Radzimski, C.; Lattwein, E.; Komorowski, L.; Stocker, W.; Schlumberger, W. Reliable Serological Testing for the Diagnosis of Emerging Infectious Diseases. *Adv. Exp. Med. Biol.* **2018**, *1062*, 19–43.

12. Razmi, N.; Hasanzadeh, M.; Willander, M.; Nur, O. Recent Progress on the Electrochemical Biosensing of *Escherichia coli* O157:H7: Material and Methods Overview. *Biosensors* **2020**, *10*, 54. [CrossRef]

13. Ali, A.A.; Altemimi, A.B.; Alhelfi, N.; Ibrahim, S.A. Application of Biosensors for Detection of Pathogenic Food Bacteria: A Review. *Biosensors* **2020**, *10*, 58. [CrossRef]

14. Linger, Y.; Knickerbocker, C.; Sipes, D.; Golova, J.; Franke, M.; Calderon, R.; Lecca, L.; Thakore, N.; Holmberg, R.; Qu, P.; et al. Genotyping Multidrug-Resistant Mycobacterium tuberculosis from Primary Sputum and Decontaminated Sediment with an Integrated Microfluidic Amplification Microarray Test. *J. Clin. Microbiol.* **2018**, *56*, e01652-17. [CrossRef]

15. Gopfert, L.; Elsner, M.; Seidel, M. Isothermal haRPA detection of blaCTX-M in bacterial isolates from water samples and comparison with qPCR. *Anal. Methods* **2021**, *13*, 552–557. [CrossRef]

16. Kuhnemund, M.; Hernandez-Neuta, I.; Sharif, M.I.; Cornaglia, M.; Gijs, M.A.; Nilsson, M. Sensitive and inexpensive digital DNA analysis by microfluidic enrichment of rolling circle amplified single-molecules. *Nucleic Acids Res.* **2017**, *45*, e59. [CrossRef]

17. Moon, J.; Kwon, H.J.; Yong, D.; Lee, I.C.; Kim, H.; Kang, H.; Lim, E.K.; Lee, K.S.; Jung, J.; Park, H.G.; et al. Colorimetric Detection of SARS-CoV-2 and Drug-Resistant pH1N1 Using CRISPR/dCas9. *ACS Sens.* **2020**, *5*, 4017–4026. [CrossRef]

18. Dreyer, V.; Utpatel, C.; Kohl, T.A.; Barilar, I.; Groschel, M.I.; Feuerriegel, S.; Niemann, S. Detection of low-frequency resistance-mediating SNPs in next-generation sequencing data of Mycobacterium tuberculosis complex strains with binoSNP. *Sci. Rep.* **2020**, *10*, 7874. [CrossRef] [PubMed]

19. Baltekin, O.; Boucharin, A.; Tano, E.; Andersson, D.I.; Elf, J. Antibiotic susceptibility testing in less than 30 min using direct single-cell imaging. *Proc. Natl. Acad. Sci. USA* **2017**, *114*, 9170–9175. [CrossRef] [PubMed]

20. Kang, D.K.; Ali, M.M.; Zhang, K.; Huang, S.S.; Peterson, E.; Digman, M.A.; Gratton, E.; Zhao, W. Rapid detection of single bacteria in unprocessed blood using Integrated Comprehensive Droplet Digital Detection. *Nat. Commun.* **2014**, *5*, 5427. [CrossRef] [PubMed]

21. Chen, J.; Li, H.; Xie, H.; Xu, D. A novel method combining aptamer-Ag10NPs based microfluidic biochip with bright field imaging for detection of KPC-2-expressing bacteria. *Anal. Chim. Acta* **2020**, *1132*, 20–27. [CrossRef] [PubMed]

22. Nguyen, H.V.; Nguyen, V.D.; Nguyen, H.Q.; Chau, T.H.; Lee, E.Y.; Seo, T.S. Nucleic acid diagnostics on the total integrated lab-on-a-disc for point-of-care testing. *Biosens. Bioelectron.* **2019**, *141*, 111466. [CrossRef]

23. Ma, Y.D.; Li, K.H.; Chen, Y.H.; Lee, Y.M.; Chou, S.T.; Lai, Y.Y.; Huang, P.C.; Ma, H.P.; Lee, G.B. A sample-to-answer, portable platform for rapid detection of pathogens with a smartphone interface. *Lab Chip* **2019**, *19*, 3804–3814. [CrossRef] [PubMed]

24. Wang, H.; Ma, Z.; Qin, J.; Shen, Z.; Liu, Q.; Chen, X.; Wang, H.; An, Z.; Liu, W.; Li, M. A versatile loop-mediated isothermal amplification microchip platform for Streptococcus pneumoniae and Mycoplasma pneumoniae testing at the point of care. *Biosens Bioelectron.* **2019**, *126*, 373–380. [CrossRef] [PubMed]

25. Chao, C.Y.; Wang, C.H.; Che, Y.J.; Kao, C.Y.; Wu, J.J.; Lee, G.B. An integrated microfluidic system for diagnosis of the resistance of Helicobacter pylori to quinolone-based antibiotics. *Biosens. Bioelectron.* **2016**, *78*, 281–289. [CrossRef] [PubMed]

26. Naderi, A.; Bhattacharjee, N.; Folch, A. Digital Manufacturing for Microfluidics. *Annu. Rev. Biomed. Eng.* **2019**, *21*, 325–364. [CrossRef]

27. Nielsen, A.V.; Beauchamp, M.J.; Nordin, G.P.; Woolley, A.T. 3D Printed Microfluidics. *Annu. Rev. Anal. Chem.* **2020**, *13*, 45–65. [CrossRef] [PubMed]

28. Unger, M.A.; Chou, H.P.; Thorsen, T.; Scherer, A.; Quake, S.R. Monolithic Microfabricated Valves and Pumps by Multilayer Soft Lithography. *Science* **2000**, *288*, 113–116. [CrossRef]

29. Eddings, M.A.; Johnson, M.A.; Gale, B.K. Determining the optimal PDMS–PDMS bonding technique for microfluidic devices. *J. Micromech. Microeng.* **2008**, *18*, 067001. [CrossRef]

30. Lee, Y.S.; Bhattacharjee, N.; Folch, A. 3D-printed Quake-style microvalves and micropumps. *Lab Chip* **2018**, *18*, 1207–1214. [CrossRef]
31. Tzivelekis, C.; Sgardelis, P.; Waldron, K.; Whalley, R.; Huo, D.; Dalgarno, K. Fabrication routes via projection stereolithography for 3D-printing of microfluidic geometries for nucleic acid amplification. *PLoS ONE* **2020**, *15*, e0240237. [CrossRef] [PubMed]

Article

OralDisk: A Chair-Side Compatible Molecular Platform Using Whole Saliva for Monitoring Oral Health at the Dental Practice

Desirée Baumgartner [1,2,*], Benita Johannsen [1], Mara Specht [1], Jan Lüddecke [1], Markus Rombach [1], Sebastian Hin [1], Nils Paust [1,2], Felix von Stetten [1,2], Roland Zengerle [1,2], Christopher Herz [3], Johannes R. Peham [3], Pune N. Paqué [4], Thomas Attin [4], Joël S. Jenzer [4], Philipp Körner [4], Patrick R. Schmidlin [4], Thomas Thurnheer [4], Florian J. Wegehaupt [4], Wendy E. Kaman [5,6], Andrew Stubbs [7], John P. Hays [5], Viorel Rusu [8], Alex Michie [9], Thomas Binsl [9], David Stejskal [10,11], Michal Karpíšek [12,13], Kai Bao [14], Nagihan Bostanci [14], Georgios N. Belibasakis [14] and Konstantinos Mitsakakis [1,2,*]

1   Hahn-Schickard, Georges-Koehler-Allee 103, 79110 Freiburg, Germany; Benita.Johannsen@Hahn-Schickard.de (B.J.); Mara.Specht@Hahn-Schickard.de (M.S.); Jan.Lueddecke@Hahn-Schickard.de (J.L.); Markus.Rombach@Hahn-Schickard.de (M.R.); sebastian.hin@iuvas.de (S.H.); Nils.Paust@Hahn-Schickard.de (N.P.); Felix.von.Stetten@Hahn-Schickard.de (F.v.S.); Roland.Zengerle@Hahn-Schickard.de (R.Z.)
2   Laboratory for MEMS Applications, IMTEK–Department of Microsystems Engineering, University of Freiburg, Georges-Koehler-Allee 103, 79110 Freiburg, Germany
3   AIT Austrian Institute of Technology, Molecular Diagnostics, Giefinggasse 4, 1210 Wien, Austria; christopher_herz@pall.com (C.H.); Johannes.Peham@ait.ac.at (J.R.P.)
4   Clinic of Conservative and Preventive Dentistry, Center of Dental Medicine, University of Zurich, Plattenstrasse 11, 8032 Zurich, Switzerland; punenina.paque@zzm.uzh.ch (P.N.P.); thomas.attin@zzm.uzh.ch (T.A.); Joel.Jenzer@icloud.com (J.S.J.); philipp.koerner@zzm.uzh.ch (P.K.); patrick.schmidlin@zzm.uzh.ch (P.R.S.); Thomas.Thurnheer@zzm.uzh.ch (T.T.); florian.wegehaupt@zzm.uzh.ch (F.J.W.)
5   Department of Medical Microbiology and Infectious Diseases, Erasmus University Medical Centre Rotterdam (Erasmus MC), 3015 CN Rotterdam, The Netherlands; w.e.kaman@acta.nl (W.E.K.); j.hays@erasmusmc.nl (J.P.H.)
6   Department of Oral Biochemistry, Academic Centre for Dentistry Amsterdam (ACTA), Free University of Amsterdam and University of Amsterdam, 1081 LA Amsterdam, The Netherlands
7   Department of Pathology and Clinical Bioinformatics, Erasmus University Medical Centre Rotterdam (Erasmus MC), 3015 CN Rotterdam, The Netherlands; a.stubbs@erasmusmc.nl
8   Magtivio B.V., Daelderweg 9, 6361 HK Nuth, The Netherlands; vru@magtivio.com
9   ClinicaGeno Ltd., 11 Station Approach, Coulsdon CR5 2NR, UK; alex@clinicageno.com (A.M.); thomas@clinicageno.com (T.B.)
10 Department of Biomedical Sciences, Faculty of Medicine, University of Ostrava, Syllabova 19, 70300 Ostrava, Czech Republic; david.stejskal@fno.cz
11 Institute of Laboratory Diagnostics, University Hospital Ostrava, 17. Listopadu 1790/5, 70800 Ostrava, Czech Republic
12 BioVendor-Laboratorní Medicína a.s., Research & Diagnostic Products Division, Karasek 1767/1, Reckovice, 62100 Brno, Czech Republic; karpisek@biovendor.com
13 Faculty of Pharmacy, Masaryk University, Palackeho trida 1946/1, 61242 Brno, Czech Republic
14 Section of Oral Health and Periodontology, Division of Oral Diseases, Department of Dental Medicine, Karolinska Institutet, 14104 Huddinge, Sweden; kai.bao@ki.se (K.B.); nagihan.bostanci@ki.se (N.B.); george.belibasakis@ki.se (G.N.B.)
*   Correspondence: Konstantinos.Mitsakakis@Hahn-Schickard.de (K.M.); Desiree.Baumgartner@Hahn-Schickard.de (D.B.); Tel.: +49-761-203-73252 (K.M.); +49-761-203-98724 (D.B.)

Citation: Baumgartner, D.; Johannsen, B.; Specht, M.; Lüddecke, J.; Rombach, M.; Hin, S.; Paust, N.; von Stetten, F.; Zengerle, R.; Herz, C.; et al. OralDisk: A Chair-Side Compatible Molecular Platform Using Whole Saliva for Monitoring Oral Health at the Dental Practice. *Biosensors* 2021, 11, 423. https://doi.org/10.3390/bios11110423

Received: 1 October 2021
Accepted: 24 October 2021
Published: 28 October 2021

Publisher's Note: MDPI stays neutral with regard to jurisdictional claims in published maps and institutional affiliations.

Abstract: Periodontitis and dental caries are two major bacterially induced, non-communicable diseases that cause the deterioration of oral health, with implications in patients' general health. Early, precise diagnosis and personalized monitoring are essential for the efficient prevention and management of these diseases. Here, we present a disk-shaped microfluidic platform (OralDisk) compatible with chair-side use that enables analysis of non-invasively collected whole saliva samples and molecular-based detection of ten bacteria: seven periodontitis-associated (*Aggregatibacter actinomycetemcomitans*, *Campylobacter rectus*, *Fusobacterium nucleatum*, *Prevotella intermedia*, *Porphyromonas gingivalis*, *Tannerella forsythia*, *Treponema denticola*) and three caries-associated (oral *Lactobacilli*, *Streptococcus mutans*, *Streptococcus sobrinus*). Each OralDisk test required 400 µL of homogenized whole saliva. The

15

automated workflow included bacterial DNA extraction, purification and hydrolysis probe real-time PCR detection of the target pathogens. All reagents were pre-stored within the disk and sample-to-answer processing took < 3 h using a compact, customized processing device. A technical feasibility study (25 OralDisks) was conducted using samples from healthy, periodontitis and caries patients. The comparison of the OralDisk with a lab-based reference method revealed a ~90% agreement amongst targets detected as positive and negative. This shows the OralDisk's potential and suitability for inclusion in larger prospective implementation studies in dental care settings.

Keywords: dental practice; point-of-care diagnostics; treatment monitoring; oral health; periodontitis; caries; saliva diagnostics

1. Introduction

Oral diseases are the most prevalent chronic diseases worldwide, accounting for almost 5 billion cases globally [1]. They are the third most expensive group of diseases to treat in the EU, following diabetes and cardiovascular diseases [2]. In addition, the overprescription of antibiotics in dentistry is a challenge [3]. Indicatively, up to 80% of prophylactic antibiotic use in the US is considered unjustified [4].

The two most prevalent oral microbial diseases are caries and periodontitis. Dental caries affects the hard tissue of the teeth, causing tooth decay. Periodontitis affects the tissues that surround and support the teeth, leading to progressive loss of the bone and soft tissue attachment and eventually tooth loss. There are different clinical manifestations and degrees of severity of periodontal disease. According to the Global Burden of Disease (GBD) study, 796 million people around the globe had severe periodontitis in 2017, ranking it within the top 10 most prevalent conditions worldwide [5]. Across Europe, 5–20% of middle-aged people and up to 40% of elderly people are affected by it [6]. In the US, 80% of the population has some form of periodontal disease [7]. Periodontitis has also been related to systemic diseases such as type 2 diabetes mellitus (T2DM), cardiovascular diseases, Alzheimer's disease and others [8–10]. Therefore, being able to detect periodontitis-causing bacteria has a much broader clinical significance than simply monitoring oral health, as it can potentially provide a signal for deteriorating systemic health [11].

Caries and periodontitis are both treatable, and the rationale around their treatment is the same, namely, the removal of the microbial biofilm which is the causative factor. The treatment of periodontal disease, upon the removal of the biofilm, requires good oral hygiene protocols and monitoring to ensure that the inflammation has subsided. For dental caries, upon removal of the infective caries tissue, the treatment mandates restoration of the lost hard tissue with appropriate restorative material (fillings). Although the treatment protocols for both diseases are well-defined, there exist urgent and unmet medical (dental) needs related to: (i) early diagnosis (even before symptoms emerge), which would assist in preventing the disease, benefiting the patients' quality of life while also saving on the costs of treatment and (ii) accurate monitoring during and after treatment (i.e., during the maintenance phase of treatment) in cases where a patient presents with advanced or aggressive disease. Accurate monitoring would allow the dentist to make the correct assessment of when to start and finish treatment: not too early, running the risk of re-emergence, and not too late, thus spending unnecessary resources.

The current state of the art for performing such diagnoses and monitoring is still largely dependent on clinical examination, patient history and radiographic imaging (X-rays). However, radiographs mainly observe damage that occurred in the past, while plain clinical examinations tend to miss the first signs of incipient dysbiosis and therefore do not contribute to early detection and prevention. Furthermore, cumulative X-ray radiation during periodontal treatment and frequent visits for follow-up monitoring poses health risks and special concerns in some patients (e.g., pregnant women). Periodontal probing is a frequently used methodology in which a probe is inserted into the gingival sulcus in order to measure pocket depths around a tooth and to assess the health status of the periodontium [12,13]. However, it is an invasive method in the gums, and as such, it should not be used in patients with T2DM as it may lead to bacteremia [14,15], thereby increasing systemic inflammation and the infection risk in already vulnerable T2DM patients. Additionally, general clinical examinations do not include probing of the whole gingival sulcus of the teeth. Other typical (simplified) measurements for basic periodontal screening include the community periodontal index (CPI) and the basic periodontal examination (BPE) [16], which are performed by looking for increased pocket depths and bleeding on probing.

Unfortunately, these current 'gold-standard' diagnostic approaches may risk missing the diagnosis of emerging periodontitis. Therefore, complementary diagnostic tools, which are largely less invasive and less tedious, could improve the diagnostic sensitivity, precision and accuracy, for example, by screening the microbial ecology of the oral cavity in order to help indicate whether a patient needs further diagnostic evaluation and treatment or to monitor patients' post-treatment status.

In this respect, the current publication proposes a rapid, molecular-based and non-invasive platform for detecting oral disease-causing bacteria, with a workflow that is compatible with point-of-care (POC) dental settings [17] and which can function as an auxiliary tool to the current gold standard diagnostic methods. The OralDisk microfluidic cartridge integrates all the biochemical reagents needed for fully automated analysis, namely: (i) customized buffers and microfluidic-optimized magnetic particles for the purification of bacterial DNA; (ii) amplification reagents in a lyophilized form; and (iii) POC-ready real-time qPCR TaqMan primers/probes for specific bacterial detection. The OralDisk is an application-specific version of the centrifugal microfluidic LabDisk platform, which has already demonstrated its utility in applications where a single infectious pathogen is to be detected [18–21]. This small-scale technical feasibility study demonstrates for the first time the platform's implementation in the field of oral health, where multiple bacteria may be present simultaneously in the oral cavity and whole saliva is used as diagnostic specimen. It presents data on the detection of three caries-associated and seven periodontitis-associated bacterial species in complex saliva samples collected from individuals that were classified into three study groups (healthy, caries and periodontitis) following an assessment by dental specialists. The patients' status assessment and the sample collection were performed in a previous clinical study [22]. The results from the OralDisk were compared with a lab-based extraction and qPCR reference method, as well as with the iai PadoTest (Institut für Angewandte Immunologie IAI AG, Zuchwil, Switzerland) commercial reference method. The OralDisk exhibited comparable (and in some cases superior) behavior while offering the additional benefit of automation.

2. Materials and Methods

2.1. Sample Collection and Ethics Permission

The samples used in this study were a sub-group ($n = 24$) of a large cohort ($n = 214$) of samples that had been collected at the Center for Dental Medicine, University of Zurich. These samples were intended for the microbial analysis of saliva with the aim of identifying oral infections in patients [22]. The 24 samples (seven healthy, nine caries and eight periodontitis samples) were selected in order to demonstrate the technical feasibility of the OralDisk, without the intention of generating clinical conclusions. The sample collection and the study protocol were approved by the local Swiss ethics committee (BASEC-no. 2016-00435) and all sample donors signed a written informed consent form prior to saliva collection. The collection contained unstimulated whole saliva that was aliquoted and stored at $-80\ ^\circ$C until further analysis. Details on the inclusion/exclusion criteria and on the collection methodology are available in Paqué et al. [22].

2.2. Selected Bacterial Panel

Quantitative shifts in the levels of multiple, rather than single, bacterial species may more accurately reflect dysbiotic changes in oral microbial ecology that are commensurate with the initiation or progression of oral disease [23]. Measurement of the species dynamics of oral polymicrobial populations is the key element that allows potential monitoring for the prevention, early diagnosis and post-treatment follow-up of oral diseases. Therefore, ten bacteria were included in the OralDisk panel: seven Gram-negative bacteria related to periodontitis (*Aggregatibacter actinomycetemcomitans*, *Campylobacter rectus*, *Fusobacterium nucleatum*, *Prevotella intermedia*, *Porphyromonas gingivalis*, *Tannerella forsythia* and *Treponema denticola*) and three Gram-positive bacteria associated with caries (oral associated *Lactobacilli*, *Streptococcus mutans* and *Streptococcus sobrinus*).

This panel was chosen based on: (i) current knowledge on the association of certain oral bacteria with caries and periodontitis [24]; (ii) feedback from experts in response to survey questionnaires; and (iii) the goal of including bacteria which were identified in subgingival [25] and supragingival [26] biofilm samples of subjects with and without periodontitis. Furthermore, the role of these bacteria as differentiators between healthy, periodontitis and caries groups was demonstrated in a preceding clinical study by Paqué et al. [22], which provided additional evidence for including these particular bacteria in the current technical feasibility study of the OralDisk. In the aforementioned clinical study, statistically significant polymicrobial differentiators were observed (i) between healthy and periodontitis groups (*C. rectus*, *T. forsythia*, *P. gingivalis*, *S. mutans*, *F. nucleatum*, *T. denticola*, *P. intermedia* and oral *Lactobacilli*); (ii) between healthy and caries groups (*S. mutans* and *T. denticola*); and (iii) between caries and periodontitis groups (*S. mutans*).

2.3. Reference Method #1: Lab-Based DNA Extraction and qPCR

Enzymatic lysis was performed on 920 µL of whole saliva using the GenElute™ Bacterial Genomic DNA Kit (Sigma-Aldrich, Saint Louis, MO, USA), followed by silica column-based extraction (Figure 1). An adjusted manufacturer's protocol was used, as described in previous work [22]. An eluate volume of 135 µL was stored at −25 °C and later used to perform qPCR (Roche LightCycler). Details on the POC-compatible qPCR assay development, primer/probe design, assay validation (including qPCR assay sensitivity and limit of detection), utilized amplification conditions and data analysis, are available in Paqué et al. [22].

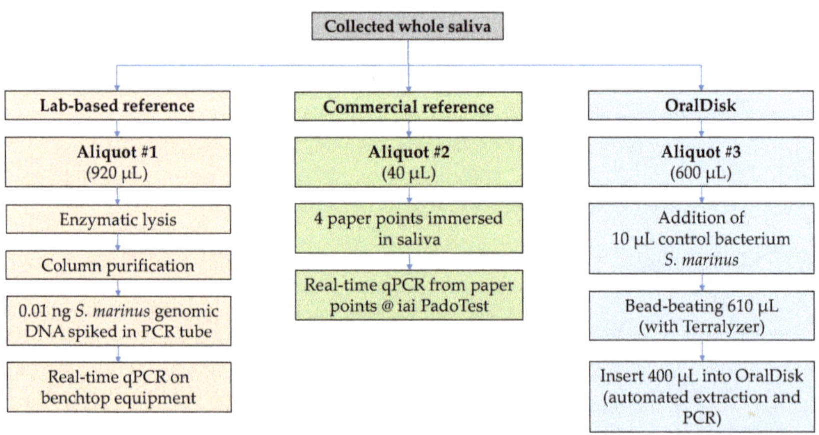

Figure 1. Experimental workflows from sample collection until analysis.

2.4. Reference Method #2: Commercial iai PadoTest

The iai PadoTest (Institut für Angewandte Immunologie IAI AG, Zuchwil, Switzerland) [27] is a commercially available test that was used as a second reference to the OralDisk (Figure 1). It performs a multiplex real-time qPCR assay that estimates bacterial cell counts based on 16S rRNA [28]. The manufacturer's collection protocol stipulates that paper points are inserted into dental pockets to collect gingival crevicular fluid (GCF). The paper points are stored in vials (one or more paper points per vial are possible, in single- or pool-mode of analysis). To make the results comparable with our saliva-based detection, the sample collection protocol was slightly modified: four paper points (Roeko Iso 55, Coltène, Altstätten, Switzerland) were immersed in 40 µL of thawed whole saliva in a tube. The tubes with the paper points were sent to iai PadoTest AG for analysis.

2.5. Mechanical Lysis and Homogenization of Saliva Samples Prior to Insertion into the OralDisk

Lysis of bacteria, with simultaneous homogenization of the saliva, was performed prior to insertion into the OralDisk by means of a mechanical bead-beating process using a hand-held device (Terralyzer, Zymo Research, Irvine, CA, USA). Notably, this was the only manual step in the protocol (Figure 1). A volume of 600 µL of whole saliva and 10 µL of 1:15 (or 1:10 for two samples) diluted Gram-positive bacterium *Serinicoccus marinus* [29] (process control [22]) was inserted into a 2-mL tube with 1.30 g of 0.2-mm steel beads (Next Advance Inc., Troy, NY, USA). The tube was then inserted into the Terralyzer. The bead-beating protocol for all samples was 2 × 10 s with a 20 s break (apart from samples GTT33 and KxTC22, for which it was 2 × 20 s with a 10 s break).

2.6. OralDisk Design and Workflow for Fully Automated Real-Time PCR

Oral bacteria were detected using the centrifugal microfluidic OralDisk, which incorporated all the microfluidic unit operations [30] required for the fully automated analysis of whole saliva samples. A volume of 400 µL of ex situ homogenized saliva (Section 2.5) was inserted into the OralDisk (Figure 2, #1) and on-disk DNA extraction and purification was based on a bind-wash-elute protocol [31]. Dedicated buffers were developed by magtivio B.V., the Netherlands, and were stored in pouches (stickpacks [32]) on the disk (Figure 2, #2a–2d). Upon centrifugation (and assisted by controlled heating), liquids were released into their respective (radially outward) chambers (Figure 2, #4a–4d). The stickpacks contained 440 µL of binding buffer (#2a); 200 µL of wash buffer 1 (#2b); 200 µL of wash buffer 2 (#2c); and 180 µL of elution buffer (#2d). Magnetic beads (MagSi-DNA mf beads, ferrimagnetic core with silica shell, cat. no. MD0200010002) were developed by magtivio B.V. especially for microfluidic use for this application and were dry-stored on the disk (Figure 2, #3). Upon magnetic bead rehydration by the binding buffer and lysate, the magnetic beads captured the DNA and were transported through the subsequent chambers (Figure 2, #4a–4d) by means of controlled continuous disk rotation and integrated magnets [33]. In chamber #4d (Figure 2), the DNA was eluted from the magnetic beads, and 160 µL of the eluate was pumped radially inwards into chamber #6, through structure #5 and by means of temperature change-rate (TCR) actuated valving [34] and centrifugo-dynamic inward pumping [35]. In chamber #6, the amplification reagents were pre-stored in the form of a lyophilized pellet (46 µL; TaqMan® Lyophilized 1-Step qPCR Master Mix; 3.5×, Thermo Fisher Scientific, USA). Upon lyopellet rehydration and thorough mixing using a dedicated microfluidic protocol to ensure homogeneity [36], the mixture was aliquoted [37] into the PCR reaction chambers (#7) where the primers/probes for each oral bacterium (plus those for the control bacterium *S. marinus*) were dry-stored. Chamber (i) was a sacrificial chamber to collect residual liquid. Upon rehydration, the thermocycling protocol for real-time PCR started: 95 °C for 3 min (initial denaturation) and 40 cycles of 95 °C for 10 s and 60 °C for 30 s.

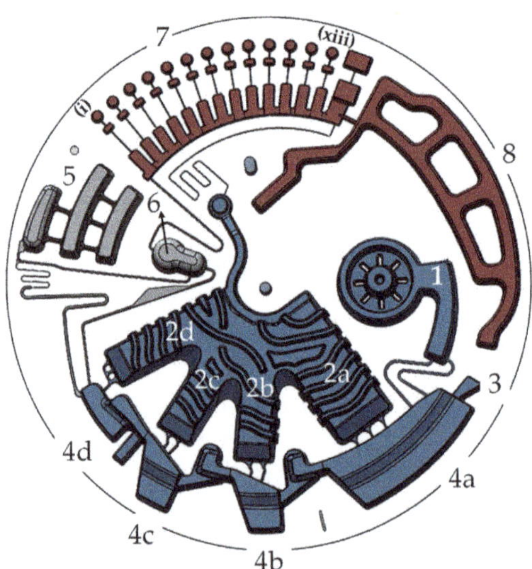

Figure 2. OralDisk design. Blue sector: magnetic bead-based extraction and purification of DNA. #1: sample inlet; #2: stickpacks for storage of buffers for: binding (2a), 1st washing (2b), 2nd washing (2c) and elution (2d); #3: pre-stored (air-dried) magnetic beads; #4: chambers for binding (4a), washing (4b, 4c) and elution of DNA from magnetic beads (4d). Grey sector: eluate transfer module, automating the inward pumping (#5) and eluate mixing with the lyopellet (#6). Red sector: amplification module, automating the preparation and execution of the real-time PCR in structure #7 in the reaction chambers labelled as (i)–(xiii). Structure #8 assists the liquid transfer from chamber #6 to the PCR structure #7.

The microfluidic protocol comprised a slightly modified version of a protocol previously published by the authors [21]. According to this slightly modified protocol, the times lapsed during the microfluidic processes in the blue-, grey- and red-marked modules in Figure 2 were: ~37 min, ~12 min and ~107 min, respectively (the latter including the PCR thermocycling). Notably, the OralDisk does not have any sample outlet port but is a closed system, as are all microfluidic cartridges in similar systems, so that the amplified DNA does not contaminate the cartridge processing instrument, thereby risking false positive results during the next measurement. Since we used human sample material, the disposal of the OralDisk was performed by autoclaving, as for other typical laboratory consumables (e.g., wells, tubes) that are used in nucleic acid amplification practices.

PCR thermocycling was performed in a customized LabDisk processing device functional model (QIAGEN Lake Constance, currently DIALUNOX GmbH, Lake Constance, Germany) (Figure 3) comprising: (i) a thermal module that enables global air heating for performing the necessary thermocycling protocols; (ii) a mechanical module for the precise positioning, acceleration and deceleration of the disks; (iii) an optical module for detection of the real-time fluorescence signal derived from the nucleic acid amplification product; and (iv) integrated magnets for bead transfer during the DNA extraction and purification. The raw data acquired with the LabDisk Player were analyzed using a RotorGene (QIAGEN, Hilden, Germany) software program to acquire Cq values.

2.7. OralDisk Fabrication

The OralDisks were fabricated by microthermoforming [38,39] of polycarbonate (PC) polymer foils (250 µm thickness, Makrofol® DE 1-1 000000, Covestro, Leverkusen, Germany) using a hot embossing machine (HEX01, Jenoptik AG, Jena, Germany) at the Hahn-Schickard Lab-on-a-Chip Foundry Service [40]. Microthermoforming technology is well-known from macro-scale blister package fabrication, which has been adapted and transferred to the micro-scale. In short, an elastomeric mold made of poly(dimethylsiloxane) (PDMS) was heated. The overlying polymer foil was heated as well, and at a specific temperature above the foil's glass transition, air was blown onto it so that it assumed the shape of the mold. Ultimately, this technology possesses the following advantages: (i) it allows monolithic fabrication of the cartridge and (ii) it is scalable and able to produce several tens of thousands of pieces when required. After the foil structuring, a Teflon coating (0.5% w/w Teflon (Teflon Amorphous Fluoropolymer, Chemours International Operations Sarl, Geneva, Switzerland) in Fluorinert™ FC-770 (art. # F3556-100ML, Sigma-Aldrich Chemie GmbH, Darmstadt, Germany)) was applied to chambers #4a–4d (Figure 2) in order to provide hydrophobic microfluidic properties in the nucleic acid extraction module. Then, 20 µL of magnetic beads with 10 µL of 250 mg/mL trehalose were pipetted into the corresponding chamber (Figure 2, #3), and 3.5 µL of each primer/probe reaction mix with 0.5 µL of 1 M trehalose (final concentration 50 mM) were pipetted into each reaction chamber (Figure 2, #7). The Teflon coating and the drying of the magnetic beads and primers/probes followed a previously published protocol (1 h at 50 °C) [20,41,42]. The lyopellets containing the amplification reagents were manually inserted into the disk. All buffers were stored in dedicated aluminum pouches (stickpacks). The sealing temperature and pressure used to prepare the stickpacks allowed their opening at a specific rotational frequency (70 Hz), whereby buffers were released into the corresponding chambers of the extraction module (Figure 2). The stickpacks were also manually inserted into the disk. The whole cartridge was sealed using a pressure-sensitive adhesive foil (9795R, 3M, Maplewood, MN, USA). The cartridge was then inserted into an aluminum pouch with a nitrogen atmosphere and desiccant bags, and was stored at room temperature until use.

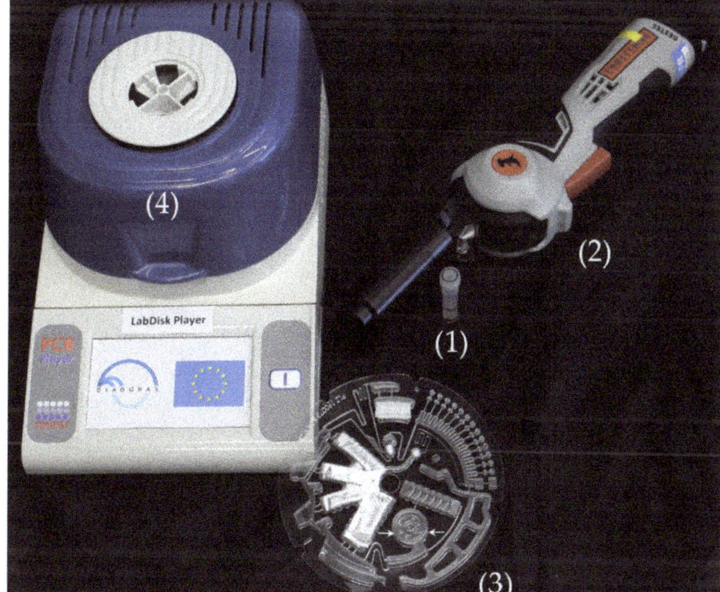

Figure 3. Image of the items which comprise the experimental setup. (**1**) Tube containing the mixture of the saliva sample, *S. marinus* control bacterium and steel beads. (**2**) Hand-held device (Terralyzer), into which the tube is inserted for performing the mechanical lysis and saliva homogenization. (**3**) The OralDisk, where the lysate is pipetted in the chamber indicated by the two white arrows. (**4**) The LabDisk Player instrument that performs the OralDisk processing and real-time PCR.

2.8. Statistics

The *p*-values for *T. forsythia* were calculated using the pairwise Wilcoxon Rank Sum Test (statistical software R [43] including the package *tidyverse* [44]) without correction for multiple testing and with a significance level of 0.05. *p*-values were also calculated for *P. gingivalis* but were higher than 0.05 with both the OralDisk and the lab-based reference method. For these two bacteria, all Cq measurements were available (i.e., no 'ND' values in Supplementary Table S1). For the other eight bacterial species, one or more Cq values were missing and were deemed to be beyond the limit of detection (marked as 'ND' in Supplementary Table S1). These datasets were not used for the calculation of *p*-values. For all species and diagnosis groups, we generated boxplots (Supplementary Figure S1) using Origin®2019 software (version 2019 (9.60)) and included a calculation of the median values and interquartile ranges (IQR) of the Cq results (Supplementary Table S2).

3. Results and Discussion

3.1. Real-Time PCR on the OralDisk

In this study, 25 disks were used to test 24 clinical samples (from seven healthy, nine caries and eight periodontitis patients) and one negative control (H_2O). Each sample was tested once per disk. Figure 4 shows representative OralDisk real-time PCR curves from which the Cq values were calculated for all bacteria that were detected in a single sample. The Cq values of all bacteria in all the samples tested with OralDisks are summarized in Supplementary Table S1. These values were used for the subsequent analysis and the comparison with the lab-based reference and commercial iai PadoTest methods. The methods were compared by means of: (i) the number of assay targets detected as positive and negative by the OralDisk, the iai PadoTest and the lab-based reference method and (ii) the scatter plots of the acquired Cq values for each clinical diagnosis group and for each bacterium for the OralDisk and the lab-based reference method. These analyses were performed in order to examine possible trends in the data.

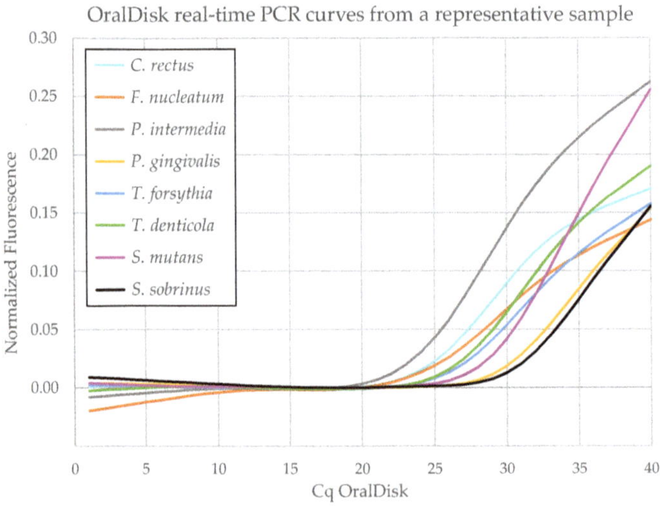

Figure 4. Representative real-time PCR curves for the oral bacteria detected with the OralDisk in one whole saliva sample.

3.2. Performance Comparison between the OralDisk and the Lab-Based Reference Method

Each saliva sample was tested using one OralDisk, which simultaneously screened for ten bacterial species by means of its geometric multiplexing configuration. None of the samples tested were found to contain all ten of the bacterial species screened for. This was in line with our previous findings using a full cohort study, in which the corresponding lab-based PCR reference method was used [22]. In order to assess the degree of qualitative agreement (i.e., bacterial presence/absence) between the OralDisk and the corresponding lab-based reference method, we divided the results (Table 1) into the following four groups:

(a) assay targets detected as positive by both the OralDisk and the lab-based reference (agreement in positive samples: 154/175 (88.0%) cases);
(b) assay targets detected as positive by the OralDisk but negative by the lab-based reference (disagreement in 7/59 (11.9%) cases);
(c) assay targets detected as negative by both the OralDisk and the lab-based reference (agreement in negative samples: 52/59 (88.1%) cases);
(d) assay targets detected as negative by the OralDisk but positive by the lab-based reference (disagreement in 21/175 (12.0%) cases).

A comparison between the OralDisk and the lab-based reference results was performed (i) for each clinical diagnosis group and (ii) for each bacterial species. Regarding the former, the highest agreement between the two methods in terms of positively detected targets (group (a)) was found in the caries samples (91.9%), followed by the healthy samples (85.4%) and, finally, the periodontitis samples (85.0%). Regarding the latter, the positive agreement between the two methods ranged from 85.7% to 100% for *C. rectus*, *P. intermedia*, *P. gingivalis*, *T. denticola*, *S. mutans* and *S. sobrinus*. Lower agreement between the OralDisk and the lab-based reference positives was observed for *F. nucleatum* (16/23 (69.6%) cases) and *Lactobacillus* spp. (7/11 (63.6%) cases).

It is important to mention that each sample was analyzed with three technical replicates with the lab-based PCR reference compared to one with the OralDisk to simulate the POC workflow. In three cases within group (d), the lab-based PCR reference did not give identical results for the triplicates. In two cases, two of three repeats for *F. nucleatum* were detected positive and in one case, one of three repeats for *Lactobacillus* spp. was detected positive. This may imply that these particular PCR assays in those samples were close to the limit of detection for the lab-based reference, which would explain why they were missed by the OralDisk. Finally, the OralDisk detected *A. actinomycetemcomitans* in only one of the four lab-based reference positive samples. In line with data from the preceding clinical study, this species was not often detected among the samples, and when detected, it was often associated with very low levels of target genome equivalents [22]. This species also did not appear to play any discriminatory role between the healthy, caries and periodontitis groups in the aforementioned study and for the recruited age groups [22]. However, *A. actinomycetemcomitans* was included in the panel because it may play a role in cases where early onset periodontitis (i.e., younger patient ages than usual) was suspected.

A possible source of disagreement between the two methods may be the different approaches for bacterial lysis, DNA extraction and purification prior to PCR amplification. The lab-based reference method used an enzymatic lysis methodology (lysozyme, mutanolysin, proteinase K enzymes [22]) with prolonged incubation times, followed by column-based purification. For the OralDisk, mechanical lysis was performed ex situ using a hand-held bead-beating device (Terralyzer, Zymo Research, USA), followed by a magnetic bead-based bind-wash-elute protocol [31] for DNA extraction and purification on the disk. Enzymatic or mechanical lysis may be more efficient, depending on the cell wall properties of certain bacterial species—the OralDisk panel included both Gram-positive and Gram-negative bacteria (Section 2.2).

Discrepancies in DNA extraction efficiency may also be expected between the different approaches (column versus bead-based), as well as between different test devices. These factors may have an impact on the detection of low abundances of bacteria (i.e., higher Cq values). Consequently, targets that are close to the limit of detection of the OralDisk (in its current configuration) may still be detectable by the lab-based reference method. This could partly account for discrepancies between the two datasets where the lab-based reference, but not the OralDisk, appeared to detect certain bacterial species more frequently.

3.3. Performance Comparison between the OralDisk and the Commercial iai PadoTest

The iai PadoTest (iai PadoTest, Institut für Angewandte Immunologie IAI AG, Zuchwil, Switzerland), a commercially available system for the detection of periodontal pathogens, was used to analyze 18 out of the 24 samples. We compared the data obtained by the iai PadoTest and the OralDisk for these 18 samples and for five species per sample (as not all ten species were shared between these two methods). To enable a direct interpretation of the results, we only compared the qualitative outcomes of the methods (i.e., presence/absence of the target bacteria), as the OralDisk in this study did not provide quantitative values for bacteria concentrations. The results from the OralDisk and iai PadoTest are summarized in Table 2. The iai PadoTest positively detected 28 of the microbial target assays, while the OralDisk detected the same 28 and 33 more, thus giving 61 in total. Possible explanations for this discrepancy in detection between the two methods may be the different molecular identification principles and/or assay protocols used [22,27]. In fact, the iai PadoTest is designed to examine gingival crevicular fluid (GCF). However, for better comparability with the lab-based reference and the OralDisk we used a modified protocol based on saliva, as described and discussed previously [22].

3.4. Comparison of Cq Performance

The performance comparison between the OralDisk and the lab-based reference (Table 1) does not consider the Cq values of the two methods. In this section, the Cq values of group (a) (Section 3.2) are therefore shown as scatter plots, in order to observe whether one of the two methods exhibited any trend in Cq for all bacteria in all samples, as well as for all bacteria in each clinical diagnosis group (Figure 5). Each data point corresponds to the detection of a specific bacterial species in a specific sample with both methods. The y-axis error bars are derived from the standard deviations of triplicate (or in some cases duplicate) measurements made with the lab-based reference PCR method [22]. The calculation of such standard deviations was not possible for the x-axis (OralDisk PCR), as each sample was tested with only one OralDisk cartridge.

The diagonal line y = x is shown in the scatter plots simply to assist the observation and assessment of whether the OralDisk or the lab-based reference method show any trend in specific Cq areas. Data points above the y = x line mean that for a specific measurement, Cq(Lab-reference) > Cq(OralDisk), while data points below the y = x line mean that Cq(OralDisk) > Cq(Lab-reference). Interestingly, it appears that for the healthy group, more data points lie below the y = x line, while for the caries and periodontitis groups and for all samples combined, the number of points above and below the x = y line appear to be balanced (Table 3). When deriving a linear fit trendline, the slope of Cq(Lab-reference)/Cq(OralDisk) is very similar between the total, healthy, caries and periodontitis groups (0.34, 0.38, 0.32, 0.36, respectively). This is an indication that the relation between the two methods is consistent and independent of the nature of the three groups included in the study. From qualitative observation, there seems to be a center of the scatterplot clusters at Cq ~ 24. Across the area where the Cq is higher than ~24 (i.e., at lower bacterial concentrations), the OralDisk method tends to generate positive signals at higher Cq values than the lab-based reference. This further supports the hypothesis expressed in Section 3.2, namely that lower bacterial concentrations that can still be detected by the lab-based reference may be missed by the OralDisk. Conversely, across the area where Cq values were lower than ~24 (i.e., at higher bacterial concentrations), increasing PCR signals tend to be observed earlier for the OralDisk.

Table 1. Comparison of the lab-based reference versus the OralDisk for positive/negative results. Each number in the table denotes the number of times an assay target was detected in the OralDisk and lab-based reference experiments. 'Ch.1–Ch10' refers to the numbering of the respective OralDisk chamber where each assay target was detected. 'Pos.' and 'Neg.' are abbreviations for 'positive' and 'negative', respectively.

	Ch.1	Ch.2	Ch.3	Ch.4	Ch.5	Ch.6	Ch.7	Ch.8	Ch.9	Ch.10	
	A. actinomycetemcomitans	*C. rectus*	*F. nucleatum*	*P. intermedia*	*P. gingivalis*	*T. forsythia*	*T. denticola*	*Lactobacillus* spp.	*S. mutans*	*S. sobrinus*	Number of Pos., Neg. over Total Assay Targets
Lab-reference Pos.	4	23 [a]	23 [a]	24	22 [a]	22 [a]	15	11	24	7	175/234
Lab-reference Neg.	20	0	0	0	0	0	9	13	0	17	59/234
(a) OralDisk Pos. and Lab-reference Pos.	1	21	16	22	22	22	15	7	22	6	154/175 (88.0%)
(b) OralDisk Pos. and Lab-reference Neg.	0	0	0	0	0	0	2	5	0	0	7/59 (11.9%)
(c) OralDisk Neg. and Lab-reference Neg.	20	0	0	0	0	0	7	8	0	17	52/59 (88.1%)
(d) OralDisk Neg. and Lab-reference Pos.	3	2	7	2	0	0	0	4	2	1	21/175 (12.0%)

[a] In one disk, signals of *C. rectus* and *F. nucleatum* where not measured, while in two (other) disks, signals of *P. gingivalis* and *T. forsythia* were not measured. For reasons of comparison, these assays were excluded from the lab-based reference.

Table 2. Comparison between the iai PadoTest and the OralDisk. 'Ch.1–Ch10' refers to the numbering of the OralDisk chambers where each bacterium was detected. Grey shading denotes the bacteria that are not included in the iai PadoTest panel. 'Pos.' and 'Neg.' are abbreviations for 'positive' and 'negative', respectively.

	Ch.1	Ch.2	Ch.3	Ch.4	Ch.5	Ch.6	Ch.7	Ch.8	Ch.9	Ch.10	
	A. actinomycetemcomitans	*C. rectus*	*F. nucleatum*	*P. intermedia*	*P. gingivalis*	*T. forsythia*	*T. denticola*	*Lactobacillus* spp.	*S. mutans*	*S. sobrinus*	Number of Pos., Neg. Assay Targets
OralDisk Pos.	1			16	16	16	12				61
iai PadoTest Pos.	0			7	8	3	10				28
OralDisk Neg.	17			2	0	0	6				25
iai PadoTest Neg.	18			11	8	13	8				58

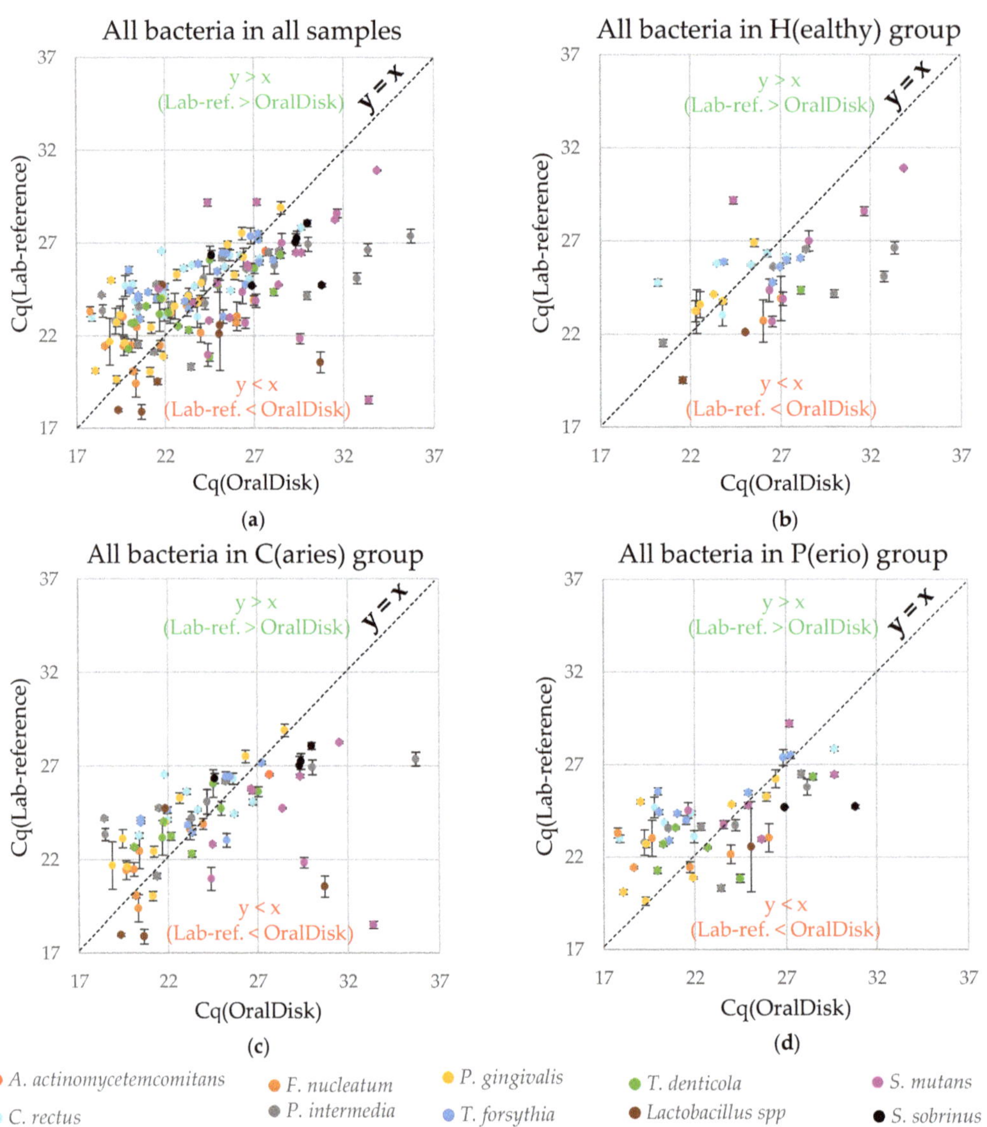

Figure 5. Scatter plots correlating the Cq values from the OralDisk and the lab-based reference for all bacteria: (**a**) in all three groups; (**b**) in the healthy group; (**c**) in the caries group; and (**d**) in the periodontitis group. Similar scatter plots for each individual bacterium are depicted in the Supplementary Figure S2.

As each OralDisk analyzes the presence/absence of ten bacteria simultaneously, the scatter plots in Figure 5 represent the entire panel and contain the collective information from all bacteria. In order to investigate whether some individual bacteria exhibit any specific trends, we composed the scatter plots for each bacterium for all clinical diagnosis groups. From Supplementary Figure S2, we observe different tendencies for some bacteria (quantitatively summarized in Table 3). For example: (i) *C. rectus*, *P. gingivalis* and *T. forsythia* tend to appear in the Cq(Lab-reference) > Cq(OralDisk) area (i.e., the OralDisk amplification curves yielded earlier Cq values than the lab-based reference); (ii) *S. mutans* tends to appear in the Cq(Lab-reference) < Cq(OralDisk) area (i.e., the OralDisk amplification curves yielded later Cq values than the lab-based reference); and (iii) there is no trend for *F. nucleatum*, *P. intermedia* and *T. denticola*.

Table 3. Number of cases where the Cq(Lab-reference) was higher or lower than the Cq(OralDisk). Calculations are based on the scatter plots in Figure 5. The bacteria *A. actinomycetemcomitans*, *Lactobacillus* spp. and *S. sobrinus* are given only indicatively as their total cases were very few.

	Number of Cases with Cq(Lab-Reference) > Cq(OralDisk)		Number of Cases with Cq(Lab-Reference) < Cq(OralDisk)		TOTAL Cases
For all bacteria in all groups	80	51.9%	74	48.1%	154
For all bacteria in healthy group	12	35.3%	22	64.7%	34
For all bacteria in caries group	41	55.9%	30	44.1%	68
For all bacteria in periodontitis group	27	57.7%	22	42.3%	52
For *A. actinomycetemcomitans* in all groups	0	0.0%	1	100.0%	1
For *C. rectus* in all groups	15	71.4%	6	28.6%	21
For *F. nucleatum* in all groups	8	53.3%	7	46.7%	15
For *P. intermedia* in all groups	10	45.5%	12	54.5%	22
For *P. gingivalis* in all groups	17	77.3%	5	22.7%	22
For *T. forsythia* in all groups	16	72.7%	6	27.3%	22
For *T. denitcola* in all groups	8	50.0%	8	50.0%	16
For *Lactobacillus* spp. in all groups	1	14.3%	6	85.7%	7
For *S. mutans* in all groups	4	18.2%	18	81.8%	22
For *S. sobrinus* in all groups	1	16.7%	5	83.3%	6

We further analyzed the Cq values from the OralDisk and the lab-based reference and mapped their distributions per bacterial species for all three clinically diagnosed groups (Supplementary Figure S1). It should be noted that due to the small number of measured samples, clinical conclusions cannot be extracted from the results per se. From the descriptive graphical representation, it can be observed that for *P. intermedia*, *P. gingivalis* and *T. forsythia* (and possibly also *T. denticola*, although N = 2 for the OralDisk), the boxplots would tend to be distinguishable (especially if/when a higher number of samples were tested), even though there are overlapping standard deviations. In cases where testing a higher number of samples would lead to a smaller standard deviation, these four bacteria would possibly be the first ones that the OralDisk would detect as differentiators between clinical diagnosis groups. Within this small-scale study, the tendency towards differentiation seems to be stronger with the OralDisk than with the lab-based reference, as the median lines of the OralDisk results for the different diagnosis groups seem to be further apart (even though not necessarily lower) than those of the lab-based reference, especially between the healthy and the caries groups (actual median values given in Supplementary Table S2). Indicative of this (and being aware of the small number of samples for thorough statistical analysis), for *T. forsythia*, the p-value between the healthy and the caries groups was 0.033 with the OralDisk and 0.29 with the lab-based reference, and between the healthy and the periodontitis groups, it was 0.057 with the OralDisk and 0.29 with the lab-based reference (Section 2.8).

3.5. Overall Evaluation of the OralDisk

The development and implementation of chair-side molecular diagnostics in the field of oral health lags behind many other fields of healthcare, including infectious diseases such as respiratory tract, bloodstream and gastrointestinal infections, for which point-of-care or near-patient systems are commercially available or at the product development stage [45]. However, the socioeconomic burden of oral diseases is high, with expenditure for the treatment of dental diseases reaching EUR 90 billion and additional productivity losses of over EUR 50 billion in EU member states in 2015 [46]. Furthermore, the documented relationship between periodontal and systemic diseases [8,47–50] has started to raise awareness of the importance of oral health and especially of early diagnosis, prevention and post-treatment monitoring.

Table 4 summarizes some existing technologies, together with the bacterial panels they detect. A commercial test for the biomarker-based detection of periodontitis is available from Dentognostics GmbH (PerioSafe®) [51]. It detects a single protein marker, namely the active matrix metalloproteinase-8 (aMMP-8) [52–54] but no bacteria per se. Furthermore, although the molecular-based detection of oral bacteria has been reported by MyPerioPath® [55], HR5™ (High Risk Pathogen Test from Direct Diagnostics [56]) and iai PadoTest [27], all these methods are not chair-side-compatible but laboratory-based. Thus, the samples need to be transported to a laboratory and the results are only available after some days, whereas the OralDisk delivers results in <3 h. The PerioSafe® and PerioPOC® [57,58] require only 5 and 20 min, respectively, due to their lateral flow configuration. However, the panels of both tests are limited: for the former, to a single protein biomarker; for the latter, to five periodontitis pathogens. Additionally, none of the methods in Table 4 detects caries-related bacteria but focus only on periodontitis-associated bacteria. Interestingly, in terms of throughput, all chair-side technologies test one sample per run. For the lab-based technologies, the number of samples tested per run may depend on the logistics of the particular laboratory. The overarching features of the OralDisk platform compared to the listed systems are that it detects ten major caries- and periodontitis-related bacteria simultaneously from non-invasively collected whole saliva using molecular-based detection and a chair-side compatible system.

In terms of specimens, some of the methods listed in Table 4 use paper points from periodontal probing (iai PadoTest, PerioPOC®). However, paper points have the inherent disadvantage that they are invasive and only reflect the local bacterial distributions of specific tooth pockets [12,13]. Saliva is an increasingly popular candidate for analysis as it is easy to collect in tubes even in milliliter volumes and its collection is non-invasive in nature [59,60]. The OralDisk requires 400 µL of saliva (compared to 3–4 drops of oral rinse for PerioSafe® and 160 µL of lysis solution for the immersion of paper points of PerioPOC®, with 20 µL of this volume being added to the lateral flow chip). For the lab-based methods of iai PadoTest, MyPerioPath® and HR5™ the required volume is not known. In any case, the availability of large amounts of saliva makes the required volume of this specimen type less critical in terms of the test to be selected. Non-oral-related diagnostics have also attempted to use saliva to detect biomarkers or nucleic acids in diseases such as COVID-19, type 2 diabetes mellitus, cardiovascular diseases and Alzheimer's disease [61–67]. The specimens used in the current study were whole saliva samples derived from recruited individuals of diverse oral health background (healthy, caries, periodontitis) during a clinical study [22]. Compared to spiked samples with known compositions, this methodology offers decisive advantages: (i) the oral flora in healthy and diseased individuals is best represented in saliva, since it already contains many of the bacteria to be examined and (ii) for POC use, it is essential to ensure that the OralDisk is compatible with the physical, biological and chemical properties of the natural matrix, i.e., whole saliva and not any artificial saliva matrix. Indeed, the compatibility of the LabDisk with this complex matrix paves the way for applicability of the platform in areas that can make use of this abundant and easy-to-acquire specimen.

Table 4. List of basic performance characteristics for a range of molecular-based methods of oral health assessment. 'LFT' stands for 'Lateral Flow Test'. Bacteria that are not included in the panels of some technologies are marked with 'X'.

Characteristics	PerioSafe®	iai PadoTest	MyPerioPath®	PerioPOC®	HR5™	OralDisk
Saliva sampling	Yes (Oral rinse)	No (Probe)	Yes	No (Probe)	Yes	Yes
Chair-side compatible	Yes	No (Requires lab)	No (Requires lab)	Yes	No (Requires lab)	Yes [a]
Quantitative capacity	Yes/No (LFT) [b]	Yes (qPCR)	Yes (qPCR)	No (LFT) [c]	Yes (qPCR)	Yes (qPCR)
Caries panel included	No	No	No	No	No	Yes
Number of bacteria in panel, tested per run	None (detects aMMP-8 biomarker)	6	11	5	5	10
Bacteria panel						
Aggregatibacter actinomycetemcomitans	-	X	X	X	X	X
Tannerella forsythia	-	X	X	X	X	X
Porphyromonas gingivalis	-	X	X	X	X	X
Treponema denticola	-	X	X	X	-	X
Prevotella intermedia	-	X	X	X	X	X
Filifactor alocis	-	X	-	-	-	-
Fusobacterium nucleatum	-	-	X	-	X	X
Campylobacter rectus	-	-	X	-	-	X
Capnocytophaga species (gingivalis, ochracea, sputigena)	-	-	X	-	-	-
Streptococcus mutans	-	-	-	-	-	X
Streptococcus sobrinus	-	-	-	-	-	X
Lactobacillus spp.	-	-	-	-	-	X
Peptostreptococcus micros	-	-	X	-	-	-
Eikenella corrodens	-	-	X	-	-	-
Eubacterium nodatum	-	-	X	-	-	-

[a] Once the mechanical lysis of bacteria is transferred from the Terralyzer into the OralDisk [68]. [b] Lateral Flow Test (protein detection) [50–53]. Visual readout provides qualitative analysis. Quantitative analysis requires a processing instrument. [c] Lateral Flow Test (nucleic acid hybridization). Visual readout provides qualitative analysis [58].

An outstanding advantage of the OralDisk is the automated sample processing and detection of the assay targets. The system is based on a microfluidic platform that integrates all the necessary biochemical reagents and operations in a protocol requiring minimal and simple hands-on work. The only short manual step is the homogenization of whole saliva using a bead-beating hand-held device (Terralyzer, Zymo Research, USA), which combines homogenization and bacterial lysis in one step (saliva bead-beating has also been previously reported as a lysis method for lab-based downstream analysis [69,70]). Even in such a configuration, the hands-on work was estimated to be only 10 min, including the ex situ saliva homogenization and mixing with the control bacterium *S. marinus*, the pipetting into the disk, and the insertion of the latter into the processing instrument. This time is far shorter than the manual sample preparation time (at least 1.5 h) that is required prior to a laboratory-based PCR. Additionally, previously published work by the authors has already demonstrated the use of an on-disk microfluidic unit operation for saliva homogenization [68] (as a pre-analytic approach for downstream protein analysis), which uses disk-integrated magnets actuated by magnets above the disk [33]. This disk-compatible approach is planned to be tested in the future as an in situ lysis and homogenization step. It will be integrated with the DNA purification module and will thus replace the only manual step (i.e., bead-beating with the mobile Terralyzer device) of the current OralDisk protocol. Further automation can be achieved if the aforementioned in situ mechanical lysis step is combined with downstream dilution and direct amplification (after biochemical optimization), without the need for a bead-based bind-wash-elute protocol. This modification is expected to reduce the total time-to-result by approximately 0.5 h. In addition, a drastic reduction in the time-to-result is also planned to be achieved by means of a new LabDisk Player that will implement contact heating using Peltier elements instead of air heating. This method is expected to reduce the PCR time from ~2 h to < 0.5 h, thereby resulting in an estimated total time-to-result of <1 h.

The major feature of automation of the OralDisk workflow leads also to a drastic reduction of hands-on work and time that is required by laboratory personnel, which can result in a reduction of the overall 'hidden costs' of laboratory workflows. Even though we cannot assess the end-user transfer price of the OralDisk (as it is currently at development stage), the actual cost-driver aspects have been identified, as well as the actions that need to be taken at the product development stage, e.g., re-assignment from thermoforming to injection molding fabrication method (for manufacturing of hundreds of thousands of OralDisk cartridges); screening for more cost-effective polymeric cartridge materials and the complete automation of the entire manufacturing workflow. In any case, the price of such a diagnostic system should not be compared with single-biomarker or single-parameter detection systems, as the OralDisk offers an increased range of information for a high number of pathogens and for both periodontitis and caries diseases. In fact, the realistic goal for the OralDisk is that the cost per pathogen tested becomes significantly lower than that using a laboratory test.

The analytics were performed on a batch of samples sufficiently representative to allow the platform to demonstrate its performance and applicability for the detection of a broad panel of ten Gram-negative and Gram-positive bacterial species. As previously mentioned, this means that both periodontitis and caries can be monitored simultaneously on a single OralDisk. Periodontitis has been mainly associated with Gram-negative anaerobic bacteria, while caries has been mainly associated with Gram-positive carbohydrate-fermenting bacteria [71]. The OralDisk contains one of the broadest panels available for bacterial detection among comparable systems (Table 4). This is achievable due to the multiple reaction chambers (#7, Figure 2) that enable geometric multiplexing. The level of multiplexing can be further increased by multiple wavelength detection in the same chamber (color multiplexing). For example, in this publication the Gram-positive bacterium *S. marinus* was included as a process control and underwent the same lysis, extraction, purification and amplification processes as the oral bacteria present in the test sample. *S. marinus* primers/probes were included together with the bacteria-specific primers/probes in each reaction chamber, providing an indication that the internal biochemical and microfluidic processes had been correctly implemented. In the disks that we tested, there was no case where the *S. marinus* was not detected, thereby confirming qualitatively that the sample-to-answer process functioned successfully. The simultaneous detection of several bacteria, including the process control, within a single sample is a major achievement of the platform, which until now had demonstrated its capability for infectious diseases that are associated with either one or two pathogens, such as sepsis [18], tropical infections [20] and respiratory tract infections [19,21].

The ~90% agreement between the OralDisk and the lab-based reference amongst targets detected as positive and negative indicates that the OralDisk platform may be suitable for accurate molecular-based oral bacteria detection. Any differences could be attributable to the different lysis, extraction and/or test device approaches utilized by the two methods. An increase in the test sensitivity of the OralDisk might be achievable using a pre-amplification step, which in a past application of the LabDisk technology was shown to detect down to a few bacteria/mL [18]. However, high sensitivity may not be particularly crucial for oral health screening purposes. For example, the oral microbiota is diverse in both healthy and diseased patients, including both commensals and opportunistic pathogens. It is the dynamic changes over time of bacteria among the oral microbiota, rather than their mere presence/absence, that drives the dysbiotic changes that lead to oral disease [72]. The fact that the OralDisk has demonstrated its capacity to analyze oral bacteria provides a basis upon which the platform can be implemented in time-course studies associated with patient monitoring, similar to the rationale reported by Paqué et al. [73]. In such future studies, we will have the opportunity to proceed to the quantification of bacteria, using OralDisk-derived calibration curves to convert Cq values into numeric bacterial loads.

The LabDisk and the corresponding customized processing device used in this work are amenable to further performance improvements (such as inclusion of the lysis step on the disk and the subsequent reduction of the time-to-result) and may be applicable as an auxiliary tool for oral microbial screening in dental POC settings. Finally, for truly holistic monitoring of oral health, bacteria-based microbiological examination in the OralDisk will be combined with protein biomarker concentration monitoring. The LabDisk has already been shown to be compatible with immunoassays by the running of a basic reaction (without detection) as a proof-of-principle demonstration [74]. A newly developed and tested bead-based immunoassay with oral biomarkers [75] will enable an immunoassay disk to be run in the same instrument as the PCR disk. This will increase the interoperability between immunological and microbiological diagnostic outputs using this platform [76]. Subsequently, by using combined computational technologies and the OralDisk device, new diagnostic predictive models of disease development and progression may be generated [73]. Such developments will enable evidence-based pre- and post-treatment monitoring of a range of oral (OralDisk) and non-oral (application-specific LabDisk) diseases.

4. Conclusions

This technical feasibility study demonstrated the ability of the OralDisk to detect a broad range of seven periodontitis- and three caries-related bacteria from whole saliva samples in < 3 h, using an automated platform for molecular-based detection. Importantly, the platform was proven to be compatible with saliva as a sample matrix for the first time, which paves the way for (i) 'modernization' of contemporary oral health by using a molecular-based diagnostic tool for saliva (instead of probe-based) close to the chair-side practice, acting supportively to 'traditional' methods and (ii) implementation in further areas of non-invasive saliva-based diagnostics. Compared to the lab-based reference test, the OralDisk results showed ~90% agreement amongst targets detected as positive and negative. We observed that higher levels of bacteria (Cq values < 24) generally resulted in lower Cq values with the OralDisk compared to the lab-based reference method. This was primarily observed for *C. rectus*, *P. gingivalis* and *T. forsythia*. On the other hand, lower bacterial levels (Cq values > 24) were mostly detected later with the OralDisk compared to the lab-based reference method. This was primarily observed for *S. mutans*.

Optimization of the platform towards automation and reduction of the time-to-result will further increase its potential for adoption at the chair-side. The future goals are to apply the OralDisk platform in larger clinical studies and to determine its clinical potential in (i) providing early detection/prevention of oral diseases before they are detectable with a conventional clinical examination or radiographic methods and (ii) monitoring progress during and after periodontal treatment by providing quantitative information on changes in bacterial load. Furthermore, such a diagnostic system could potentially contribute to more informed decision making regarding the prescription of oral antibiotics, while also acting as an early warning system for underlying systemic diseases [77] such as diabetes and cardiovascular diseases, which have previously been correlated with periodontitis.

Supplementary Materials: The following are available online at https://www.mdpi.com/article/10.3390/bios11110423/s1, Table S1: Raw experimental data (Cq values) from the OralDisk and the lab-based reference method, as well as qualitative data from the iai PadoTest. Figure S1: Boxplots of the Cq values for each bacterium detected by the OralDisk and the lab-based reference. Table S2: Median and interquartile range (IQR, describing the middle 50% of values when ordered from lowest to highest) of the Cq values from the OralDisk and lab-based reference measurements. Figure S2: Scatter plots correlating the Cq values from the lab-based reference and the OralDisk for all bacteria in the healthy, caries and periodontitis groups.

Author Contributions: Conceptualization, D.B., B.J., R.Z., C.H., J.R.P., P.N.P., F.J.W., N.B., G.N.B. and K.M.; methodology, D.B., B.J., M.S., J.L., R.Z., C.H., J.R.P., P.N.P., V.R., N.B., G.N.B. and K.M.; software, S.H., C.H., J.S.J., A.S., A.M. and T.B.; validation, D.B., P.N.P., T.A., J.S.J., P.K., P.R.S., T.T., F.J.W. and V.R.; formal analysis, D.B., C.H., P.N.P., J.S.J., A.S., V.R., K.B., G.N.B. and K.M.; investigation, D.B., B.J., M.S., J.L., M.R., S.H., C.H., P.N.P., J.S.J., W.E.K. and K.B.; resources, M.R., R.Z., P.N.P., T.A., J.S.J., P.K., P.R.S., T.T., F.J.W., V.R., D.S. and M.K.; data curation, C.H., J.S.J. and A.S.; writing—original draft preparation, D.B., G.N.B. and K.M.; writing—review and editing, D.B., B.J., M.S., J.L., S.H., N.P., F.v.S., R.Z., C.H., J.R.P., P.N.P., T.A., J.S.J., P.K., P.R.S., T.T., F.J.W., W.E.K., J.P.H., V.R., A.M., T.B., M.K., K.B., N.B., G.N.B. and K.M.; visualization, D.B., S.H., C.H., J.R.P., W.E.K. and K.M.; supervision, N.P., F.v.S., R.Z., J.R.P., P.N.P., T.A., P.R.S., T.T., J.P.H., D.S., N.B., G.N.B. and K.M.; project administration, R.Z., J.R.P., P.N.P., T.A., T.T., D.S., M.K., N.B., G.N.B. and K.M.; funding acquisition, J.R.P., P.N.P., T.A., F.J.W., J.P.H., A.M., M.K., N.B., G.N.B. and K.M. All authors have read and agreed to the published version of the manuscript.

Funding: This project has received funding from the European Union's Horizon 2020 research and innovation programme under grant agreement No 633780 ('DIAGORAS' project), as well as the KI/SLL Styrgruppen för Odontologisk Forskning (SOF) Dnr. 4-823/2019. The article processing charge was funded by the Baden-Wuerttemberg Ministry of Science, Research and Art and the University of Freiburg in the funding programme Open Access Publishing.

Institutional Review Board Statement: The study was conducted according to the guidelines of the Declaration of Helsinki and approved by the local Swiss Ethics Committee with BASEC-No. 2016-00435 and the date of approval: 9 January 2016.

Informed Consent Statement: Informed consent was obtained from all the subjects involved in the study.

Data Availability Statement: Data is contained within the article or in the supplementary material.

Acknowledgments: The authors would like to acknowledge the Hahn-Schickard Lab-on-a-Chip Foundry Service for LabDisk production. We would like to thank Helga Lüthi-Schaller for her excellent laboratory and technical support during the experiments of the clinical study. We would also like to thank Jennifer Berkman and Claudia Steffen from Thermo Fisher Scientific, who greatly assisted this research by providing access and tailoring a custom TaqMan® Lyophilized 1-Step qPCR Master Mix towards the needs of this project. We are immensely grateful for their constructive and prompt support.

Conflicts of Interest: The authors declare no conflict of interest. The funders had no role in the design of the study; in the collection, analyses, or interpretation of data; in the writing of the manuscript, or in the decision to publish the results.

References

1. Dental Diseases and Oral Health. Available online: www.who.int/oral_health/publications/en/orh_fact_sheet.pdf (accessed on 30 September 2021).
2. Listl, S.; Grytten, J.I.; Birch, S. What is health economics? *Community Dent. Health* **2019**, *36*, 262–274. [PubMed]
3. Belibasakis, G.N.; Lund, B.; Krüger Weiner, C.; Johannsen, B.; Baumgartner, B.; Manoil, C.; Hultin, M.; Mitsakakis, K. Healthcare Challenges and Solutions in Dental Practice: Assessing Oral Antibiotic Resistances by Contemporary Point-of-Care Solutions. *Antibiotics* **2020**, *9*, 810. [CrossRef] [PubMed]
4. Suda, K.J.; Calip, G.S.; Zhou, J.F.; Rowan, S.; Gross, A.E.; Hershow, R.C.; Perez, R.I.; McGregor, J.C.; Evans, C.T. Assessment of the appropriateness of antibiotic prescriptions for infection prophylaxis before dental procedures, 2011 to 2015. *JAMA Netw. Open* **2019**, *2*, e193909. [CrossRef]
5. Bernabe, E.; Marcenes, W.; Hernandez, C.R.; Bailey, J.; Abreu, L.G.; Alipour, V.; Amini, S.; Arabloo, J.; Arefi, Z.; Arora, A.; et al. Global, Regional, and National Levels and Trends in Burden of Oral Conditions from 1990 to 2017: A Systematic Analysis for the Global Burden of Disease 2017 Study. *J. Dent. Res.* **2020**, *99*, 362–373. [PubMed]
6. WHO, Regional Office for Europe, Oral Health, Data and Statistics. Available online: http://www.euro.who.int/en/health-topics/disease-prevention/oral-health/data-and-statistics (accessed on 30 September 2021).
7. NIDCR, NIH, Periodontal (Hum) Disease. Available online: http://www.nidcr.nih.gov/OralHealth/Topics/GumDiseases/PeriodontalGumDisease.htm (accessed on 30 September 2021).
8. Seitz, M.W.; Listl, S.; Bartols, A.; Schubert, I.; Blaschke, K.; Haux, C.; Van Der Zande, M.M. Current Knowledge on Correlations Between Highly Prevalent Dental Conditions and Chronic Diseases: An Umbrella Review. *Prev. Chronic. Dis.* **2019**, *16*, 180641. [CrossRef]
9. D'Aiuto, F.; Gable, D.; Syed, Z.; Allen, Y.; Wanyonyi, K.L.; White, S.; Gallagher, J.E. Evidence summary: The relationship between oral diseases and diabetes. *Br. Dent. J.* **2017**, *222*, 944–948. [CrossRef]
10. Bui, F.Q.; Almeida-da-Silva, C.L.C.; Huynh, B.; Trinh, A.; Liu, J.; Woodward, J.; Asadi, H.; Ojcius, D.M. Association between periodontal pathogens and systemic disease. *Biomed. J.* **2019**, *42*, 27–35. [CrossRef]
11. Petersen, P.E. Global policy for improvement of oral health in the 21st century–implications to oral health research of World Health Assembly 2007, World Health Organization. *Community Dent. Oral Epidemiol.* **2009**, *37*, 1–8. [CrossRef]
12. Listgarten, M.A. Periodontal probing: What does it mean? *J. Clin. Periodontol.* **1980**, *7*, 165–176. [CrossRef]
13. Hefti, A.F. Periodontal probing. *Crit. Rev. Oral. Biol. Med.* **1997**, *8*, 336–356. [CrossRef]
14. Olsen, I. Update on bacteraemia related to dental procedures. *Transfus. Apher. Sci.* **2008**, *39*, 173–178. [CrossRef]
15. Daly, C.G.; Mitchell, D.H.; Highfield, J.E.; Grossberg, D.E.; Stewart, D. Bacteremia due to periodontal probing: A clinical and microbiological investigation. *J. Periodontol.* **2001**, *72*, 210–214. [CrossRef]
16. Preshaw, P.M. Detection and diagnosis of periodontal conditions amenable to prevention. *BMC Oral Health* **2015**, *15*, S5. [CrossRef] [PubMed]
17. Mitsakakis, K.; Stumpf, F.; Strohmeier, O.; Klein, V.; Mark, D.; von Stetten, F.; Peham, J.R.; Herz, C.; Tawakoli, P.N.; Wegehaupt, F.; et al. Chair/bedside diagnosis of oral and respiratory tract infections, and identification of antibiotic resistances for personalised monitoring and treatment. *Stud. Health Technol. Inform.* **2016**, *224*, 61–66. [PubMed]
18. Czilwik, G.; Messinger, T.; Strohmeier, O.; Wadle, S.; von Stetten, F.; Paust, N.; Roth, G.; Zengerle, R.; Saarinen, P.; Niittymaki, J.; et al. Rapid and fully automated bacterial pathogen detection on a centrifugal-microfluidic LabDisk using highly sensitive nested PCR with integrated sample preparation. *Lab Chip* **2015**, *15*, 3749–3759. [CrossRef] [PubMed]
19. Stumpf, F.; Schwemmer, F.; Hutzenlaub, T.; Baumann, D.; Strohmeier, O.; Dingemanns, G.; Simons, G.; Sager, C.; Plobner, L.; von Stetten, F.; et al. LabDisk with complete reagent prestorage for sample-to-answer nucleic acid based detection of respiratory pathogens verified with influenza A H3N2 virus. *Lab Chip* **2016**, *16*, 199–207. [CrossRef] [PubMed]
20. Hin, S.; Lopez-Jimena, B.; Bakheit, M.; Klein, V.; Stack, S.; Fall, C.; Sall, A.; Enan, K.; Mustafa, M.; Liz Gillies, L.; et al. Fully automated point-of-care differential diagnosis of acute febrile illness. *PLoS Negl. Trop. Dis.* **2021**, *15*, e0009177. [CrossRef]
21. Rombach, M.; Hin, S.; Specht, M.; Johannsen, B.; Lüddecke, J.; Paust, N.; Zengerle, R.; Roux, L.; Sutcliffe, T.; Peham, J.R.; et al. RespiDisk: A Point-of-Care platform for fully automated detection of respiratory tract infection pathogens in clinical samples. *Analyst* **2020**, *145*, 7040–7047. [CrossRef]
22. Paqué, P.N.; Herz, C.; Jenzer, J.S.; Wiedemeier, D.; Attin, T.; Bostanci, N.; Belibasakis, G.N.; Bao, K.; Körner, P.; Fritz, T.; et al. Microbial analysis of saliva to identify oral diseases using a point-of-care compatible qPCR assay. *J. Clin. Med.* **2020**, *9*, 2945. [CrossRef]
23. Lamont, R.J.; Hajishengallis, G. Polymicrobial synergy and dysbiosis in inflammatory disease. *Trends Mol. Med.* **2015**, *21*, 172–183. [CrossRef]
24. Bostanci, N.; Bao, K.; Greenwood, D.; Silbereisen, A.; Belibasakis, G.N. Periodontal disease: From the lenses of light microscopy to the specs of proteomics and next-generation sequencing. *Adv. Clin. Chem.* **2019**, *93*, 263–290. [PubMed]

25. Socransky, S.S.; Haffajee, A.D.; Cugini, M.A.; Smith, C.; Kent Jr, R.L. Microbial complexes in subgingival plaque. *J. Clin. Periodontol.* **1998**, *25*, 134–144. [CrossRef] [PubMed]
26. Haffajee, A.D.; Socransky, S.S.; Patel, M.R.; Song, X. Microbial complexes in supragingival plaque. *Oral Microbiol. Immunol.* **2008**, *23*, 196–205. [CrossRef] [PubMed]
27. Website of iai PadoTest. Available online: https://www.padotest.ch/en (accessed on 30 September 2021).
28. Belibasakis, G.N.; Schmidlin, P.R.; Sahrmann, P. Molecular microbiological evaluation of subgingival biofilm sampling by paper point and curette. *APMIS* **2014**, *122*, 347–352. [CrossRef]
29. Yi, H.; Schumann, P.; Sohn, K.; Chun, J. Serinicoccus marinus gen. nov., sp. nov., a novel actinomycete with L-ornithine and L-serine in the peptidoglycan. *Int. J. Syst. Evol. Microbiol.* **2004**, *54*, 1585–1589. [CrossRef]
30. Strohmeier, O.; Keller, M.; Schwemmer, F.; Zehnle, S.; Mark, D.; von Stetten, F.; Zengerle, R.; Paust, N. Centrifugal microfluidic platforms: Advanced unit operations and applications. *Chem. Soc. Rev.* **2015**, *44*, 6187–6229. [CrossRef]
31. Boom, R.; Sol, M.M.; Jansen, C.L.; Wertheim-van Dillen, P.M.; van der Noordaa, J. Rapid and simple method for purification of nucleic acids. *J. Clin. Microbiol.* **1990**, *28*, 495–503. [CrossRef]
32. Van Oordt, T.; Barb, Y.; Smetana, J.; Zengerle, R.; von Stetten, F. Miniature stick-packaging—An industrial technology for pre-storage and release of reagents in lab-on-a-chip systems. *Lab Chip* **2013**, *13*, 2888–2892. [CrossRef]
33. Hin, S.; Paust, N.; Rombach, M.; Lüddecke, J.; Specht, M.; Zengerle, R.; Mitsakakis, M. Minimizing ethanol carry-over in centrifugal microfluidic nucleic acid extraction by advanced bead handling and management of diffusive mass transfer. In Proceedings of the 20th International Conference on Solid-State Sensors, Actuators and Microsystems & Eurosensors XXXIII, Berlin, Germany, 23–27 June 2019. [CrossRef]
34. Keller, M.; Czilwik, G.; Schott, J.; Schwarz, I.; Dormanns, K.; von Stetten, F.; Zengerle, R.; Paust, N. Robust temperature change rate actuated valving and switching for highly integrated centrifugal microfluidics. *Lab Chip* **2017**, *17*, 864–875. [CrossRef]
35. Zehnle, S.; Schwemmer, F.; Roth, G.; von Stetten, F.; Zengerle, R.; Paust, N. Centrifugo-dynamic inward pumping of liquids on a centrifugal microfluidic platform. *Lab Chip* **2012**, *12*, 5142–5145. [CrossRef]
36. Hin, S.; Paust, N.; Keller, M.; Rombach, M.; Strohmeier, O.; Zengerle, R.; Mitsakakis, K. Temperature change rate actuated bubble mixing for homogeneous rehydration of dry pre-stored reagents in centrifugal microfluidics. *Lab Chip* **2018**, *18*, 362–370. [CrossRef] [PubMed]
37. Mark, D.; Weber, P.; Lutz, S.; Focke, M.; Zengerle, R.; von Stetten, F. Aliquoting on the centrifugal microfluidic platform based on centrifugo-pneumatic valves. *Microfluid. Nanofluid.* **2011**, *10*, 1279–1288. [CrossRef]
38. Focke, M.; Stumpf, F.; Faltin, B.; Reith, P.; Bamarni, D.; Wadle, S.; Müller, C.; Reinecke, H.; Schrenzel, J.; Francois, P. Microstructuring of polymer films for sensitive genotyping by real-time PCR on a centrifugal microfluidic platform. *Lab Chip* **2010**, *10*, 2519–2526. [CrossRef]
39. Focke, M.; Kosse, D.; Al-Bamerni, D.; Lutz, S.; Müller, C.; Reinecke, H.; Zengerle, R.; von Stetten, F. Microthermoforming of microfluidic substrates by soft lithography (µTSL): Optimization using design of experiments. *J. Micromech. Microeng.* **2011**, *21*, 115002. [CrossRef]
40. Hahn-Schickard, Lab-on-a-Chip Foundry Service. Available online: https://www.hahn-schickard.de/en/production/lab-on-a-chip-foundry (accessed on 30 September 2021).
41. Rombach, M.; Kosse, D.; Faltin, B.; Wadle, S.; Roth, G.; Zengerle, R.; von Stetten, F. Real-time stability testing of air-dried primers and fluorogenic hydrolysis probes stabilized by trehalose and xanthan. *BioTechniques* **2014**, *57*, 151–155. [CrossRef]
42. Hin, S.; Baumgartner, D.; Specht, M.; Lüddecke, J.; Arjmand, E.M.; Johannsen, B.; Schiedel, L.; Rombach, M.; Paust, N.; von Stetten, F.; et al. VectorDisk: A Microfluidic Platform Integrating Mosquito Vector Markers for Evidence Based Control Applications. *Processes* **2020**, *8*, 1677. [CrossRef]
43. R Core Team. *R: A Language and Environment for Statistical Computing*; R Foundation for Statistical Computing: Vienna, Austria, 2015. Available online: https://www.R-project.org/ (accessed on 23 October 2021).
44. Wickham, H.; Averick, M.; Bryan, J.; Chang, W.; D'Agostino McGowan, L.; François, R.; Grolemund, G.; Hayes, A.; Henry, L.; Hester, J.; et al. Welcome to the tidyverse. *J. Open Source Softw.* **2019**, *4*, 1686. [CrossRef]
45. Mitsakakis, K.; D'Acremont, V.; Hin, S.; von Stetten, F.; Zengerle, R. Diagnostic tools for tackling febrile illness and enhancing patient management. *Microelectron. Eng.* **2018**, *201*, 26–59. [CrossRef]
46. Righolt, A.; Jevdejvic, M.; Marcenes, W.; Listl, S. Global-, Regional-, and Country-Level Economic Impacts of Dental Diseases in 2015. *J. Dent. Res.* **2018**, *97*, 501–507. [CrossRef]
47. Cullinan, M.P.; Ford, P.J.; Seymour, G.J. Periodontal disease and systemic health: Current status. *Aust. Dent. J.* **2009**, *54* Suppl. 1, S62–S69. [CrossRef]
48. Seymour, G.J.; Ford, P.J.; Cullinan, M.P.; Leishman, S.; Yamazaki, K. Relationship between periodontal infections and systemic disease. *Clin. Microbiol. Infect.* **2007**, *13*, 3–10. [CrossRef]
49. Grigoriadis, A.; Sorsa, T.; Raisanen, I.; Parnanen, P.; Tervahartiala, T.; Sakellari, D. Prediabetes/Diabetes Can Be Screened at the Dental Office by a Low-Cost and Fast Chair-Side/Point-of-Care aMMP-8 Immunotest. *Diagnostics* **2019**, *9*, 151. [CrossRef]
50. Smits, K.P.J.; Listl, S.; Plachokova, A.S.; Van der Galien, O.; Kalmus, O. Effect of periodontal treatment on diabetes-related healthcare costs: A retrospective study. *BMJ Open Diab. Res. Care* **2020**, *8*, e001666. [CrossRef]
51. Website of Dentognostics GmbH. Available online: https://www.dentognostics.de/en/ (accessed on 30 September 2021).

52. Leppilahti, J.M.; Ahonen, M.M.; Hernandez, M.; Munjal, S.; Netuschil, L.; Uitto, V.J.; Sorsa, T.; Mantyla, P. Oral rinse MMP-8 point-of-care immuno test identifies patients with strong periodontal inflammatory burden. *Oral Dis.* **2011**, *17*, 115–122. [CrossRef] [PubMed]

53. Borujeni, S.I.; Mayer, M.; Eickholz, P. Activated matrix metalloproteinase-8 in saliva as diagnostic test for periodontal disease? A case-control study. *Med. Microbiol. Immunol.* **2015**, *204*, 665–672. [CrossRef]

54. Al-Majid, A.; Alassiri, S.; Rathnayake, N.; Tervahartiala, T.; Gieselmann, D.R.; Sorsa, T. Matrix Metalloproteinase-8 as an Inflammatory and Prevention Biomarker in Periodontal and Peri-Implant Diseases. *Int. J. Dent.* **2018**, *2018*, 7891323. [CrossRef]

55. Website of OralDNA Labs, MyPerioPath®. Available online: https://www.oraldna.com/test/myperiopath/ (accessed on 30 September 2021).

56. Website of Direct Diagnostics. Available online: https://www.directdiagnostics.com/hr5 (accessed on 30 September 2021).

57. Website of PerioPOC®. Available online: https://en.periopoc.com/ (accessed on 30 September 2021).

58. Arweiler, N.B.; Marx, V.K.; Laugisch, O.; Sculean, A.; Auschill, T.M. Clinical evaluation of a newly developed chairside test to determine periodontal pathogens. *J. Periodontol.* **2019**, *91*, 387–395. [CrossRef]

59. Bellagambi, F.G.; Lomonaco, T.; Salvo, P.; Vivaldi, F.; Hangouet, M.; Ghimenti, S.; Biagini, D.; Di Francesco, F.; Fuoco, R.; Errachid, A. Saliva sampling: Methods and devices. An overview. *TrAC Trends Anal. Chem.* **2020**, *124*, 115781. [CrossRef]

60. Khurshid, Z.; Zohaib, S.; Najeeb, S.; Zafar, M.S.; Slowey, P.D.; Almas, K. Human Saliva Collection Devices for Proteomics: An Update. *Int. J. Mol. Sci.* **2016**, *17*, 846. [CrossRef]

61. Khanna, P.; Walt, D.R. Salivary diagnostics using a portable point-of-service platform: A Review. *Clin. Ther.* **2015**, *37*, 498–504. [CrossRef] [PubMed]

62. Williams, E.; Bond, K.; Zhang, B.; Putland, M.; Williamson, D.A. Saliva as a Noninvasive Specimen for Detection of SARS-CoV-2. *J. Clin. Microbiol.* **2020**, *58*, e00776-20. [CrossRef] [PubMed]

63. To, K.K.-W.; Tsang, O.T.-Y.; Yip, C.C.-Y.; Chan, K.-H.; Wu, T.-C.; Chan, J.M.-C.; Leung, W.-S.; Chik, T.S.-H.; Choi, C.Y.-C.; Kandamby, D.H.; et al. Consistent Detection of 2019 Novel Coronavirus in Saliva. *Clin. Infect. Dis.* **2020**, *71*, 841–843. [CrossRef] [PubMed]

64. Jacobs, R.; Maasdorp, E.; Malherbe, S.; Loxton, A.G.; Stanley, K.; van der Spuy, G.; Walzl, G.; Chegou, N.N. Diagnostic Potential of Novel Salivary Host Biomarkers as Candidates for the Immunological Diagnosis of Tuberculosis Disease and Monitoring of Tuberculosis Treatment Response. *PLoS ONE* **2016**, *11*, e0160546. [CrossRef] [PubMed]

65. Khan, R.S.; Khurshid, Z.; Asiri, F.Y.I. Advancing Point-of-Care (PoC) Testing Using Human Saliva as Liquid Biopsy. *Diagnostics* **2017**, *7*, 39. [CrossRef]

66. Rathnayake, N.; Gieselmann, D.R.; Heikkinen, A.M.; Tervahartiala, T.; Sorsa, T. Salivary Diagnostics: Point-of-Care diagnostics of MMP-8 in dentistry and medicine. *Diagnostics* **2017**, *7*, 7. [CrossRef] [PubMed]

67. Ji, S.; Choi, Y. Point-of-care diagnosis of periodontitis using saliva: Technically feasible but still a challenge. *Front. Cell. Infect. Microbiol.* **2015**, *5*, 65. [CrossRef]

68. Johannsen, B.; Müller, L.; Baumgartner, D.; Karkossa, L.; Fruh, S.M.; Bostanci, N.; Karpisek, M.; Zengerle, R.; Paust, N.; Mitsakakis, K. Automated Pre-Analytic Processing of Whole Saliva Using Magnet-Beating for Point-of-Care Protein Biomarker Analysis. *Micromachines* **2019**, *10*, 833. [CrossRef]

69. De Boer, R.; Peters, R.; Gierveld, S.; Schuurman, T.; Kooistra-Smid, M.; Savelkoul, P. Improved detection of microbial DNA after bead-beating before DNA isolation. *J. Microbiol. Methods* **2010**, *80*, 209–211. [CrossRef]

70. Li, X.L.; Bosch-Tijhof, C.J.; Wei, X.; de Soet, J.J.; Crielaard, W.; van Loveren, C.; Deng, D.M. Efficiency of chemical versus mechanical disruption methods of DNA extraction for the identification of oral Gram-positive and Gram-negative bacteria. *J. Int. Med. Res.* **2020**, *48*, 0300060520925594.

71. Larsen, T.; Fiehn, N.E. Dental biofilm infections—An update. *Apmis* **2017**, *125*, 376–384. [CrossRef]

72. Belibasakis, G.N.; Bostanci, N.; Marsh, P.D.; Zaura, E. Applications of the oral microbiome in personalized dentistry. *Arch. Oral Biol.* **2019**, *104*, 7–12. [CrossRef]

73. Paqué, P.N.; Herz, C.; Wiedemeier, D.B.; Mitsakakis, K.; Attin, T.; Bao, K.; Belibasakis, G.N.; Hays, J.P.; Jenzer, J.S.; Kaman, W.E.; et al. Salivary Biomarkers for Dental Caries Detection and Personalized Monitoring. *J. Pers. Med.* **2021**, *11*, 235. [CrossRef]

74. Zhao, Y.; Czilwik, G.; Klein, V.; Mitsakakis, K.; Zengerle, R.; Paust, N. C-reactive protein and interleukin 6 microfluidic immunoassays with on-chip pre-stored reagents and centrifugo-pneumatic liquid control. *Lab Chip* **2017**, *17*, 1666–1677. [CrossRef]

75. Johannsen, J.; Karpíšek, M.; Baumgartner, D.; Klein, V.; Bostanci, N.; Paust, N.; Früh, S.M.; Zengerle, R.; Mitsakakis, K. One-step, wash-free, bead-based immunoassay employing bound-free phase detection. *Anal. Chim. Acta* **2021**, *1153*, 338280. [CrossRef]

76. Mitsakakis, K. Novel lab-on-a-disk platforms: A powerful tool for molecular fingerprinting of oral and respiratory tract infections. *Expert Rev. Mol. Diagn.* **2021**, *21*, 523–526. [CrossRef] [PubMed]

77. Seitz, M.W.; Haux, C.; Smits, K.P.J.; Kalmus, O.; Van Der Zande, M.M.; Lutyj, J.; Listl, S. Development and evaluation of a mobile patient application to enhance medical-dental integration for the treatment of periodontitis and diabetes. *Int. J. Med. Inform.* **2021**, *152*, 104495. [CrossRef] [PubMed]

Article

Investigating the Regulation of Neural Differentiation and Injury in PC12 Cells Using Microstructure Topographic Cues

Xindi Sun [1], Wei Li [1], Xiuqing Gong [1], Guohui Hu [2], Junyi Ge [1], Jinbo Wu [1,*] and Xinghua Gao [1,*]

[1] Materials Genome Institute, Shanghai University, Shanghai 200444, China; sunxindi@shu.edu.cn (X.S.); liwei19981126@163.com (W.L.); gongxiuqing@shu.edu.cn (X.G.); junyi_ge@t.shu.edu.cn (J.G.)

[2] Shanghai Key Laboratory of Mechanics in Energy Engineering, Shanghai Institute of Applied Mathematics and Mechanics, School of Mechanics and Engineering Science, Shanghai University, Shanghai 200072, China; ghhu@staff.shu.edu.cn

* Correspondence: jinbowu@t.shu.edu.cn (J.W.); gaoxinghua@t.shu.edu.cn (X.G.)

Abstract: In this study, we designed and manufactured a series of different microstructure topographical cues for inducing neuronal differentiation of cells in vitro, with different topography, sizes, and structural complexities. We cultured PC12 cells in these microstructure cues and then induced neural differentiation using nerve growth factor (NGF). The pheochromocytoma cell line PC12 is a validated neuronal cell model that is widely used to study neuronal differentiation. Relevant markers of neural differentiation and cytoskeletal F-actin were characterized. Cellular immunofluorescence detection and axon length analysis showed that the differentiation of PC12 cells was significantly different under different isotropic and anisotropic topographic cues. The expression differences of the growth cone marker growth-associated protein 43 (GAP-43) and sympathetic nerve marker tyrosine hydroxylase (TH) genes were also studied in different topographic cues. Our results revealed that the physical environment has an important influence on the differentiation of neuronal cells, and 3D constraints could be used to guide axon extension. In addition, the neurotoxin 6-hydroxydopamine (6-OHDA) was used to detect the differentiation and injury of PC12 cells under different topographic cues. Finally, we discussed the feasibility of combining the topographic cues and the microfluidic chip for neural differentiation research.

Keywords: microfluidics; microstructure topography; PC12 cells; neural differentiation

Citation: Sun, X.; Li, W.; Gong, X.; Hu, G.; Ge, J.; Wu, J.; Gao, X. Investigating the Regulation of Neural Differentiation and Injury in PC12 Cells Using Microstructure Topographic Cues. *Biosensors* **2021**, *11*, 399. https://doi.org/10.3390/bios11100399

Received: 6 September 2021
Accepted: 14 October 2021
Published: 16 October 2021

Publisher's Note: MDPI stays neutral with regard to jurisdictional claims in published maps and institutional affiliations.

1. Introduction

The nervous system consists of the peripheral nervous system (PNS), supporting communication between the body and the brain, and the central nervous system (CNS), integrating information from various organ systems [1]. The CNS contains a variety of neuronal cell types connected by axons and dendrites that together build complex neural networks through migration, differentiation, and synapse formation, and realize complex brain functions including transmission, storage, and processing of information, thereby controlling behavior. The main causes of CNS diseases are neurodegeneration and injury [2]. The most common neurodegenerative diseases include Parkinson's disease (PD), Alzheimer's disease (AD), and Huntington disease (HD), and their prevalence increases with age [3–6]. At present, there is no effective treatment for these disorders and their prognosis is poor, seriously impacting the patient's quality of life and posing a significant socioeconomic burden to society [7,8]. In order to promote neuroregeneration and recover nervous system function, scientists are studying nerve cell growth in vitro. A common strategy is to create structures that support nerve cell growth and axon extension. However, the nervous system has a complex 3D environment consisting of distinct physical, chemical, and biological properties [9,10]. Topographic cues in the physical microenvironment can influence nerve cells, so it is critical to simulate nerve regeneration in vitro [11]. When neurons are injured, the effective connections maintaining the normal function of these cells are lost, and external topographic cues promoting the growth of neurites are necessary to enable the reconstruction of neural connections. Simulation of topographical cues in vitro allows neurons to be placed in a simplified, controlled environment to study the response of neurons to different molecular and physical guidance cues. The different responses of nerve cells to topographical cues include cell adhesion, neurite outgrowth, and cell morphology changes [12–14]. These responses can further affect the growth, differentiation, and proliferation of cells in the process of neuroregeneration.

At present, 2D cultures are commonly used although they cannot fully simulate the complex 3D microenvironment of the brain [15,16]. Microfluidics is an emerging fluid control technology that integrates basic operating units involved in the traditional biochemical laboratory on a small chip and thereby enables many analytical functions on a tiny platform. It offers the advantages of convenient operation, savings on reagent costs, and increased analysis speed and has been widely used in many interdisciplinary fields such as biology, chemistry, materials, medicine [17–19]. It is worth noting that due to the design of the specific channels of the microfluidic chip, the physical separation of the soma and axon of neurons can be realized for the study of neuronal behavior in a 3D environment [20]. Under the specific microstructure, the cell bodies of nerve cells were fixed in a certain position, and axon behavior could be observed after axons separated the cell bodies. It was convenient to study the behavior of neurons in a specific environment. Francisco et al. studied the ability of embryonic chicken primary sensory neurons to extend axons in a variety of environments characterized by 3D physical constraints in the form of "rooms" and corridor "walls" [21]. Kywe Moe et al. designed a multi-architecture chip (MARC) with different aspect ratios and hierarchical structures, nanometer-to-micrometer sizes, and different structural complexity of the topography. MARCs were then replicated in polydimethylsiloxane (PDMS) to study the effects of different geometric sizes on neural differentiation in primary mouse neural progenitor cells (MNPC) [22]. These studies suggest that microfluidic chips may aid neurological disease models by simulating topographical clues for drug screening or toxicology studies.

Given the above, we designed and manufactured a series of different microstructure topographical cues for inducing neuronal differentiation of cells in vitro, with different topography, sizes, and structural complexities. The different topological microstructures were replicated in Polydimethylsiloxane (PDMS). Replicated microstructures were treated with Poly-L-lysine (PLL). PC12 cells were cultured in their microstructures and then induced neural differentiation using nerve growth factor (NGF). PC12 cells were commonly used for in vitro models of neurodegenerative diseases [23]. Cytoskeletal F-actin and β-tubulin III of neural differentiation were characterized. The results of the cell immunofluorescence experiment and axon length analysis showed that PC12 cells differentiated significantly under different topographic cues. We also explored whether there were differences in the expression of sympathetic nerve marker tyrosine hydroxylase (TH) and growth cone marker growth-associated protein 43 (GAP-43) genes under different topographic cues. Our experimental results showed that the physical environment had an important influence on the differentiation of neuronal cells, and three-dimensional constraints could be used to guide the extension of axons. The neurotoxin 6-hydroxydopamine (6-OHDA) was used to detect the differentiation and injury of PC12 cells under different topographic cues. Finally, we discussed the feasibility of combining the topographic cues and the microfluidic chip for neural differentiation research.

2. Materials and Methods

2.1. Design and Preparation of Topological Microstructures

Topological microstructures of different patterns were designed using CAD software. In the field area of 8 mm × 2.5 mm, several patterned topographies were designed. Anisotropic topographies include: (1) a straight line of a width of 10 μm, spacing of 10 μm (named S); (2) 45° polyline with a width of 10 μm, spacing of 10 μm (named C45); (3) 90° polyline with a width of 10 μm, spacing of 10 μm (named C90); (4) 135° polyline with a width of 10 μm, spacing of 10 μm (named C135). Isotropic topographies include: (5) 60° distribution micropillar array with a column diameter of 10 μm (named P60); (6) 90° distribution micropillar array with a diameter of 10 μm (named P90). All chambers were 10 μm high. The microstructures were manufactured in silicon using dry deep etching, courtesy of the Shanghai Institute of Microsystems and Information Technology, Chinese Academy of Sciences.

Different topographic microstructures were replicated using PDMS in subsequent experiments as the substrate for cell adhesion or the bottom layer of the microfluidic chip. The process of replication was as follows: firstly, the silicon template was treated with plasma to remove surface contaminants; then, the silicon template was treated for 1 min using 1H,1H,2H,2H-perfluorooctanyl triethoxy silane (Sigma Aldrich, Burlington, MA, USA) and then heated in a 120 °C circulating blast oven for 2 h. Secondly, PDMS and curing agents were mixed to a ratio of 10:1. The mixture was poured into the silicon template, degassed in a vacuum, and cured in an oven at 65 °C for at least 2 h. Lastly, the PDMS replicas were carefully peeled off the silicon template and sterilized. In order to verify the fidelity of PDMS replicas, they were placed in an oven at 65 °C for 48 h to ensure that the PDMS surface was dry. The obtained PDMS replicas were characterized by SEM (SU8230, Hitachi, Japan).

Microfluidic chip materials include cyclic olefin copolymer (COC), Polymethylmethacrylate (PMMA), PDMS among others [24]. PDMS as a microfluidic chip material shows certain characteristics: high thermal stability, high biocompatibility, good optical properties, chemically inert, and so on [25,26]. Considering the advantages of PDMS, we finally used PDMS as the microfluidic chip material in this experiment. Similarly, there are many types of processing methods for microfluidic systems, such as injection molding, laser ablation, molding, and 3D printing, etc. [27,28]. Each method has its own advantages and disadvantages. For example, 3D printing is flexible and fast in the chip processing process. However, the choice of materials for 3D printing is very limited [29].

2.2. Design and Production of the Upper Layer of the Microchip

The upper layer of the microchip was fabricated using standard soft lithography etching [30]. The upper layer comprised two independent and identical symmetric cell culture channels, each 1000 μm wide and 300 μm high (Figure 1H). The interval between the two cell culture channels was 500 μm, thus forming a cell culture channel limiting nerve cell PC12 to one side to observe the axonal growth. The gap between the two cell culture channels was the main area in which axonal differentiation is observed. SU-8 photoresist was used to fabricate the upper layer of the chip structure. The chip was made of polydimethylsiloxane (PDMS, Sylgard 184, Dow Corning, Midland, MI, USA). The obtained upper layer could be sealed with the patterned bottom layer topological microstructure using oxygen plasma.

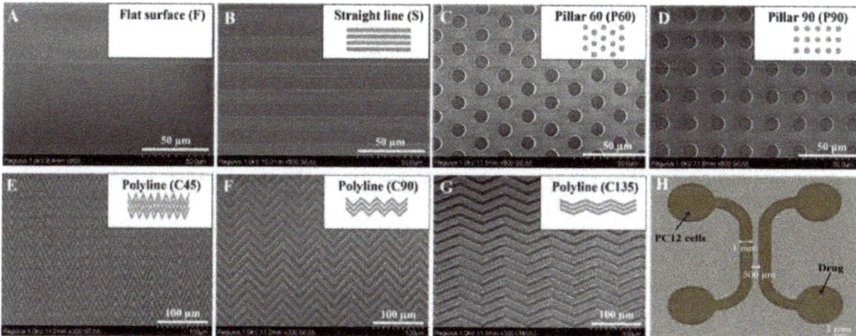

Figure 1. SEM images of PDMS replicas with different topological microstructures and images of the upper layer with the symmetric cell culture channels. (**A**) flat surface (F). (**B**) straight channel (S). (**C**) 60° distribution micropillar array (P60). (**D**) 90° distribution micropillar array (P90). (**E**) 45° polyline (C45). (**F**) 90° polyline (C90). (**G**) 135° polyline (C135). Scale bar = 50 μm or 100 μm. (**H**) Images of the upper layer mold with the symmetric cell culture channels. Scale bar = 1 mm.

2.3. Cell Culture and Seeding

Rat adrenal medullary pheochromocytoma PC12 cells were used in this study. The PC12 cell line was purchased from the Cell Bank/Stem Cell Bank of the Chinese Academy of Sciences Cell (ATCC source). Cells were cultured in Roswell Park Memorial Institute Medium 1640 (1640, Gibco, Waltham, MA, USA) with 10% horse serum (HS, Gibco), 5% fetal bovine serum (FBS, Gibco), and 1% penicillin-streptomycin double antibody (Gibco). Cells were passaged once or twice a week and used between 4 and 6 generations. The obtained PDMS replicas with microstructures were placed in a 6-well plastic culture dish, coated overnight with 0.1% polylysine (PLL, Sigma), and rinsed with phosphate buffer solution (PBS) 3 times. PC12 cells were seeded at a density of 5×10^6 cells/mL and allowed to settle and attach to the surface. After the overnight culture, cells were rinsed with medium to remove any unattached cells. PC12 cells were then cultured in a fresh medium supplemented with 100 ng/mL NGF (BBI) at 37 °C and 5% CO_2 for 5 days. The medium was changed every 2–3 days. The experimental process of 6-OHDA group was PC12 cells cultured in normal fresh medium for 4 days and treated with 20 ng/mL 6-OHDA for 24 h, and the experimental process of NGF+6-OHDA group was PC12 cells were cultured in fresh medium supplemented with 100 ng/mL NGF for 4 days, then treated with 6-OHDA at a concentration of 20 ng/mL for 24 h.

2.4. Immunofluorescence and Imaging

Alexa Fluor® 488 phalloidin (Life Invitrogen, Waltham, MA, USA) was used to characterize cellular F-actin. Staining was performed according to the manufacturer's instructions. In addition, the tubulin expression of PC12 cells was detected using immunofluorescence staining. Cells on PDMS were washed carefully with phosphate-buffered saline (PBS) and fixed with 4% paraformaldehyde (Sigma) for 20 min. After the addition of goat blocking buffer (Abcam, San Jose, CA, USA), the culture was incubated at room temperature for 1 h. The primary antibody (Tubulin, Abcam), diluted to a ratio of 1:100 (Abcam), was added. The sample was left overnight at 4 °C. After washing with PBS, the corresponding secondary antibody marked with Alexa Fluor 568 was added (goat anti-rabbit IgG). The sample was incubated at room temperature for 1 h. After the addition of 4′,6-diamidino-2-phenylindole (DAPI, Life Invitrogen) to counterstain nuclei, the sample was incubated at room temperature for 10 min. After washing with PBS three times, fluorescence images were acquired.

2.5. Total RNA Isolate and RT-qPCR

We separated total RNA from the F group, S group, and P 90 group of PC12 after treated NGF 5 days with TRIzol® reagent according to the manufacturer's instructions (Invitrogen). We used a One-Step RT-qPCR Kit (Sangon Biotech, Shanghai, China) to perform cDNA synthesis and PCR reaction on total RNA and used Lightcycle 96 (Roche, Basel, Switzerland) for detection. Average CT values were normalized to GAPDH expression levels. Primers synthesized by Sangon Biotech (Shanghai) Co., Ltd. were as follows:

GAP-43: 5′-CGACAGGATGAGGGTAAAGAA-3′ (forward), 5′-GACAGGAGAGGAA ACTTCAGAG-3′ (reverse);

TH: 5′-GTGAACCAATTCCCCATGTG-3′ (forward), 5′-CAGTACCGTTCCAGAAGCTG-3′ (reverse);

GAPDH: 5′-TCCAGTATGACTCTACCCACG-3′ (forward), 5′-CACGACATACTCAGC ACCAG-3′ (reverse).

2.6. Data Analysis

Image J was used for image analysis and cell counting. CellSens Standard software was used for the axon length measurement. The standard errors in the experimental data were obtained from Student's t-test under multiple groups of data ($n \geq 3$).

3. Results and Discussion

3.1. Characterization of Topological Microstructures in PDMS Replicas

The PDMS replicas with topological microstructures were obtained from the silicone templates and patterns included isotropic pillars of different sizes and anisotropic micrometer gratings of different angles. The patterning groups could be divided into anisotropic and isotropic topographies, of which anisotropic topographies were the straight channel (S), 45° polyline (C45), 90° polyline (C90), and 135° polyline (C135), and isotropic topographies were the 60° distribution micropillar array (P60), and 90° distribution micropillar array (P90). All patterns were characterized by SEM as shown in Figure 1A–G. There are a total of 6 patterned topographic microstructures and flat surfaces (F). Topological microstructures of the PDMS replicas exhibited good fidelity with a smooth surface and clear structure. Figure 1H showed the optical photo of the upper layer mold.

3.2. Differentiation of PC12 Cells in Different Topological Microstructures

The pheochromocytoma cell line (PC12) is a validated neuronal cell model that is widely used to study neuronal differentiation [22]. PC12 cells rely on nerve growth factor (NGF) and respond to this neurotrophic factor with nerve cell differentiation, showing a typical neural phenotype and neurite outgrowth. In this experiment, PC12 cells were cultured in different topological microstructures together with nerve growth factor NGF at a concentration of 100 ng/mL to establish a microchip neuronal differentiation model. In response to NGF, PC12 cells proliferated and showed obvious neurite growth, suggesting that the in vitro differentiation model was successfully established. Cellular immunofluorescence images visualizing F-actin and neuronal-associated protein β-tubulin III are shown in Figures 2 and 3.

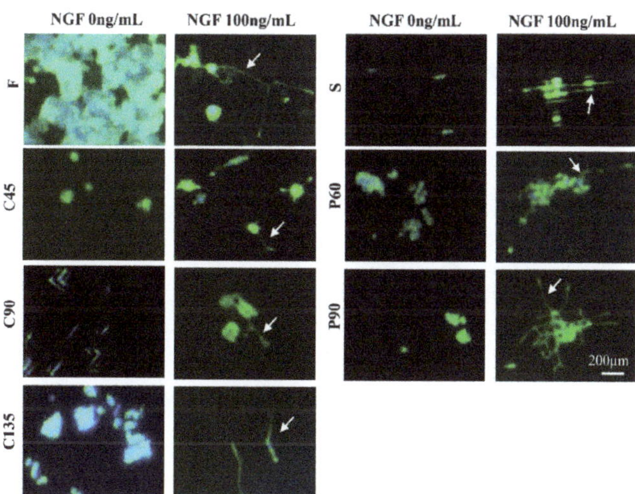

Figure 2. Fluorescence images of F-actin expression of PC12 cells under different conditions. Green, F-actin; blue, cell nucleus; NGF:100 ng/mL; Scale bar = 200 μm.

We analyzed the axonal length of the PC12 cells on different micro topologies, similar to previous work by Spillane et al. [31]. The histogram of the axon length of PC12 cells on different micro topologies at day 5 was shown in Figure 4A,B. The results show that when PC12 cells were not treated with NGF, the average axon length under different conditions was all less than 10 μm, which was much smaller than the result of PC12 cells treated with NGF. And then the morphology of PC12 cells on different micro topologies changed significantly after applying NGF treatment. This indicated that NGF played a key role in the axon elongation and neural differentiation of PC12 cells. And the results also show that PC12 cells grown on different microstructures could differentiate into neurons and exhibited increased cell body size, sprouting neurites, and the formation of a neural network. In this, compared with the flat group (F group), the axon length of PC12 cells increased significantly in the S, P60, P90, and C135 groups. The C45 and C90 groups may be affected by the bending angle of the polyline microstructure, the axon length did not become significantly longer, and C45 was also significantly shorter and only 20.3 ± 2.8 μm. This revealed that the axon could rotate around the corner within the range of $45°$ to $180°$ of the measured angles on anisotropic patterns; the larger this angle, the more pronounced the axon growth was (S: 84.0 ± 2.9 μm > C135: 54.8 ± 3.8 μm > C90: 48.1 ± 2.5 μm > C45: 20.3 ± 2.8 μm).

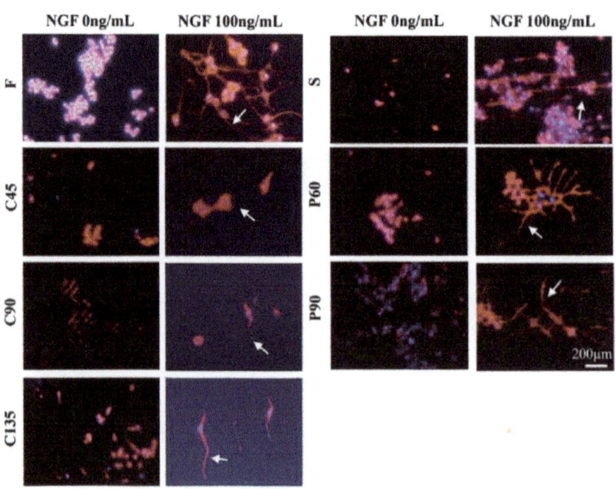

Figure 3. Fluorescence images of β-tubulin expression of PC12 cells under different conditions. Red, β-tubulin; blue, cell nucleus; NGF:100 ng/mL; Scale bar = 200 μm.

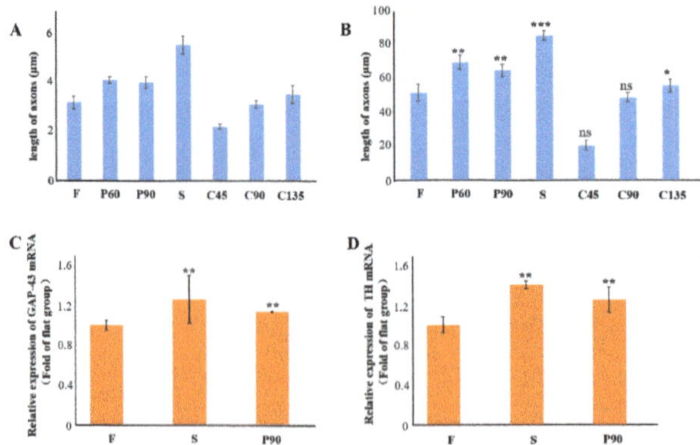

Figure 4. (**A**) Histogram of axon length of PC12 cell on different topographies after 5 days. (**B**) Histogram of axon length of PC12 cell on different topographies after NGF treatment for 5 days. (**C**) Gene expression of GAP-43 of PC12 cell on different topographies after NGF treatment for 3 days. (**D**) Gene expression of TH of PC12 cell on different topographies after NGF treatment for 3 days. NGF: 100 ng/Ml, GAPDH refers to the internal parameter, * $p < 0.05$, ** $p < 0.01$, *** $p < 0.001$. The data is the mean value of each parameter for each experimental condition.

In addition, PC12 cell differentiation also showed a certain cell orientation and the ability to form neural networks. The axon extension on C45, C90, C135, and S structures tended to occur in the direction of the topography itself. Conversely, the neurite extension on P60 and P90 structures was relatively more random, although most of the axons turned and extended around a pillar. The immunofluorescence staining results of β-tubulin (Figure 3) and F-actin were similar. Among them, the cell axons of the S group showed the orientation of growing along the straight microstructure, while in the P60 and P90 group the cells were similar to the F group and could form the neural network. It might be because there was much space between the micropillars in the P60 and P90 groups the physical constraints have little effect on the direction of cell axon growth. This was also reflected in the similar axon length of P60 (68.3 ± 4.1 μm) and P90 (63.8 ± 3.6 μm) and suggested that the axon length was not affected by the angle between the micropillars.

To further prove the relationship between microstructural cues and PC12 neural differentiation, the expressions of growth cone marker GAP-43 and sympathetic marker TH were used to identify the sympathetic characteristics of NGF-treated PC12 cells in different topographies, similar to previous work by Li-Ying Zhong et al. [32]. According to the previous experimental results, we chose the anisotropic topography S and the isotropic topography P90. The results were shown in Figure 4C,D. Compared with the control group (F group), the mRNA expression of the GAP-43 gene and TH gene in S and P90 topographies were significantly increased ($p < 0.01$), and the increase in S topography was more significant. This indicated that some special microstructure cues could guide neurons to grow in microenvironments of different terrain, size, and structural complexity, and specific three-dimensional constraints can effectively guide the extension of axons.

3.3. Cellular Neurotoxin Injury of PC12 Cells on Different Topological Microstructures

6-OHDA is a neurotoxin with a similar structure to dopamine which can selectively elicit cell death of dopaminergic neurons [33]. It is widely used in animal models of Parkinson's disease. Figure 5 shows immunofluorescence images of β-tubulin III of PC12 cells treated with 6-OHDA (20 ng/mL) under different conditions. PC12 cells treated with 6-OHDA on different patterned surfaces exhibited substantial cellular apoptosis and almost no axonal differentiation. When the PC12 cells were pretreated with NGF for 5 days to obtain differentiated cells, and then 6-OHDA was added, the cells still exhibited a certain degree of apoptosis, but some axons could have remained. We also calculated the axon length as shown in Figure 6. The effect of 6-OHDA on cells was huge. The axons were hardly generated, and the generated axons also disappear as the cell dies. Among all groups, after being treated by NGF and 6-OHDA, the remaining axon length of the S group was the longest. This result may be helpful for us to construct a nerve injury model or a toxicological drug evaluation model.

3.4. Analysis on the Trend of Neuronal Axon Differentiation on a Microchip

Based on the above research, we also discussed the feasibility of combining the topographic cues and the microfluidic chip for neural differentiation research. The design of the combined microchip is shown in Figure 7A. The microchip was contained two parts; one was the upper layer containing two independent channels and the other was the patterned bottom layer with topological microstructure like the S group. PC12 cells were seeded on the left side microchannel and the drug or growth factor was added to the right microchannel. After mixing the drug with the medium and after mixing the growth factor with the medium, pipetting guns were used to slowly add the right side of the microchannel, and the above steps were repeated every 8 h. The microchannels on the left and right sides were connected to each other by the straight microstructure. The results of immunofluorescence and axon length were shown in Figure 7B,C. The results indicated that there was significant differentiation of PC12 cells treated with NGF, with axon lengths reaching 105.8 ± 6.1 μm. This result was even greater than the axon length of PC12 on the S microstructure cues with NGF (84.0 ± 2.9 μm). In addition, the axon orientation of the cells was obvious, and they all extend along the microstructure topography to the microchannel on the other side. On the other hand, the neurotoxin experiment with 6-OHDA added also showed similar results. PC12 cells died to a certain extent after 6-OHDA treatment. PC12 cells undergo 6-OHDA treatment after differentiation, some axons of PC12 cells remained, but there was serious damage. We believe that the combined model of the microchip with microstructure cues may be more conducive to the observation of cell neural differentiation and the measurement of axon size. It can simulate related neurological disease models in vitro.

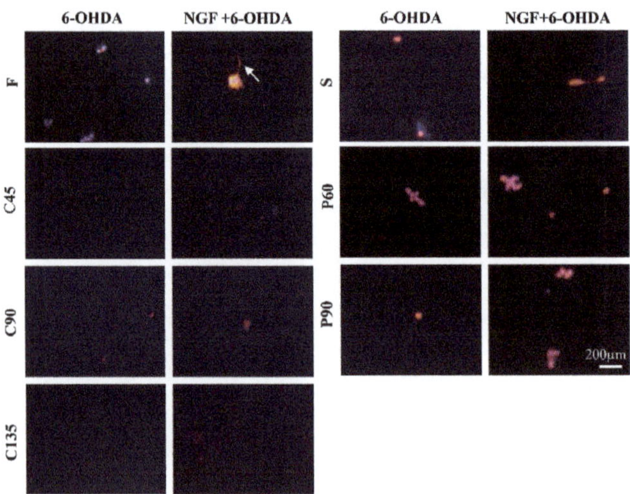

Figure 5. Fluorescence images of β-tubulin expression of PC12 cells under different conditions after 6-OHDA treatment. Red, β-tubulin; blue, cell nucleus; NGF: 100 ng/mL; 6-OHDA: 20 ng/mL. Scale bar = 200 μm.

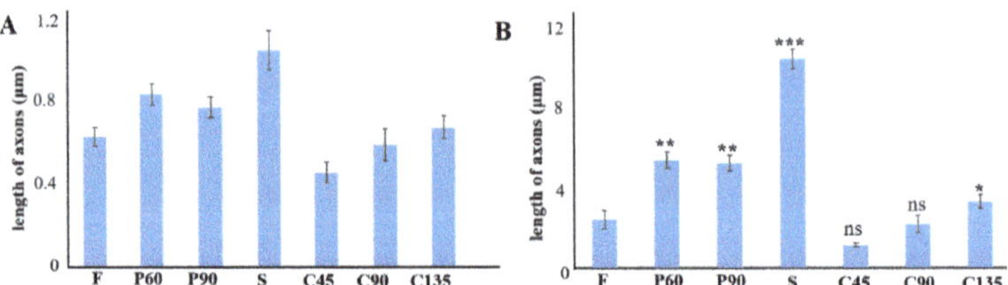

Figure 6. (**A**) Histogram of axon length of PC12 cell on different topographies after 6-OHDA treatment without NGF. (**B**) Histogram of axon length of PC12 cell on different topographies after 6-OHDA and NGF treatment. NGF: 100 ng/mL. 6-OHDA: 20 ng/mL. * $p < 0.05$, ** $p < 0.01$, *** $p < 0.001$. The data was the mean value of each parameter for each experimental condition.

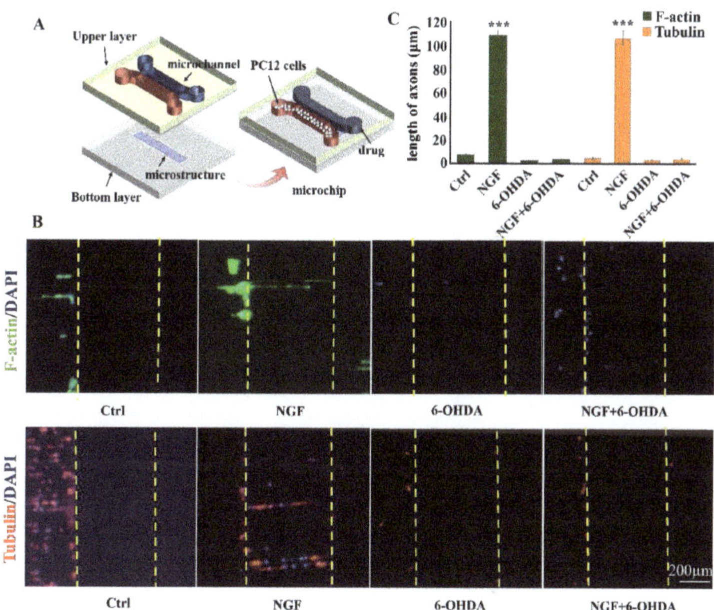

Figure 7. (**A**) Chip design. (**B**) Fluorescence images of F-actin and β-tubulin expression of PC12 cells on the combined chip after 5 days. Red, β-tubulin; blue, cell nucleus; Scale bar = 200 μm. C: Histogram of axon length of PC12 cell on the combined chip after 5 days. NGF: 100 ng/mL; 6-OHDA: 20 ng/mL. *** $p < 0.001$. The data is the mean value of each parameter for each experimental condition.

4. Conclusions

To summarize, here we describe the design and manufacture of a series of different topographical cues for the culture of PC12 cells in vitro. The structures enabled guidance of neuronal growth in distinct 3D microenvironments with different topographies, sizes, and structural complexity. We studied the neuronal differentiation of PC12 cells in these topologies using PDMS replicas and assessed the influence of biophysical factors on the differentiation of nerve cells. Using these topologies, we can effectively simulate the 3D environment of the human body in vitro and show that 3D constraints could be used to guide axonal extension. This research using a microfluidic chip platform is a small part of the potential application of neuronal network reconstruction in vitro. Future work may increase the complexity of the microchannel, involving more complex networks of more neuronal subtypes and provide the basis for the in vitro study of Alzheimer's disease, Parkinson's disease, and other neurological disorders.

Author Contributions: Conceptualization, X.G. (Xinghua Gao), J.W. and X.S.; methodology, X.G. (Xinghua Gao) and X.S.; software, G.H.; validation, X.G. (Xinghua Gao) and J.W.; formal analysis, J.G. and W.L.; investigation, X.S.; resources, X.G. (Xinghua Gao) and X.G. (Xiuqing Gong); data curation, X.S. and J.G.; writing—original draft preparation, X.S. and W.L.; writing—review and editing, X.G. (Xiuqing Gong) and G.H.; visualization, X.S.; supervision, J.W.; project administration, X.G. (Xinghua Gao); funding acquisition, X.G. (Xiuqing Gong), J.W. and X.G. (Xinghua Gao) All authors have read and agreed to the published version of the manuscript.

Funding: This work was supported by the National Natural Science Foundation of China (Grant No. 31800848 and No. 21775101).

Institutional Review Board Statement: Not applicable.

Informed Consent Statement: Not applicable.

Conflicts of Interest: The authors declare no conflict of interest.

References

1. Yang, C.Y.; Huang, W.Y.; Chen, L.H.; Liang, N.W.; Wang, H.C.; Lu, J.; Wang, X.; Wang, T.W. Neural tissue engineering: The influence of scaffold surface topography and extracellular matrix microenvironment. *J. Mater. Chem. B* **2021**, *9*, 567–584. [CrossRef]
2. Kovacs, G.G. Concepts and classification of neurodegenerative diseases. *Handb. Clin. Neurol.* **2017**, *145*, 301–307.
3. Fraser, K.B.; Rawlins, A.B.; Clark, R.G.; Alcalay, R.N.; Standaert, D.G.; Liu, N.; Parkinson's Disease Biomarker Program Consortium; West, A.B. Ser(P)-1292 LRRK2 in Urinary Exosomes Is Elevated in Idiopathic Parkinson's Disease. *Mov. Disord.* **2016**, *31*, 1543–1550. [CrossRef]
4. Nisbet, R.M.; Gotz, J. Amyloid-β and Tau in Alzheimer's Disease: Novel Pathomechanisms and Non-Pharmacological Treatment Strategies. *J. Alzheimers Dis.* **2018**, *64*, S517–S527. [CrossRef]
5. Ricco, M.; Vezzosi, L.; Balzarini, F.; Gualerzi, G.; Ranzieri, S. Prevalence of Huntington Disease in Italy: A systematic review and meta-analysis. *Acta Biomed.* **2020**, *91*, 119–127. [PubMed]
6. Sveinbjornsdottir, S.; Neurochem, J. The clinical symptoms of Parkinson's disease. *J. Neurochem.* **2016**, *139* (Suppl. 1), 318–324. [CrossRef] [PubMed]
7. Liljegren, M.; Naasan, G.; Temlett, J.; Perry, D.C.; Rankin, K.P.; Merrilees, J.; Grinberg, L.T.; Seeley, W.W.; Englund, E.; Miller, B.L. Criminal Behavior in Frontotemporal Dementia and Alzheimer Disease. *JAMA Neurol.* **2015**, *72*, 295–300. [CrossRef] [PubMed]
8. Thies, W.; Bleiler, L. Alzheimer's Association Report 2011 Alzheimer's disease facts and figures Alzheimer's Association. *Alzheimers Dement* **2011**, *7*, 208–244. [CrossRef] [PubMed]
9. Kothapalli, C.R.; Kamm, R.D. 3D matrix microenvironment for targeted differentiation of embryonic stem cells into neural and glial lineages. *Biomaterials* **2013**, *34*, 5995–6007. [CrossRef] [PubMed]
10. Puckert, C.; Tomaskovic-Crook, E.; Gambhir, S.; Wallace, G.G.; Crook, J.M.; Higgins, M.J. Molecular interactions and forces of adhesion between single human neural stem cells and gelatin methacrylate hydrogels of varying stiffness. *Acta Biomater.* **2020**, *106*, 156–169. [CrossRef] [PubMed]
11. Lim, J.Y.; Donahue, H.J. Cell Sensing and Response to Micro- and Nanostructured Surfaces Produced by Chemical and Topographic Patterning. *Tissue Eng.* **2007**, *13*, 1879–1891. [CrossRef]
12. Zamani, F.; Amani-Tehran, M.; Latifi, M.; Shokrgozar, M.A. The influence of surface nanoroughness of electrospun PLGA nanofibrous scaffold on nerve cell adhesion and proliferation. *J. Mater. Sci. Mater. Med.* **2013**, *24*, 1551–1560. [CrossRef]
13. Kulangara, K.; Adler, A.F.; Wang, H.; Chellappan, M.; Hammett, E.; Yasuda, R.; Leong, K.W. The effect of substrate topography on direct reprogramming of fibroblasts to induced neurons. *Biomaterials* **2014**, *35*, 5327–5336. [CrossRef]
14. Xia, H.; Chen, Q.; Fang, Y.; Liu, D.; Zhong, D.; Wu, H.; Xia, Y.; Yan, Y.; Tang, W.; Sun, X. Directed neurite growth of rat dorsal root ganglion neurons and increased colocalization with Schwann cells on aligned poly (methyl methacrylate) electrospun nanofifibers. *Brain Res.* **2014**, *1565*, 18–27. [CrossRef]
15. Yang, X.; Li, K.; Zhang, X.; Liu, C.; Guo, B.; Wen, W.; Gao, X. Nanofiber Membrane supported lung-on-a-chip Microdevice for Anti-cancer Drug Testing. *Lab Chip* **2018**, *18*, 486–495. [CrossRef]
16. Yao, K.; Li, W.; Li, K.; Wu, Q.; Gu, Y.; Zhao, L.; Zhang, Y.; Gao, X. Simple Fabrication of Multicomponent Heterogeneous Fibers for Cell Co-Culture via Microfluidic Spinning. *Macromol. Biosci.* **2020**, *20*, e1900395. [CrossRef] [PubMed]
17. Li, Z.; Bai, Y.; You, M.; Hu, J.; Yao, C.; Cao, L.; Xu, F. Fully integrated microfluidic devices for qualitative, quantitative and digital nucleic acids testing at point of care. *Biosens. Bioelectron.* **2021**, *177*, 112952. [CrossRef] [PubMed]
18. Lee, W.B.; Chien, C.C.; You, H.L.; Kuo, F.C.; Lee, M.S.; Lee, G.B. Rapid antimicrobial susceptibility tests on an integrated microfluidic device for precision medicine of antibiotics. *Biosens. Bioelectron.* **2021**, *176*, 112890. [CrossRef]
19. Wu, R.; Kim, T. Review of microfluidic approaches for fabricating intelligent fiber devices: Importance of shape characteristics. *Lab Chip* **2021**, *21*, 1217–1240. [CrossRef] [PubMed]
20. Peyrin, J.; Deleglise, B.; Saias, L.; Vignes, M.; Gougis, P.; Magnifico, S.; Betuing, S.; Pietri, M.; Caboche, J.; Vanhoutte, P.; et al. Axon diodes for the reconstruction of oriented neuronal networks in microfluidic chambers. *Lab Chip* **2011**, *11*, 3663–3673. [CrossRef]
21. Francisco, H.; Yellen, B.B.; Halverson, D.S.; Friedman, G.; Gallo, G. Regulation of axon guidance and extension by three-dimensional constraints. *Biomaterials* **2007**, *28*, 3398–3407. [CrossRef]
22. Moe, A.A.; Suryana, M.; Marcy, G.; Lim, S.K.; Ankam, S.; Goh, J.Z.; Jin, J.; Teo, B.K.; Law, J.B.; Low, H.Y.; et al. Microarray with Micro- and Nano-topographies Enables Identifification of the Optimal Topography for Directing the Differentiation of Primary Murine Neural Progenitor Cells. *Small* **2012**, *8*, 3050–3061. [CrossRef]
23. Li, H.; Tang, Z.; Chu, P.; Song, Y.; Yang, Y.; Sun, B.; Niu, M.; Qaed, E.; Shopit, A.; Han, G.; et al. Neuroprotective effect of Phosphocreatine on oxidative stress and mitochondrial dysfunction induced apoptosis in vitro and in vivo: Involvement of dual PI3K/Akt and Nrf2/HO-1 pathways. *Free Radic. Biol. Med.* **2018**, *120*, 228–238. [CrossRef]
24. Nguyen, T.; Ngo, T.A.; Dang, D.B.; Wolff, A. Optimising the supercritical angle fluorescence structures in polymer microfluidic biochips for highly sensitive pathogen detection: A case study on *Escherichia coli*. *Lab Chip* **2019**, *19*, 3825–3833. [CrossRef]
25. Arora, A.; Simone, G.; Salieb-Beugelaar, G.B.; Kim, J.T.; Manz, A. Latest developments in micro total analysis systems. *Lab Chip* **2010**, *82*, 4830–4847. [CrossRef] [PubMed]
26. Hulme, J.P.; An, S.; Goddard, N.; Miyahara, Y.; Oki, A. Fabrication of a flexible multi-referenced surface plasmon sensor using room temperature nanoimprint lithography. *Curr. Appl. Phys.* **2019**, *2*, e185–e188. [CrossRef]

27. Owens, C.E.; Hart, A.J. High-precision modular microfluidics by micromilling of interlocking injection-molded blocks. *Lab Chip* **2018**, *18*, 890–901. [CrossRef] [PubMed]
28. Peter, L.; Marcus, S.G.; Anurag, M.; Mae, R.W.; Healy, K.E.; Raghavan, R. μOrgano: A lego-like plug & play system for modular multi-organ-chips. *PLoS ONE* **2015**, *10*, e0139587.
29. Ji, Q.; Zhang, J.M.; Liu, Y.; Li, X.; Lv, P.; Jin, D.; Duan, H. A modular microfluidic device via multimaterial 3D printing for emulsion generation. *Sci. Rep.* **2018**, *8*, 4791. [CrossRef] [PubMed]
30. Yang, X.; Xu, X.; Zhang, Y.; Wen, W.; Gao, X. 3D Microstructure Inhibits Mesenchymal Stem Cells Homing to the Site of Liver Cancer Cells on a Microchip. *Genes* **2017**, *8*, 218. [CrossRef]
31. Spillane, M.; Ketschek, A.; Merianda, T.T.; Twiss, J.L.; Gallo, G. Mitochondria Coordinate Sites of Axon Branching through Localized Intra-axonal Protein Synthesis. *Cell Rep.* **2013**, *5*, 1564–1575. [CrossRef] [PubMed]
32. Zhong, L.Y.; Fan, X.R.; Shi, Z.J.; Fan, Z.C.; Luo, J.; Lin, N.; Liu, C.Y.; Wu, L.; Zeng, X.R.; Cao, J.M.; et al. Hyperpolarization-Activated Cyclic Nucleotide-Gated Ion (HCN) Channels Regulate PC12 Cell Differentiation toward Sympathetic Neuron. *Front. Cell. Neurosci.* **2019**, *13*, 415. [CrossRef] [PubMed]
33. Ayaka, S.; Toko, M.; Ko, F. Ko Fujimori. Protection of 6-OHDA neurotoxicity by PGF2α through FP-ERK-Nrf2 signaling in SH-SY5Y cells. *Toxicology* **2021**, *450*, 152686.

Article

Microfluidic Lab-on-a-Chip Based on UHF-Dielectrophoresis for Stemness Phenotype Characterization and Discrimination among Glioblastoma Cells

Elisa Lambert [1,†], Rémi Manczak [1], Elodie Barthout [2,†], Sofiane Saada [2], Elena Porcù [3,4], Francesca Maule [5], Barbara Bessette [2], Giampietro Viola [3,4], Luca Persano [3,4], Claire Dalmay [1], Fabrice Lalloué [2] and Arnaud Pothier [1,*]

[1] XLIM-UMR 7252, University of Limoges/CNRS, 87060 Limoges, France; elisa.lambert@xlim.fr (E.L.); remi.manczak@xlim.fr (R.M.); claire.dalmay@xlim.fr (C.D.)
[2] CAPTuR-EA 3842, University of Limoges, 87025 Limoges, France; elodie.barthout@unilim.fr (E.B.); sofiane.saada@unilim.fr (S.S.); barbara.bessette@unilim.fr (B.B.); fabrice.lalloue@unilim.fr (F.L.)
[3] Department of Women's and Children's Health (DSB), University of Padova, 35128 Padova, Italy; elena.porcu@unipd.it (E.P.); giampietro.viola.1@unipd.it (G.V.); luca.persano@unipd.it (L.P.)
[4] Institute of Pediatric Research (IRP), 35127 Padova, Italy
[5] Arnie Charbonneau Cancer Institute, Department of Biochemistry and Molecular Biology, University of Calgary, Calgary, AB T2N 1N4, Canada; francesca.maule@ucalgary.ca
* Correspondence: arnaud.pothier@unilim.fr
† These authors contributed equally to this work.

Citation: Lambert, E.; Manczak, R.; Barthout, E.; Saada, S.; Porcù, E.; Maule, F.; Bessette, B.; Viola, G.; Persano, L.; Dalmay, C.; et al. Microfluidic Lab-on-a-Chip Based on UHF-Dielectrophoresis for Stemness Phenotype Characterization and Discrimination among Glioblastoma Cells. *Biosensors* **2021**, *11*, 388. https://doi.org/10.3390/bios11100388

Received: 15 September 2021
Accepted: 12 October 2021
Published: 13 October 2021

Publisher's Note: MDPI stays neutral with regard to jurisdictional claims in published maps and institutional affiliations.

Abstract: Glioblastoma (GBM) is one of the most aggressive solid tumors, particularly due to the presence of cancer stem cells (CSCs). Nowadays, the characterization of this cell type with an efficient, fast and low-cost method remains an issue. Hence, we have developed a microfluidic lab-on-a-chip based on dielectrophoresis (DEP) single cell electro-manipulation to measure the two crossover frequencies: f_{x01} in the low-frequency range (below 500 kHz) and f_{x02} in the ultra-high-frequency range (UHF, above 50 MHz). First, in vitro conditions were investigated. An U87-MG cell line was cultured in different conditions in order to induce an undifferentiated phenotype. Then, ex vivo GBM cells from patients' primary cell culture were passed through the developed microfluidic system and characterized in order to reflect clinical conditions. This article demonstrates that the usual exploitation of low-frequency range DEP does not allow the discrimination of the undifferentiated GBM cells from the differentiated one. However, the presented study highlights the use of UHF-DEP as a relevant discriminant parameter. The proposed microfluidic lab-on-a-chip is able to follow the kinetics of U87-MG phenotype transformation in a CSC enrichment medium and the cancer stem cells phenotype acquirement.

Keywords: high-frequency dielectrophoresis; glioblastoma cells; single cell manipulation; microfluidic point-of-care device; cancer stem cells

1. Introduction

Glioblastoma (GBM) is the most frequent and highly malignant brain tumor in adulthood classified as a high-grade glioma (grade IV), considered as the most aggressive tumors of the central nervous system. Worldwide, 240,000 brain tumors are diagnosed each year, the majority of which are GBM [1]. GBM is associated with a poor prognosis with a mean survival of 12 months. Indeed, standard treatment such as surgery and combined radio–chemotherapy [2] fails to improve patients' care [3]. Despite recent advances in targeting therapies and immunotherapies, the current treatments do not allow the improvement of the mean survival. This very poor prognosis is mainly due to frequent relapses, despite the regression or disappearance of the tumor upon the golden standard treatment, i.e., the Stupp protocol [2]. Consequently, this pathology is very difficult to handle.

The high recurrence of GBM can be explained by the high heterogeneity of cellular, genetic and morphological patterns of the cell populations present in the tumor [4], which alters the efficiency of conventional therapies. This cell heterogeneity mainly results from a small cell subpopulation, called cancer stem cells (CSCs) [4], which are now considered as one of the major factors responsible for tumor progression and relapse [5]. In fact, CSCs display an immature and undifferentiated phenotype associated with specific self-renewal features. They also are likely to regenerate the entire tumor. Moreover, due to their quiescent properties, CSCs are resistant to radio and chemotherapy targeting proliferating cells [1]. Thus, CSCs detection in solid tumor could afford a prognosis value to evaluate tumor aggressiveness and prevent recurrence risk. However, CSCs are a rare cell subpopulation difficult to characterize with specific and usual tumor biomarkers within the tumor. CSC detection currently represents a challenge to improve the management of certain solid cancers, in particular GBM. Thus, new approaches for effectively discriminating CSCs from differentiated tumor cells are regularly investigated. However, one of the main encountered difficulties in characterizing CSCs is the lack of specific markers for these cells.

In this objective, biologists commonly analyze a panel of biomarkers by using conventional approaches for characterizing CSCs such as immunofluorescence, flow cytometry and protein array analysis. These different methodologies require systematic immunolabeling to identify CSCs, thereby require multiple steps which increases both expenses and inconvenience. Indeed, these methodologies are time-consuming and additional costs are incurred for the purchase of specific antibodies against CSC biomarkers. In addition, immunolabeling can influence and modify cell behavior and differentiation mechanisms. These changes might affect cell cultures and limit further analyses [6]. In order to overcome these issues and avoid immunolabeling, researchers started to develop alternative label-free methods for cell characterization. These innovative technologies aim to characterize CSCs based on their specific physical properties. Thus, most of the recent techniques rely on an external force coupled with microfluidics. This allows the reduction of the cost and analysis time with the possibility of parallelization and the reduction of sample volume as only few μL are needed. It also prevents cells from damage by limiting mechanical stress. Among the label-free method, dielectrophoresis (DEP) presents an important interest as it allows the investigation of biological cell behavior according to its intrinsic dielectric properties. For this study, we use DEP electro-manipulation at low frequency (below 500 kHz) and at ultra-high-frequency (UHF) range (above 50 MHz), highlighting the relevance of using UHF-DEP phenomenon to discriminate undifferentiated cells (CSC) from differentiated tumor cells. Implementation of the proposed microfluidic lab-on-a-chip is performed on BiCMOS technology and allows the screening of the intracellular properties of GBM cells.

1.1. Basic DEP Theory

Dielectrophoresis is a physical phenomenon, which leads to the motion of a polarizable particle such as biological cells, in a non-uniform electric field due to the interaction between the induced dipole of the particle and the field gradient. The polarization phenomenon redistributes of the charges at the interface between the particle and the suspension medium. The dielectrophoretic force exerted on the polarized particle in a non-uniform electric field is expressed as follows [7]:

$$F_{DEP} = 2\pi r^3 \varepsilon_m Re[f_{CM}(\omega)] \nabla E^2 \tag{1}$$

where r is the radius of the particle, ε_m the permittivity of the suspension medium, $Re[f_{CM}(\omega)]$ the real part of the Clausius–Mossotti (CM) factor, E the applied electric field.

The Clausius–Mossotti factor describes the polarization state of a particle in a suspension medium. It depends on the dielectric properties (permittivity and conductivity) of the medium and the particle [7]:

$$f_{CM}(\omega) = \frac{\varepsilon_p^* - \varepsilon_m^*}{\varepsilon_p^* + 2\varepsilon_m^*} \tag{2}$$

where ε_p^* and ε_m^* are the complex permittivity of the particle and the medium, respectively. The complex permittivity can be defined as:

$$\varepsilon^* = \varepsilon - j\frac{\sigma}{\omega} \tag{3}$$

where ε is the absolute permittivity ($\varepsilon = \varepsilon_r^* \varepsilon_0$, with ε_r is the relative permittivity and ε_0 is the vacuum permittivity, of which the value is 8.854 F·m^{-1}), σ the conductivity and ω the angular frequency of the electric field. The sign of the real part of the CM factor determines the orientation of the DEP force. In Figure 1a, when Re[$f_{CM}(\omega)$] is positive, the DEP force attracts the particle to the strong field areas. This phenomenon is called positive DEP (pDEP). In Figure 1b, when Re[$f_{CM}(\omega)$] is negative, the DEP force is then repulsive and the particle is repelled towards the weak electric field areas. This is called negative DEP (nDEP).

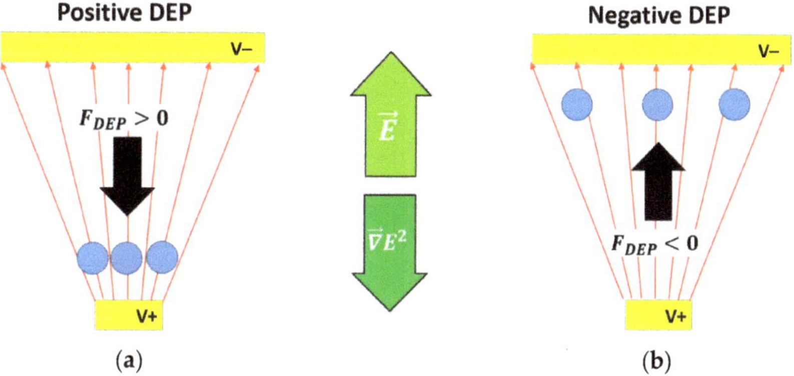

Figure 1. Particles' interaction in a non-uniform applied field: (**a**) the DEP force is positive, i.e., collinear to the electric field gradient ∇E^2, and the particles are attracted to areas of high field intensity (positive DEP); (**b**) the DEP force is negative, i.e., opposite to the electric field gradient ∇E^2, and the particles are repelled toward areas of low field intensity (negative DEP).

From a physic point of view, a biological cell can be modelized as a spherical dielectric particle that is submitted to the DEP force. A cell is a complex biological object, but it can be properly modeled into a simpler single-shell model with a reduced number of dielectric parameters associated to each component of the cell.

1.2. From a Biological Cell to a Single-Shell Model

In order to predict cells' behavior with an applied electric field, it is helpful to have a simplified model of a biological cell. The Figure 2 presents a schematic of a cell and its commonly used associated single-shell model associated where the cell membrane and cytoplasm are represented by a shell and a core with their own complex permittivity [8–10].

Indeed, the single-shell model considers the cytoplasm and its content as a homogeneous dielectric sphere enveloped by the plasma membrane. This model limits the number of dielectric parameters to take into account only the complex permittivity of the intracellular content, the cell membrane and the suspension medium. Actually, the Clausius–Mossotti factor depends on these parameters as well as the frequency of the applied field. This single-shell model will be considered in this paper.

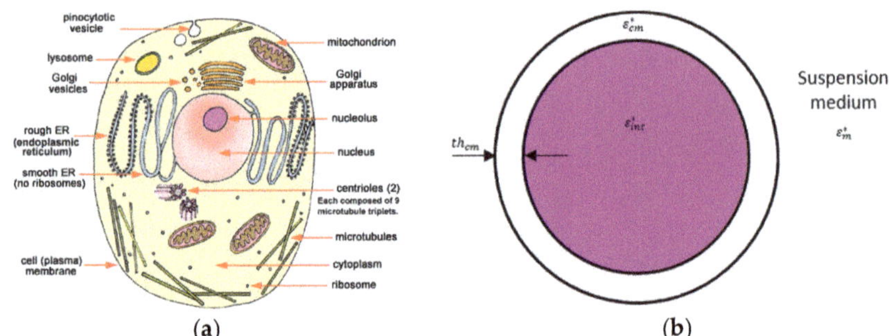

Figure 2. (a) Representation of a biological cell in a suspension medium; (b) its single-shell model with ε_{int}^{*} the complex permittivity of the cellular content, ε_{cm}^{*} the complex permittivity of the cell membrane and th_{cm} the thickness of the cell membrane and ε_{m}^{*} the complex permittivity of the suspension medium.

1.3. Effect of the Cellular Dielectric Properties on the Clausius–Mossotti Factor

Figure 3 illustrates the frequency-dependent cell behavior through the real part of the CM factor (Figure 3a). The dielectric parameters and cell geometric parameters are reported in Table 1. The real part plot in Figure 3a is computed thanks to the myDEP software [11]. nDEP behavior can be observed at very low frequency (lower than 400 kHz) and at high frequency (at least higher than 150 MHz), and pDEP behavior can be seen at medium range frequency (between 500 kHz and 100 MHz). The plot of the CM factor hence presents alternations between a repulsive state (nDEP) and an attractive state (pDEP). Two crossover frequencies f_{x01} and f_{x02} appear where the real part of the CM factor becomes null. f_{x01} occurs at low frequency, whereas f_{x02} occurs at higher frequency.

Moreover, from Figure 3a, 100 kHz and 1, 20 and 500 MHz frequencies were selected in order to study the dielectric response of the cells. Indeed, these frequencies correspond to the two different DEP behaviors but at low frequency (frequencies n°1 and n°2) and at high-frequency regime (frequencies n°3 and n°4). COMSOL Multiphysics® computations were performed with the AC/DC electric current module in Figure 3b. The parameters from Table 1 were used for the simulation in such a way that the results correspond to the curve of the real part of the CM factor. As said before, the cell is represented by the single-shell model with the core: its intracellular content, and the shell: its plasma membrane. The cell is here considered to be suspended in a low-conductivity medium. The electric potential is 1 Vpp. The shown colors represent the electric field intensity in V/m from dark blue (the field intensity is 0 V/m) to dark red (the field intensity is maximum). One can notice that for the nDEP behavior (at 100 kHz and 500 MHz), the electric field lines (black streamlines) bypass the cell, whereas for the pDEP behavior (at 1 and 20 MHz), the electric field lines seem attracted inside the cell. This is mainly due to the reorientation of the charges at the interface between the cell membrane and the medium [7]. One can also notice that at low frequency, no field can reach the cell content. The electric field is at maximum inside the plasma membrane, which hence acts as an insulator. As a result, at low frequency, the electromagnetic field will be more sensitive to the physical and dielectric properties of the cell membrane. The more the frequency of the applied signal increases, the more the electric field can penetrate inside the cell and starts to interact with and so probe the cellular content. Consequently, at high frequency, the electromagnetic wave can be deeply sensitive to the intracellular content.

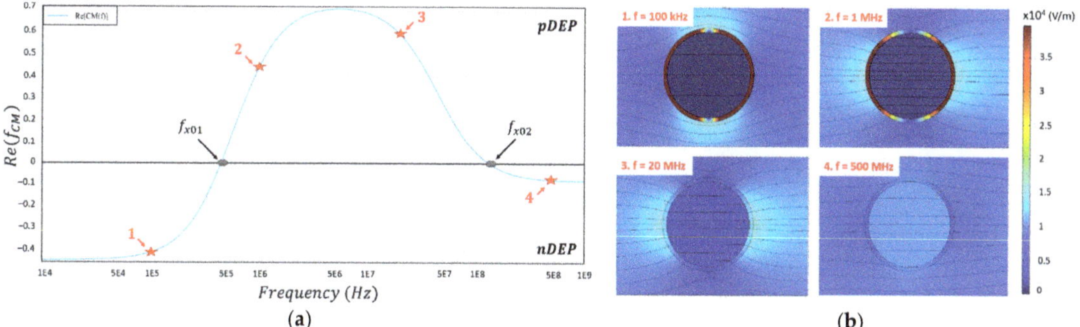

Figure 3. Numerical simulation of a cell dielectric behavior in function of the frequency with the parameters from Table 1. The design used for the biological cell is the single-shell model. (**a**) Numerical simulation of the real part of the Clausius–Mossotti factor. The red stars correspond to the chosen frequencies for the COMSOL simulation; (**b**) COMSOL simulation of the single-shell model for different frequencies (100 kHz; 1 MHz; 20 MHz; 500 MHz) which correspond to the curve of the CM factor. The color scale corresponds to the electric field intensity (V/m) and the black lines correspond to the electric streamlines.

Table 1. Values of the different dielectric and cellular parameters used in the COMSOL Multiphysics simulation.

Parameter	Value
Particle radius	11.5 µm
Membrane thickness	700 nm [1]
Intracellular relative permittivity	50
Intracellular conductivity	0.5 S/m
Membrane relative permittivity	100^2
Membrane conductivity	1.43×10^4 S/m [2]
Medium relative permittivity	78
Medium conductivity	0.02 S/m

[1] Membrane thickness was increased by 100 in order to avoid mesh issues during the computation. [2] Data were modified proportionally due the modification of the membrane thickness in order to respect the cell dielectric behavior.

Moreover, it is possible to change the value of the dielectric parameters in order to study the evolution of the real part of the CM factor and the parameter dependency of the two crossover frequencies as has been carried out in [12–14]. A first approximation of the crossover frequency f_{x01} can be expressed as [7]:

$$f_{x01} = \sigma_m \frac{th_{cm}}{\sqrt{2\pi}\, r\, \varepsilon_{cm}} \tag{4}$$

The crossover frequency f_{x01} depends mostly on the dielectric parameters of the plasma membrane, but also the particle radius. Hence, f_{x01} is more sensitive to the cell shape, its morphology and to the plasma membrane properties. It has been widely used to separate cells or polystyrene particles of different size [15,16] and to separate living from non-viable cells [17]. The second crossover frequency f_{x02} can be approximated with the assumption that the conductivity of the suspending medium (20 mS/m, see Section 2) is significantly below the intracellular value expression [18]:

$$f_{x02} = \frac{\sigma_{int}}{2\pi} \sqrt{\frac{1}{2\varepsilon_m^2 - \varepsilon_{int}\varepsilon_m - \varepsilon_{int}^2}} \tag{5}$$

The crossover frequency f_{x02} depends on the dielectric parameters of the intracellular content [19]. Hence, cells can be individually electro-manipulated by the DEP force motion according to their own dielectric properties of their cytoplasm. As an example, UHF-DEP has already been successfully used in order to discriminate differentiated from undifferentiated medulloblastoma cells [20,21].

This paper aims to show the relevance of the identification of the crossover frequency f_{x02} as the DEP signature and then using it as an appropriate discriminant biomarker to detect CSCs within tumor cell population. Therefore, the two crossover frequencies f_{x01} and f_{x02} have been measured for each investigated GBM cell and compared.

2. Materials and Methods

2.1. Cell Line Culture

Human GBM cell line U87-MG was purchased from American Type Culture Collection (ATCC). Cells were grown in different culture conditions (see below) at 37 °C in a humidified atmosphere of 5% CO_2–95% air. Cancer stem cells enrichment was obtained submitting cells to stringent culture conditions with the Define Medium (DM). Two culture conditions were used for the cells' DEP characterization:

- Normal Normoxia Medium (NM): induces normal differentiation in DMEM supplemented by 10% FBS, 2 mM glutamine and 1% penicillin/streptomycin.
- Define Normoxia Medium (DM): the starvation of 10% Fetal Bovine Serum (FBS) in this medium induces stringent conditions. DM is supplemented in two specific growth factors: EGF (Epidermal Growth Factors) and bFGF2 (basic Fibroblast Growth Factors) required for clonal expansion and the formation of glioma spheres which are composed of several thousand aggregated cells. DM composition consists in DMEM/F12 supplemented by 0.6% glucose, 1% sodium bicarbonate, 1% MEM non-essential amino acids, 5 mM HEPES, 9.6 µg/mL putrescine, 10 µg/mL ITSS, 0.063 µg/mL progesterone, 60 µg/mL N-acetyl-L-cysteine, 2 µg/mL heparin, 0.1 mg/mL penicillin/streptomycin, 50X B-27 supplement without vitamin A, 20 ng/mL EGF, 20 ng/mL bFG.

Hence, it is expected that the Define Medium will select undifferentiated cells and also promote the emergence of CSCs, whereas the Normal Medium will induce a differentiation of cells and so will result in a very low ratio of cancer stem cells. For the differentiated cells, they were cultured for 6 days in NM before the DEP characterization. For the undifferentiated cells, they were cultured for 5 days in DM or maintained during 21 days in DM in order to study the kinetics of CSC appearance and the emergence of stem cell characteristics.

2.2. Primary GBM Cell Isolations, Culturing and Separation by Flow Cytometry

For some validation experiments, four different primary cell cultures have been isolated and established in vitro from GBM patients to characterize any potential difference in term of DEP signature between bulk cells and their relative CSCs counterparts. Written informed consent for the donation of adult tumor brain tissues was obtained from patients before tissue collection under the auspices of the protocol for the acquisition of human brain tissues obtained from the Ethical Committee of the Padova University Hospital (2462P). In particular, GBM cells were isolated from tumors during surgery as previously described [22]. Briefly, resected GBM samples were dissociated, and single cell suspensions were grown in Define Medium under hypoxic conditions (DH). Indeed, GBM cells were maintained in an atmosphere of 2% oxygen, 5% carbon dioxide and balanced nitrogen in a H35 hypoxic cabinet (Don Whitley Scientific Ltd., Shipley, UK) to achieve a proper expansion of the CSC subpopulation in hypoxic conditions mimicking the GBM microenvironment [23].

After proper expansion in vitro, primary GBM cells were incubated with a PE mouse anti-CD133 (AC133) antibody (Miltenyi Biotec, Bergisch Gladbach, Germany) according to manufacturer's indications and then separated into a CD133+ and a CD133− subpopulation by means of Fluorescence Activated Cell Sorting (FACS) in a MoFlo XDP cell sorter (Beckman Coulter, Brea, CA, USA). A CD133 versus side scatter dot plot revealed the populations of interest characterized by the expression (or not) of the CD133 marker. Cell fractions were selected by setting appropriate sorting gates as previously described [24]. Upon sorting, isolated GBM cell subpopulations (CD133+ and CD133−) were washed in Phosphate Buffered Saline (PBS), frozen in medium containing 10% DMSO and then cryopreserved in liquid nitrogen for subsequent thawing and DEP characterization.

2.3. DEP Suspension Medium

Few minutes from the DEP characterization, U87-MG cells were suspended in an osmotic medium adapted for DEP electro-manipulation made from deionized water (ion free) supplemented by sucrose. The conductivity of the DEP medium is 26 mS/m and the pH is 7.4. The crossover frequency measurements were performed at room temperature.

2.4. Comparative Transcriptomic Analysis (mRNA Levels) of the Stemness Phenotype

In order to confirm the enrichment of undifferentiated cells in Define Medium (DM), a comparative transcriptomic analysis of the stemness phenotype was performed.

The extraction of the total RNA from U87-MG cell line was carried out using the RNeasy kit (Qiagen) on 1 million cells according to the manufacturer's recommendations. Quantitative Polymerase Chain Reaction (qPCR) was performed on 50 ng of cDNA using Taqman probes on *GAPDH* and *HPRT* as reference genes. The CSC markers used are *CD133*, *Nanog*, *Sox2* and *Oct4*. The analysis is performed using QuantStudio3 (Thermo Fisher) and relative expressions are estimated by $\Delta\Delta$Ct method using the average of the two reference genes as endogenous control.

2.5. Crossover Frequency Experiment

2.5.1. DEP Sensor Design and Experimental Setup

In order to identify the cells' DEP signature, we use a specific BiCMOS RF-sensor implemented on a microfluidic chip, as presented in Figure 4. The developed UHF-DEP lab-on-chip allows the electro-manipulation of one single cell. Its structure is made of four electrodes to generate a non-uniform electric field. They are set at 90° across the microfluidic channel. In order not to disturb the fluid flow and not to obstruct the channel with the passage of cells, the two electrodes parallel to the channel (in dark gray in Figure 4b) are very thin and 0.45 µm high. The other pair of electrodes perpendicular to the channel are thicker: 9 µm high, in order to ensure a sufficiently strong field over the height of the channel. The implemented gaps between the electrodes are 40 µm wide to generate dielectrophoretic force with a low applied voltage to trap efficiently biological cells. The two pairs of electrodes are biased with a high-frequency continuous wave (CW) signal. The fabrication process of the chip is detailed in [25]. The microfluidic channel is molded in a polydimethylsiloxane (PDMS) cap to drive the cell suspension to the sensor array. The channel is 150 µm wide and 50 µm high.

The experimental setup for the crossover frequency measurement is shown in Figure 5. Once the cells were suspended in the DEP medium, the Eppendorf was linked to the UHF-DEP lab-on-chip thanks to capillary tubes. The cell suspension is injected in the chip by external flow controllers. They apply input and output pressures in order to regulate the speed and the motion of the cells in the microfluidic channel.

The UHF signal is produced thanks to a radio-frequency signal generator (whose frequency range is adjustable from 10 MHz to 1.1 GHz) which is then amplified. The signal generated can reach a magnitude of 10 Vpp while keeping a high purity continuous wave (CW) signal. During the crossover frequency measurement, the signal voltage is set between 2 and 4 Vpp. The applied signal is then directed to a power divider in order to bias the pair of thick electrodes simultaneously with the same signal, while the thin electrodes are grounded. The DEP signal is propagated until the quadrupole sensor thanks to 50 Ω microstrip transmission lines which are connected to RF probes. The switch driver allows switching between the high-frequency applied signal to the low-frequency applied signal. To measure the first crossover frequency f_{x01}, a low-frequency signal can be generated by a second generator whose frequency range can be set from 1 µHz to 80 MHz. Then, the low frequency applied signal is set between 2 and 4 Vpp and propagated through the power divider to the RF probes.

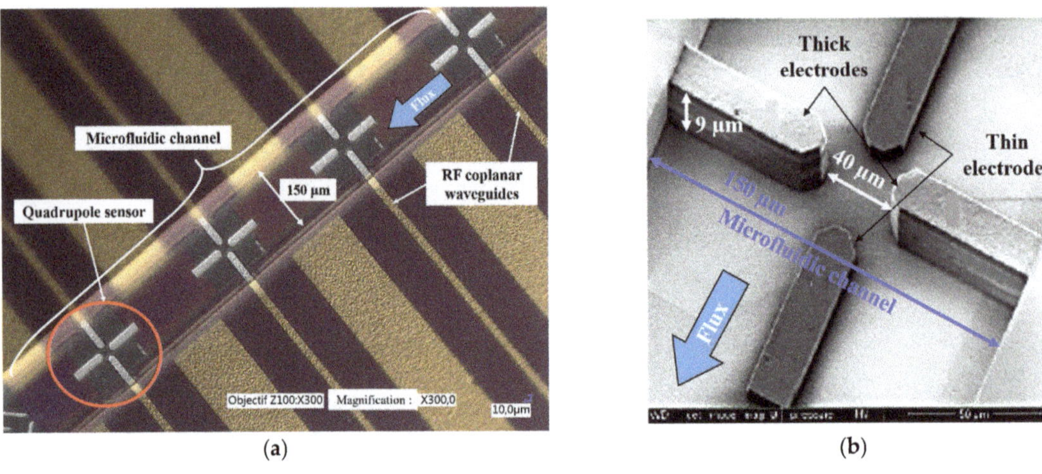

(a)   (b)

Figure 4. (**a**) Quadrupole microelectrodes sensor implemented with a microfluidic channel in BiCMOS; (**b**) Scanning Electron Microscopy (SEM) picture of a quadrupole sensor.

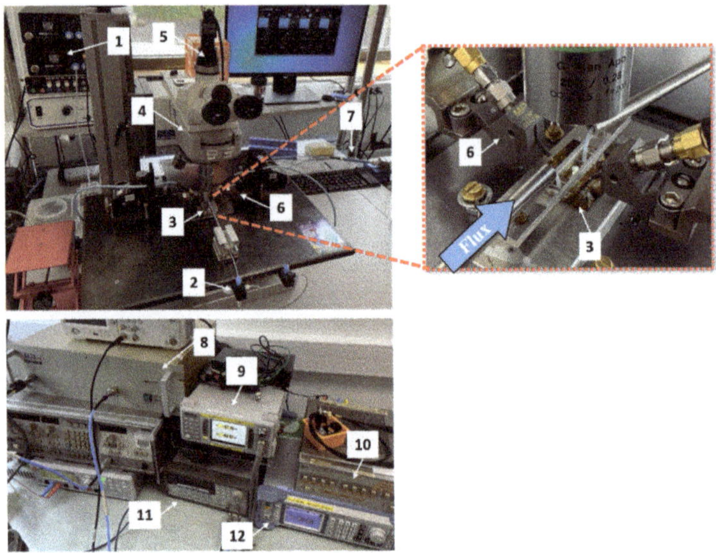

Figure 5. Experimental bench where the crossover frequencies measurements are performed. The labeled parts refer to: (**1**) external flow controllers Elveflow OB1; (**2**) cell suspension; (**3**) UHF-DEP microfluidic chip, a zoomed picture is shown in the red dotted box; (**4**) Scope.A1 Zeiss Microscope; (**5**) camera Axiocam 105 color Zeiss; (**6**) RF probes MPI TITAN T26P-GSG-150; (**7**) power divider; (**8**) power amplifier Bonn Elecktrik BLWA 100-5M; (**9**) power meter Anritsu ML2496A; (**10**) attenuator/switch driver 11713A HP; (**11**) low-frequency generator Agilent 33250A; (**12**) RF signal generator SMB 100A from Rhode & Schwarz.

2.5.2. f_{x01} and f_{x02} Crossover Frequencies Measurements

The aim of this article is to show that we can take benefit from the second crossover frequency f_{x02} to discriminate differentiated cells from undifferentiated cells in GBM cell line and patient GBM cells from primary culture, whereas f_{x01} cannot emphasize this discrimination. To do so, we used our lab-on-chip to measure the crossover frequencies and to characterize cells' DEP signature according to their different culture conditions (Normal Medium NM and Define Medium DM). First, the cells are brought to the characterization area, i.e., the quadrupole sensor, thanks to the external flow controllers. Once a single cell is present in the center of the quadrupole such as in the first picture of Figure 6b, the flow is temporarily cut off and stabilized in order to proceed to the crossover frequency measurement and the electromagnetic signal is switched on. The flow is stopped during the DEP characterization to avoid the competition between forces, so that the cell is only submitted to the dielectrophoretic force and the natural gravity.

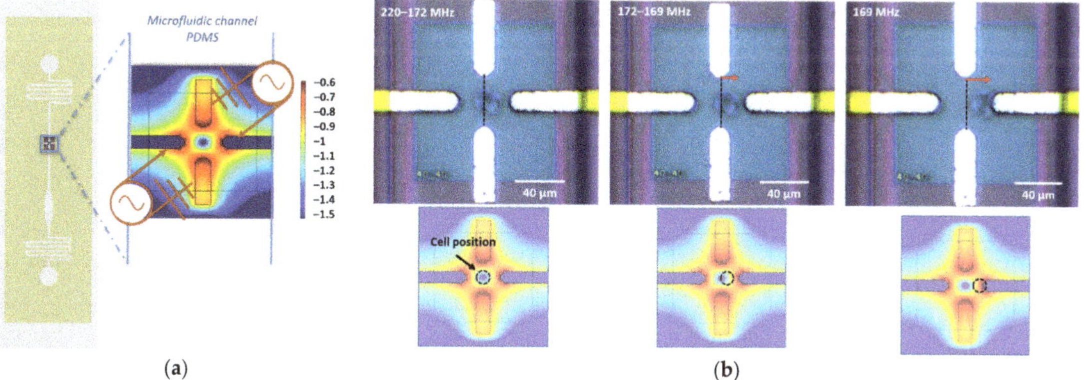

Figure 6. Principle of the crossover frequency measurement: (**a**) schematic of the microfluidic chip with a zoom in on one quadrupole sensor. Computation of the biased sensor in a non-uniform electric field (COMSOL Multiphysics®). The scale color corresponds to the normalized electromagnetic field intensity; (**b**) dielectrophoretic response of a single U87-MG NM cell under an UHF applied signal for frequencies between 220 and 169 MHz. The second crossover frequency f_{x02} is measured at 169 MHz.

The Figure 6a shows the quadrupole biased whatever the investigated frequency from the low- to high-frequency range. One can notice that the areas of strong electric field intensity (in orange/red) are located at the different edges of the electrodes. As said before, these zones are related to the pDEP cell behavior, whereas the area of weak field intensity (in dark blue) is located at the center of the electrodes, which is assimilated to the nDEP cell behavior. The DEP sensor is biased firstly with the UHF generated signal at 500 MHz. At this frequency range (Figure 3a), we expect the cell to present nDEP behavior, and it is far from its crossover frequency. The dielectrophoretic force is thus repulsive and the cell is trapped within the central electrical cage created by the quadrupole, as shown in the first picture of Figure 6b. Then, we decrease the frequency of the applied signal. The DEP force starts to become attractive and we can observe the first movement of the cell (second picture in Figure 6b). Finally, the cell is pulled toward the edge of one of the lateral electrodes, which is the pDEP area (last picture in Figure 6b). Hence, we can tune the frequency of the signal from a repulsive state in the center of the sensor to an attractive state. The crossover frequency f_{x02} can be determined from the motion of the cell from the nDEP behavior to the pDEP behavior, which can be observed optically under a microscope. In order to precisely identify f_{x02}, we first decrease the frequency of the applied signal by steps of 10 MHz in order to approach the crossover frequency. Then, we slowly scan the frequency by steps of 1 MHz to observe the cell motion. This operation is repeated once again in order to accurately determine f_{x02}. Then, we increase the applied frequency to place the cell in the center of the quadrupole. We turn off the UHF signal generator and use the switch driver in order to inject the low-frequency signal in the lab-on-a-chip and to determine the first crossover frequency f_{x01} of the same cell. The same procedure for the characterization of f_{x02} is used for the measurement of f_{x01}. We turn on the generator and we apply a sinusoidal signal at 10 kHz in order to place the cell in its nDEP behavior. Next, we increase the frequency by steps of 10 kHz until we observe the cell motion. Then, we scan slowly the frequency by steps of 1 kHz to have an accurate value of the crossover frequency. This characterization process is duplicated to confirm the measured value. Finally, the electric signal is turned off and the flow pressure at the chip inlet is increased to release the characterized cell and renew the cell suspension in the microfluidic channel. A new cell is next trapped and fully characterized following the same method.

Hence, the resolution of the two measured crossover frequencies f_{x01} and f_{x02} is, respectively, 1 kHz and 1 MHz. One should notice that due to the natural biological heterogeneity occurring among a cell population, the crossover frequencies might spread out on a more or less large frequency range. Nevertheless, the repeatability and reproducibility of the crossover frequency measurements allow us to consolidate the collected data. Afterwards, the comparison of different crossover frequencies recorded from distinct cells or conditions is validated with statistical analyses. Hence, we consider that the identification of the DEP signature (collection of crossover frequencies from distinct tumor cells) is representative of the whole cell population.

2.6. Statistical Analysis

Statistical analysis was performed using PAST software. Comparisons between groups were analyzed by ANOVA test. $p < 0.005$ was considered significant (* $p < 0.05$; ** $p < 0.01$; *** $p < 0.001$)

3. Results

3.1. Enrichment of CSC in the Define Medium

In order to enrich the tumor cell populations in undifferentiated cells related to CSC, U87-MG cells were cultured in Define Medium for 5 days. Morphological changes are observed macroscopically in these stringent culture conditions. As expected, the morphology of U87-MG NM vs. U87-MG DM is completely different (Figure 7a). In Normal Medium, cells are spread out in the petri dish, whereas in Define Medium, cells develop the ability to form glioma spheres due to the presence of specific growth factors (EGF and bFGF-2). It is known that neural stem cells cultured in vitro have the capability to generate clonal structures called "neurospheres" [26]. Glioma spheres are composed of a wealth of aggregated cells. However, just before the DEP characterization, cells are resuspended in the DEP medium and glioma spheres are mechanically broken with the action of a micropipette. When the cell suspension is injected in the lab-on-a-chip, the cells cultured in different conditions present a round shape, and no significant difference in morphology can be observed under an optical microscope (Figure 7b).

To confirm the enrichment of cancer stem cells from total cell population, we achieved a transcriptomic analysis in order to assess the changes of mRNA expression levels related to CSC biomarkers (*CD133*, *Nanog*, *Sox2* and *Oct4*) when U87-MG cells were cultured either in the Normal Medium or in define medium. mRNA relative quantification of U87-MG cultured in Define Medium were normalized, respectively, to the gene expression of U87-MG culture in the Normal Medium (dotted line) (Figure 8). As expected, CSC transcripts were overexpressed in define medium cultured cells, confirming the enrichment of CSCs in cell subpopulation.

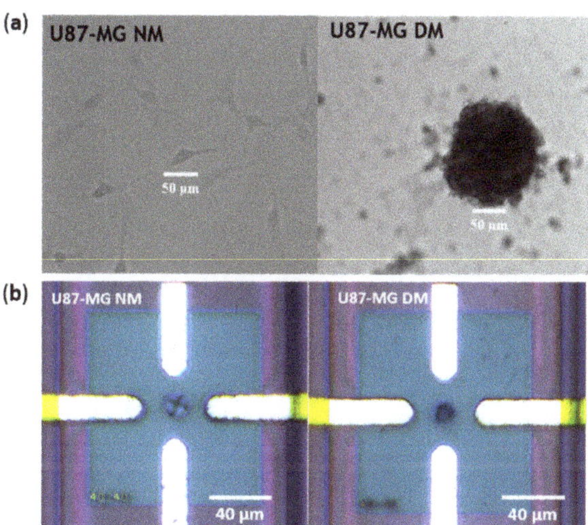

Figure 7. (**a**) Microscope view of the U87-MG cell line cultured in two different conditions: Normal Medium (NM) and after 2 days in define medium (DM); (**b**) microscope view of characterized U87-MG cell line trapped in our quadrupole sensor.

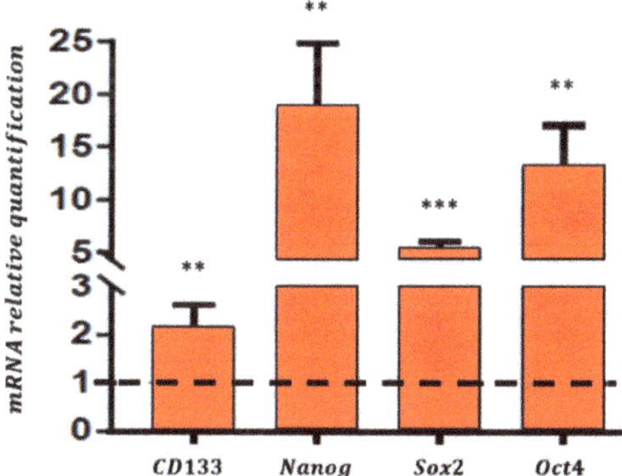

Figure 8. Comparative analysis of gene expression of four undifferentiated markers: *CD*133, *Nanog*, *Sox*2t and *Oct*4 among U87-MG cell line, cultured in Normal Medium (dotted line), or in define medium, measured by Real Time PCR. *** represents p-value < 0.001, ** represents p-value < 0.01.

3.2. Dielectrophoretic Signatures f_{x01} and f_{x02} of U87-MG Cell Line

The U87-MG cell line has been characterized using the microfluidic lab-on-a-chip using the method previously described. The first crossover frequency f_{x01} and the second crossover frequency f_{x02} of the same trapped cells have been successively measured.

The measurement results for both culture conditions summarized in the violin plot thereby illustrate the distribution over frequency of the data (Figure 9). Violin plots are very similar to box plots, except that they additionally show the probability density curve of the different data. The small white dot marker labels the median value of the dataset (small white dot). Moreover, as for the box plot, the first and fourth quartiles of the dataset are represented by the thin black line, and 50% of the whole cell population is concentrated in the thick black line. The extreme peaks correspond to the minimum and maximum values. In Figure 9, one should notice that the scales for the two crossover frequencies are the same except that the unit are kHz and MHz for f_{x01} and f_{x02}, respectively.

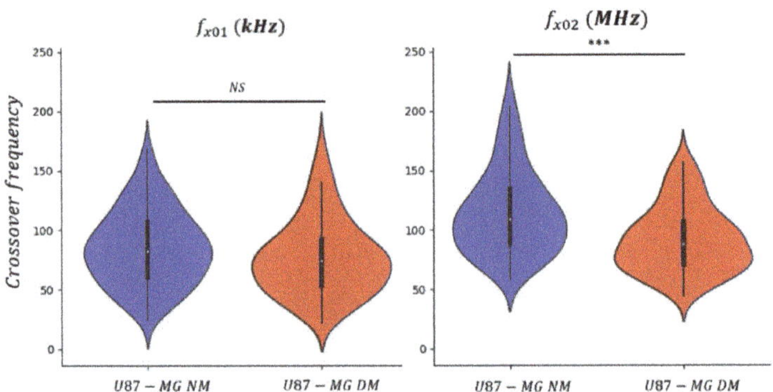

Figure 9. Graphic violin plot representation of U87-MG cells crossover frequencies f_{x01} (left graph) and f_{x02} (right graph), cultured in two different conditions: Normal Medium (NM) and define medium (DM). *** represents *p*-value < 0.0001.

Descriptive statistics including number of cells, median value and standard deviation of crossover frequency for distinct culture conditions are reported in Table 2. One can notice, for both crossover frequencies datasets on Figure 9, the wide-ranging dispersion of measured values, which is also reflected in the value of the standard deviation. This observation is due to the normal heterogeneity of the cells among the GBM cell population. At low frequency, f_{x01} is influenced by the small difference of the cell size and morphology within the cell culture, while at high frequency, f_{x02} might be dependent of intracellular changes or alterations. For instance, during the cell cycle, the nucleocytoplasmic ratio might differ from a cell to another as they are not synchronized during the culture [27]. However, the violin plot highlights the fact that cell's crossover frequencies are gathered around their respective median value.

Table 2. Values regarding the crossover frequency measurements of the U87-MG cell line.

Cell Culture Conditions	Crossover Frequency	Number of Cells Measured	Median Value	SD
Normal Medium (NN)	f_{x01}	139	82 kHz	31.5 kHz
Define medium (DN)		134	74 kHz	32.1 kHz
Normal Medium (NN)	f_{x02}	139	109 MHz	35.2 MHz
Define medium (DN)		134	88 MHz	27.9 MHz

A representative number of cells have been individually characterized to statistically consolidate the collected dataset and make the established signatures significant. One can notice that the distribution of the first crossover frequency f_{x01} is mostly the same for the two culture conditions, NM vs. DM. The median value of f_{x01} for the undifferentiated enriched population (DM) shows no significant difference with the normal conditions: 74 and 82 kHz, respectively. Indeed, as shown previously in Figure 8, both U87-MG NM and U87-MG DM present the same round shape morphology. In contrast, the distribution of the second crossover frequency f_{x02} exhibits a significant difference despite an overlap in frequency. This can be explained by the fact that GBM cell population cultured in normal conditions include a majority of differentiated cells but also few undifferentiated cells. However, for the DM conditions, there are more undifferentiated cells, since presenting stringent survival conditions, DM is more selective. Thus, we can observe a decrease in the crossover frequency f_{x02} with the presence of the DM cell pool. The median values for U87-MG NM and U87-MG DM are, respectively, 109 and 88 MHz. The decrease in f_{x02} shows a significant difference between the two cells' phenotypes, as the p-value is lower than 10^{-3}.

This demonstrates that undifferentiated cells compared to differentiated cells own different intracellular dielectric properties. Despite displaying two crossover frequencies, only the one in the UHF range, f_{x02}, is sufficiently meaningful to be exploited for identifying cells presenting an undifferentiated state or a stemness-like phenotype. These results show how promising UHF-DEP cell profile analysis might be for the discrimination of cell subpopulation within the tumor. Next, we will focus on the second crossover frequency f_{x02} and strengthen the relevance of using UHF-DEP as a discriminant parameter through a kinetic study of the evolution of the stemness phenotype.

3.3. Kinetic Evolution of the Stemness Phenotype

As the enrichment of this cell population is accomplished by seeding normal U87-MG cells in Define Medium, either cells already exerting an undifferentiated profile can survive, or other cells must acquire phenomenon is a process that requires several division cycles and we specific features related to stem cells to survive in Define Medium. With the acquirement of cell undifferentiated status, cell aggressiveness is increased and is related to tumor aggressiveness. The undifferentiation proposed to follow its kinetics by UHF-DEP to demonstrate the potential of the UHF-DEP microsystem developed.

To do so, the U87-MG cell line was cultured within three different conditions: (i) 6 days in Normal Medium (NM); (ii) 5 days in Define Medium (DM); (iii) maintained during 21 days in Define Medium (DM+). The results of the measured crossover frequency f_{x02} are presented in the violin plot chart in Figure 10. In addition to the data already collected previously for U87-MG NM and U87-MG DM cells, we remeasured about hundred more crossover frequencies f_{x02} in order to consolidate the previously obtained DEP signatures and to improve the statistical strength of our analysis.

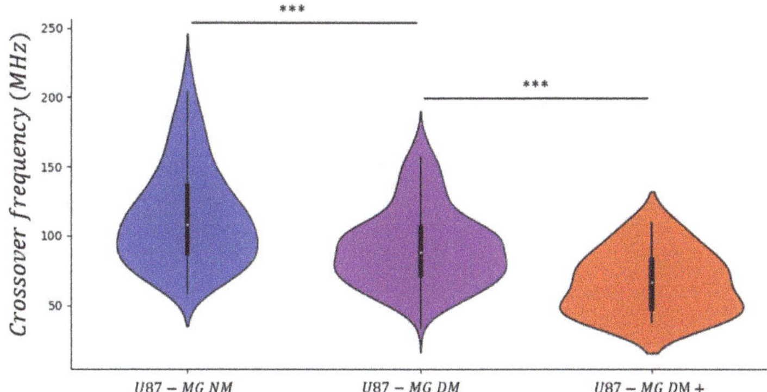

Figure 10. Graphic violin plot representation of U87-MG cells crossover frequency f_{x02}, cultured in three different conditions: (i) 5 days in Normal Medium (NM); (ii) 5 days in define medium (DM); (iii) 21 days in define medium (DM+). *** represents p-value < 0.001.

Statistics results related to the collected data are reported in Table 3, including the median value and the standard deviation of f_{x02}. As said before, the distribution of the second crossover frequency values is dispersed. This observation is highlighted by the high values of the standard deviation. However, the standard deviation seems to decrease the more the cells are maintained in Define Medium. Indeed, as the GBM cell line presents a high biological heterogeneity, the Define Medium tends to select only cells which are able to survive under such stringent culture conditions, i.e., cells with an undifferentiated phenotype. Despite the data dispersion, the violin plot shows that the cell's crossover frequency values are mostly gathered around their respective median value.

Table 3. Values regarding the crossover frequency measurements of the U87-MG cell line.

Culture Condition	Median Value	SD
Normal Medium (NM)	108 MHz	36.2 MHz
Define Medium (DM)	88 MHz	27.9 MHz
Define Medium (DM+)	67 MHz	22.1 MHz

In Normal Medium, the median value of the second crossover frequency is 108 MHz. After 5 days in Define Medium, the median DEP signature decreases to 88 MHz and after 16 additional days in Define Medium, it is 67 MHz. The two successive decreases in f_{x02} observed between the three cell populations present significant differences, as the p-value is lower than 10^{-3}. In correlation with the previous experimentation, a lower dielectric signature seems to potentially characterize cells presenting a stemness-like phenotype.

One can also notice that although more cells were characterized in the NM and DM conditions for this measurement campaign, the median value of the crossover frequency is not affected and remains the same compared to the first campaign (related in Table 2). It shows the robustness and the reproducibility of our method to measure the DEP signatures of cells.

These two experiments demonstrate the ability of the developed UHF-DEP lab-on-chip to successfully extract information about the potential stemness status of U87-MG cells by the measurement of the second crossover frequency f_{x02}.

3.4. Dielectrophoretic Signatures f_{x01} and f_{x02} of GBM Primary Cultures

We previously measured (in Section 3.2) the low- and high-frequency DEP signatures of the U87-MG cell line, cultured in two different conditions in order to induce an undifferentiation corresponding to the CSC subpopulation. From the obtained crossover frequency, f_{x01} did not show any difference between Normal Medium and Define Medium. However, we demonstrated that f_{x02} presents a significant difference and can be a relevant discriminant parameter to identify the CSC subpopulation. These results of the DEP signatures were obtained from the in vitro cell line.

To go further, we proposed in the last section of this paper to repeat this experiment on ex vivo GBM primary cultures to demonstrate the potential clinical applications of our approach.

The two crossover frequencies f_{x01} and f_{x02}, respectively were characterized in GBM primary culture cells derived from four patients. These cells were collected after surgery on patients suffering from glioblastoma. Once extracted from the tumor samples, cells were put in culture according to the procedure indicated in Materials and Methods. As for the U87-MG cell line, a few minutes before DEP characterization, the ex vivo GBM cells were resuspended in the DEP medium. Then, for each investigated cell, their two crossover frequencies f_{x01} and f_{x02} were measured with the previously described protocol.

Before DEP characterizations, the GBM cell population was first separated into two subpopulations: CD133− and CD133+. CD133 is a transmembrane protein expressed in human hematopoietic stem cells and progenitor cells [28]. As said before, CD133 is a biomarker associated with stem-like cells, and thus with tumor regeneration. It is possible to mark cells with monoclonal antibodies anti-AC133 coupled with a fluorochrome to detect the presence of the peptide CD133 on the cell surface [29]. Nevertheless, CD133 can be also expressed in differentiated cancer cells, so the whole cell population will present a fluorescent intensity gradient [30]. Hence, we impose a threshold of the fluorescent intensity during the passage of cells in the flow cytometer to define two subpopulations. The CD133+ population is the cell population that overexpresses the marker and thus is enriched in CSCs, while the CD133− population is the population of differentiated cells [28]. Therefore, we separate and isolate the CSC cell population from the differentiated one thanks to a fluorescent marker, before DEP characterizations of these populations. The results of the measured crossover frequencies f_{x01} and f_{x02} for both populations are presented in the violin plot (Figure 11).

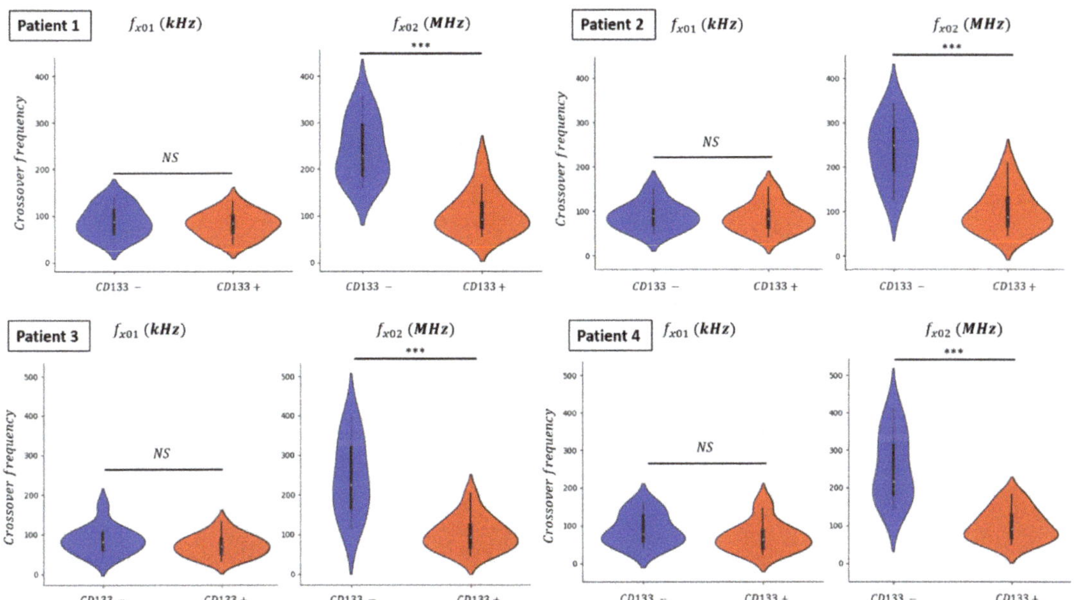

Figure 11. Graphic violin plot representation of crossover frequencies f_{x01} and f_{x02} of GBM primary cells collected from four different patients. *** represents *p*-value < 0.001.

One should notice that the scale for the two crossover frequencies is the same except that the units are kHz and MHz for f_{x01} and f_{x02}, respectively. The data distribution of crossover frequencies f_{x01} and f_{x02} from the characterized glioblastoma cell shows similar violin plot shapes although cells were derived from four different patients (Figure 11). Indeed, we can expect from ex vivo GBM cells to be even more genetically heterogeneous from one patient to another than the immortalized U87-MG cell line. Median values of the crossover frequencies measured for each isolated population are reported in the Table 4. The obtained results highlight that our DEP cell analyzer lab-on-chip is a relevant and reliable tool to study and analyze either in vitro or ex vivo dissociated samples.

The distribution of the first crossover frequency f_{x01} is mostly the same for the two isolated populations, CD133− vs. CD133+. The median value of f_{x01} for the undifferentiated population (CD133+) shows no significant changes with the differentiated one. At low frequency, the largest signature dissimilarity between the two conditions corresponds to patient 4, where f_{x01} displays a change of 15 kHz between CD133− and CD133+ conditions (from 78 to 63 kHz). However, this difference is not statistically significant to discriminate the subpopulation of CSCs overrepresented in the CD133+ cells. The distribution of the second crossover frequency f_{x02} exhibits a difference despite an overlap among the measured values. This can be explained by the method of enrichment through flow cytometry. Indeed, the CD133 biomarker is not a binary label as the whole cell population might present different fluorescent intensities. With this fluorescent gradient, we choose a threshold to separate GBM cells into two subpopulations and to be selective toward CSC population. Thus, we can observe a decrease in the crossover frequency f_{x02} with the presence of the CD133+ cell pool. The median values of the second crossover frequency display the smallest change for patient 4, which is 125 MHz (from 216 to 91 MHz). The decrease in f_{x02} shows a significant difference between the two cell populations, as the p-value is lower than 10^{-3}. Moreover, one can notice in Figure 11 that at high frequency, the CD133+ cell population displays a stoutness around the median value compared to the CD133− cell, for whom the values are gathered around their median.

Table 4. Values of the crossover frequency measurements of the GBM primary culture extracted from tumor samples of four different patients.

	Cell Population	Crossover Frequency	Median Value
Patient 1	CD133−	f_{x01}	88 kHz
	CD133+		83 kHz
	CD133−	f_{x02}	229 MHz
	CD133+		92 MHz
Patient 2	CD133−	f_{x01}	89 kHz
	CD133+		83 kHz
	CD133−	f_{x02}	248 MHz
	CD133+		86 MHz
Patient 3	CD133−	f_{x01}	81 kHz
	CD133+		70 kHz
	CD133−	f_{x02}	225 MHz
	CD133+		92 MHz
Patient 4	CD133−	f_{x01}	78 kHz
	CD133+		63 kHz
	CD133−	f_{x02}	216 MHz
	CD133+		91 MHz

Such results validate that we can exploit the intracellular dielectric properties differences between differentiated and undifferentiated cells by measuring the second crossover frequency f_{x02}. The cells extracted from patients' GBM tumor samples show similar behavior as observed with an in vitro GBM cell line. The f_{x02} median value of ex vivo cells are not the same as in vitro results, but we can extrapolate that an ex vivo cell with a low f_{x02} could be a stem-like cell.

4. Conclusions

In this article, we used an innovative microfluidic device based on high-frequency dielectrophoresis for single-cell characterization in order to discriminate and identify the cancer stem cell subpopulation. First, we evaluated the discrimination capabilities of our microfluidic device in vitro on a glioblastoma cell line. U87-MG cells were cultured in two distinct conditions: one inducing differentiation and the second selecting immature and undifferentiated cells. Our results suggest that the expression of biological CSC markers and the measurement of the UHF crossover frequency f_{x02} are closely linked. At this frequency range, our lab-on-chip is able to interact with the intracellular content, which is more representative of the undifferentiated features of cells, making UHF-DEP greatly relevant to investigate the stemness status of cancer cells. As a first step towards clinical experiments, some GBM cells were extracted and cultured from patients' tumors. These GBM primary cells have been sorted into two subpopulations according to their expression level of CSC biomarker CD133. Whatever the considered patient, observed DEP signatures display the same profile. As previously identified, f_{x02} shows a more significant and more important difference between the two cell phenotypes than f_{x01} and so is confirmed to be a relevant CSC discriminant parameter. As primary culture cells are more representative of tumor than cell lines, we believe that it might be possible to transpose this capability of UHF-DEP cell characterization for recognizing "stemness" features from tumor cells derived to a broad range of GBM patients.

UHF-DEP is a very promising tool with great potential to discriminate cells according to their internal biological properties. Hence, from the identification of UHF-DEP signatures, we can see the perspective to develop a cell sorting device for isolating cancer stem cells [31]. In the future, the early detection of CSC subpopulation in a glioma tumor with a UHF-DEP approach could have a prognosis value on therapeutic response and might allow adaption of a therapeutic strategy following diagnosis.

Author Contributions: Conceptualization, E.L. and E.B.; methodology, R.M. and S.S.; software, E.L. and E.B.; formal analysis, E.L. and E.B.; investigation, E.L., E.B., R.M., S.S., F.M. and E.P.; data curation, A.P.; writing—original draft preparation, E.L.; writing—review and editing, E.L., C.D., F.L., A.P. and L.P.; supervision, A.P., C.D., B.B., F.L., L.P. and G.V.; project administration, A.P. and F.L.; funding acquisition, A.P., F.L., L.P. and G.V. All authors have read and agreed to the published version of the manuscript.

Funding: This research was funded by the European Union's Horizon 2020 research and innovation program under grant number 737164, and by the Nouvelle Aquitaine Council with the Oncosome-track project and by funds from the Pediatric Research Institute Foundation (IRP 18/06). F.M. was supported by a fellowship from the Italian Association for Cancer Research (AIRC) (ID 19575). E.P. was supported by a fellowship from the Umberto Veronesi Foundation (#1142).

Institutional Review Board Statement: The study was conducted according to the guidelines of the Declaration of Helsinki, and approved by the committee of the Sperimentazione Clinic of Padova Provincia (n°2462P, 29 November 2016).

Informed Consent Statement: Informed consent was obtained from all subjects involved in the study as described in the Materials and Methods paragraph.

Acknowledgments: We are grateful to Chiara Frasson (Pediatric Research Institute, Padova, Italy) for technical assistance in flow cytometry procedures.

Conflicts of Interest: The authors declare no conflict of interest.

References

1. Cheray, M.; Bégaud, G.; Deluche, E.; Nivet, A.; Battu, S.; Lalloué, F.; Verdier, M.; Bessette, B. Cancer Stem-Like Cells in Glioblastoma. *Exon. Publ.* **2017**, 59–71. [CrossRef]
2. Stupp, R.; Mason, W.P.; Bent, M.V.D.; Weller, M.; Fisher, B.; Taphoorn, M.J.; Belanger, K.; Brandes, A.; Marosi, C.; Bogdahn, U.; et al. Radiotherapy plus Concomitant and Adjuvant Temozolomide for Glioblastoma. *N. Engl. J. Med.* **2005**, *352*, 987–996. [CrossRef] [PubMed]
3. Goldsmith, H.S. Potential Improvement of Survival Statistics for Glioblastoma Multiforme (WHO IV). *Surg. Neurol. Int.* **2019**, *10*, 123. [CrossRef] [PubMed]
4. Tabatabai, G.; Weller, M. Glioblastoma Stem Cells. *Cell Tissue Res.* **2011**, *343*, 459–465. [CrossRef] [PubMed]
5. Chen, J.; Li, Y.; Yu, T.-S.; McKay, R.M.; Burns, D.K.; Kernie, S.G.; Parada, L.F. A Restricted Cell Population Propagates Glioblastoma Growth after Chemotherapy. *Nature* **2012**, *488*, 522–526. [CrossRef]
6. Gossett, D.R.; Weaver, W.M.; Mach, A.J.; Hur, S.C.; Tse, H.T.K.; Lee, W.; Amini, H.; Di Carlo, D. Label-Free Cell Separation and Sorting in Microfluidic Systems. *Anal. Bioanal. Chem.* **2010**, *397*, 3249–3267. [CrossRef]
7. Pethig, R.R. *Dielectrophoresis: Theory, Methodology and Biological Applications*; John Wiley & Sons: Hoboken, NJ, USA, 2017; ISBN 978-1-118-67145-0.
8. Cottet, J.; Fabregue, O.; Berger, C.; Buret, F.; Renaud, P.; Frénéa-Robin, M. MyDEP: A New Computational Tool for Dielectric Modeling of Particles and Cells. *Biophys. J.* **2018**, *116*, 12–18. [CrossRef] [PubMed]
9. Afshar, S.; Fazelkhah, A.; Braasch, K.; Salimi, E.; Butler, M.; Thomson, D.J.; Bridges, G.E. Full Beta-Dispersion Region Dielectric Spectra and Dielectric Models of Viable and Non-Viable CHO Cells. *IEEE J. Electromagn. RF Microw. Med. Biol.* **2020**, *5*, 70–77. [CrossRef]
10. Fazelkhah, A.; Afshar, S.; Braasch, K.; Butler, M.; Salimi, E.; Bridges, G.; Thomson, D. Cytoplasmic conductivity as a marker for bioprocess monitoring: Study of Chinese hamster ovary cells under nutrient deprivation and reintroduction. *Biotechnol. Bioeng.* **2019**, *116*, 2896–2905. [CrossRef]
11. Cottet, J.; Fabregue, O.; Berger, C.; Buret, F.; Renaud, P.; Frénéa-Robin, M. MyDEP: A New Computational Tool for Dielectric Modeling of Particles and Cells. *Zenodo* **2019**. [CrossRef]
12. Pethig, R.; Menachery, A.; Pells, S.; De Sousa, P. Dielectrophoresis: A Review of Applications for Stem Cell Research. *J. Biomed. Biotechnol.* **2010**, *2010*, 182581. [CrossRef] [PubMed]
13. Sato, N.; Yao, J.; Sugawara, M.; Takei, M. Numerical Study of Particle-Fluid Flow Under AC Electrokinetics in Elec-trode-Multilayered Microfluidic Device. *IEEE Trans. Biomed. Eng.* **2018**, *66*, 453–463. [CrossRef] [PubMed]
14. Du, X.; Ma, X.; Li, H.; Li, L.; Cheng, X.; Hwang, J.C.M. Validation of Clausius–Mossotti Function in Wideband Single-Cell Dielec-trophoresis. *IEEE J. Electromagn. RF Microw. Med. Biol.* **2019**, *3*, 127–133. [CrossRef]

15. Alazzam, A.; Stiharu, I.; Bhat, R.; Meguerditchian, A.-N. Interdigitated Comb-like Electrodes for Continuous Separation of Malignant Cells from Blood Using Dielectrophoresis. *Electrophoresis* **2011**, *32*, 1327–1336. [CrossRef] [PubMed]
16. Choi, S.; Park, J.-K. Microfluidic system for dielectrophoretic separation based on a trapezoidal electrode array. *Lab Chip* **2005**, *5*, 1161–1167. [CrossRef] [PubMed]
17. Vahey, M.D.; Voldman, J. An Equilibrium Method for Continuous-Flow Cell Sorting Using Dielectrophoresis. *Anal. Chem.* **2008**, *80*, 3135–3143. [CrossRef]
18. Chung, C.; Pethig, R.; Smith, S.; Waterfall, M. Intracellular Potassium under Osmotic Stress Determines the Dielectrophoresis Cross-over Frequency of Murine Myeloma Cells in the MHz Range. *Electrophoresis* **2018**, *39*, 989–997. [CrossRef] [PubMed]
19. Chung, C.; Waterfall, M.; Pells, S.; Menachery, A.; Smith, S.; Pethig, R. Dielectrophoretic Characterisation of Mammalian Cells above 100 MHz. *J. Electr. Bioimpedance* **2011**, *2*, 64–71. [CrossRef]
20. Manczak, R.; Saada, S.; Tanori, M.; Casciati, A.; Dalmay, C.; Bessette, B.; Begaud, G.; Battu, S.; Blondy, P.; Jauberteau, M.O.; et al. High-Frequency Dielectrophoresis Characterization of Differentiated vs Undifferentiated Medulloblastoma Cells. In Proceedings of the 2018 EMF-Med 1st World Conference on Biomedical Applications of Electromagnetic Fields (EMF-Med); 2018; pp. 1–2. [CrossRef]
21. Casciati, A.; Tanori, M.; Manczak, R.; Saada, S.; Tanno, B.; Giardullo, P.; Porcù, E.; Rampazzo, E.; Persano, L.; Viola, G.; et al. Human Medulloblastoma Cell Lines: Investigating on Cancer Stem Cell-Like Phenotype. *Cancers* **2020**, *12*, 226. [CrossRef]
22. Pistollato, F.; Persano, L.; Della Puppa, A.; Rampazzo, E.; Basso, G. Isolation and Expansion of Regionally Defined Human Glio-blastoma Cells In Vitro. *Curr. Protoc. Stem Cell Biol.* **2011**, *17*, 3.4.1–3.4.10. [CrossRef]
23. Pistollato, F.; Abbadi, S.; Rampazzo, E.; Persano, L.; Della Puppa, A.; Frasson, C.; Sarto, E.; Scienza, R.; D'Avella, D.; Basso, G. Intratumoral Hypoxic Gradient Drives Stem Cells Distribution and MGMT Expression in Glioblastoma. *Stem Cells* **2010**, *28*, 851–862. [CrossRef]
24. Frasson, C.; Rampazzo, E.; Accordi, B.; Beggio, G.; Pistollato, F.; Basso, G.; Persano, L. Inhibition of PI3K Signalling Selectively Affects Medulloblastoma Cancer Stem Cells. *BioMed Res. Int.* **2015**, *2015*, 973912. [CrossRef]
25. Hjeij, F.; Dalmay, C.; Bessaudou, A.; Blondy, P.; Pothier, A.; Bessette, B.; Bégaud, G.; Jauberteau, M.O.; Lalloué, F.; Kaynak, C.B.; et al. UHF Dielectrophoretic Handling of Individual Biological Cells Using BiCMOS Microfluidic RF-Sensors. In Proceedings of the 2016 46th European Microwave Conference (EuMC), London, UK, 4–6 October 2016; pp. 265–268. [CrossRef]
26. Suslov, O.N.; Kukekov, V.G.; Ignatova, T.N.; Steindler, D.A. Neural stem cell heterogeneity demonstrated by molecular phenotyping of clonal neurospheres. *Proc. Natl. Acad. Sci. USA* **2002**, *99*, 14506–14511. [CrossRef]
27. Pethig, R.; Bressler, V.; Carswell-Crumpton, C.; Chen, Y.; Foster-Haje, L.; García-Ojeda, M.E.; Lee, R.S.; Lock, G.M.; Talary, M.S.; Tate, K.M. Dielectrophoretic Studies of the Activation of Human T Lymphocytes Using a Newly Developed Cell Pro-filing System. *Electrophoresis* **2002**, *23*, 2057–2063. [CrossRef]
28. Li, Z. CD133: A Stem Cell Biomarker and Beyond. *Exp. Hematol. Oncol.* **2013**, *2*, 17. [CrossRef] [PubMed]
29. Ren, F. CD133: A cancer stem cells marker, is used in colorectal cancers. *World J. Gastroenterol.* **2013**, *19*, 2603–2611. [CrossRef] [PubMed]
30. Kemper, K.; Sprick, M.; De Bree, M.; Scopelliti, A.; Vermeulen, L.; Hoek, M.; Zeilstra, J.; Pals, S.T.; Mehmet, H.; Stassi, G.; et al. The AC133 Epitope, but not the CD133 Protein, Is Lost upon Cancer Stem Cell Differentiation. *Cancer Res.* **2010**, *70*, 719–729. [CrossRef] [PubMed]
31. Provent, T.; Mauvy, A.; Manczak, R.; Saada, S.; Dalmay, C.; Bessette, B.; Lalloue, F.; Pothier, A. A High Frequency Dielectrophoresis Cytometer for Continuous Flow Biological Cells Refinement. In Proceedings of the 2020 50th European Microwave Conference (EuMC), Utrecht, The Netherlands, 12–14 January 2021; pp. 921–924. [CrossRef]

Article

VEGF Detection via Simplified FLISA Using a 3D Microfluidic Disk Platform

Dong Hee Kang [1], Na Kyong Kim [1], Sang-Woo Park [2,*] and Hyun Wook Kang [1,*]

[1] Department of Mechanical Engineering, Chonnam National University, 77 Youngbong-ro, Buk-Gu, Gwangju 61186, Korea; kdh05010@gmail.com (D.H.K.); naky0607@gmail.com (N.K.K.)

[2] Department of Ophthalmology, Chonnam National University Medical School and Hospital, Baekseo-ro, Dong-Gu, Gwangju 61469, Korea

* Correspondence: exo70@naver.com (S.-W.P.); kanghw@chonnam.ac.kr (H.W.K.); Tel.: +82-62-530-1662 (H.W.K.); Fax: +82-62-560-1689 (H.W.K.)

Abstract: Fluorescence-linked immunosorbent assay (FLISA) is a commonly used, quantitative technique for detecting biochemical changes based on antigen–antibody binding reactions using a well-plate platform. As the manufacturing technology of microfluidic system evolves, FLISA can be implemented onto microfluidic disk platforms which allows the detection of trace biochemical reactions with high resolutions. Herein, we propose a novel microfluidic system comprising a disk with a three-dimensional incubation chamber, which can reduce the amount of the reagents to 1/10 and the required time for the entire process to less than an hour. The incubation process achieves an antigen–antibody binding reaction as well as the binding of fluorogenic substrates to target proteins. The FLISA protocol in the 3D incubation chamber necessitates performing the antibody-conjugated microbeads' movement during each step in order to ensure sufficient binding reactions. Vascular endothelial growth factor as concentration with ng mL^{-1} is detected sequentially using a benchtop process employing this 3D microfluidic disk. The 3D microfluidic disk works without requiring manual intervention or additional procedures for liquid control. During the incubation process, microbead movement is controlled by centrifugal force from the rotating disk and the sedimentation by gravitational force at the tilted floor of the chamber.

Keywords: fluorescence-linked immunosorbent assay; lab-on-a-disk; vascular endothelial growth factor; 3D microstructure

Citation: Kang, D.H.; Kim, N.K.; Park, S.-W.; Kang, H.W. VEGF Detection via Simplified FLISA Using a 3D Microfluidic Disk Platform. *Biosensors* **2021**, *11*, 270. https://doi.org/10.3390/bios11080270

Received: 12 July 2021
Accepted: 6 August 2021
Published: 11 August 2021

Publisher's Note: MDPI stays neutral with regard to jurisdictional claims in published maps and institutional affiliations.

1. Introduction

As a consequence of recent technological and medical developments, the human average life expectancy has increased by interventions which have focused on age-related disabilities. Analysis of blood biomarkers is essential, which reveals the specificity of biological aging of individuals, such as immune aging, physical function, and anabolism [1]. Among the biomarkers, vascular endothelial growth factor (VEGF) is a signaling molecule to promote the formation of new vessel branches within tumors and progression and metastasis [2]. In particular, VEGF is a pathognomonic biomarker candidate for the diagnosis criteria of age-related macular degradation incidence, which is most related to ischemic eye disease found in the vitreous and aqueous humor in proportion to the VEGF concentration. Observing the variation of VEGF is essential for predicting the effectiveness of therapy and eye disorder prognoses [3–6]. Research works reported that the increase in the prevalence rate of ocular diseases is correlated with concentrations of VEGF [7–12]. Several ophthalmic disorders associated with VEGF concentration, such as pre-proliferative retinopathy [13], ocular ischemic syndrome [14], and retinal vein occlusions [15], can cause retinal ischemia, which can result in irreversible changes in the ocular structures, such as function and anatomy. However, the low concentration of

VEGF and limited aqueous humor sample collection make it difficult in clinical practice to measure the VEGF concentration variation.

Fluorescence-linked immunosorbent assay (FLISA) is a plate-based assay technique for quantifying proteins (antibody and antigen), even in the picogram range per milliliter [16–19]. The protocol using the well-plate platform for FLISA requires the aqueous humor sample to be more than 100 μL in order to detect protein with pg mL^{-1} levels. The amount of the sample that can be extracted for the reliable quantitative analysis is limited to 50–100 μL considering the change in intraocular pressure [20,21]. Moreover, in general, two or more incubation steps are required for 1–2 h each for the binding reaction between proteins in the FLISA protocol [22–25]. Furthermore, repeating washing steps between incubations is necessary to remove remaining reagents in a liquid. The FLISA protocol is a complicated assay procedure and time-consuming. Consequently, innovation in diagnostic testing uses only small samples and enables high-precision measurements [26–32].

In several studies focusing on an FLISA using a lab-on-a-disk platform detecting bio-chemicals, the time requirements for the assay protocol could be notably reduced to under 1 h by using microbeads. Lee et al. reported a fully automated immunoassay on a disk platform using whole blood; this entire process was terminated within 30 min through a fully automated disk for infectious disease (antibody of Hepatitis B) detection [31]. Walsh et al. showed that fluorescence intensity with a 1 ng mL^{-1} resolution could be acquired even though all reagents are simultaneously loaded into the chamber, requiring 10 min in the incubation step. FLISA protocol simplification helps improve the quantitative analysis efficiency [20]. These results show that FLISA using microbeads can decrease the protocol time due to the high specific volume of binding reagents. This is the advantage of the microfluidic platforms; only one-tenth of the volume of the reagent is required compared with traditional plate-based FLISA. Moreover, the washing and detection steps are controlled sequentially under the centrifugal force acting on the disk and microbead sedimentation via density difference, without requiring manual intervention. However, implementation of the FLISA protocol on a microfluidic disk-based platform requires the reduction in the reagent volumes as well as precise control. Furthermore, the microfluidic system is segregated for precise fluid control at each step of FLISA; thereby, it has increased the manufacturing costs and complexity of the system. Poly(methyl methacrylate) (PMMA) sheets are commonly used for manufacturing microfluidic chips at a low cost, since they enable the fabrication of thin and transparent chips [33–36]. However, processing of the PMMA sheet is improper to fabricate the precise geometry of a microfluidic channel to apply the one-step FLISA protocol. For sufficient mixing in the incubation process, novel valve components are tried in the microfluidic system [31,37]. The wax valve components in the disk would physically isolate each incubation process in the protocol. A fully automated detection system is required for the complicated manufacturing process and increases the unit cost to fabricate a microfluidic disk.

From this point of view, this research suggested a method that employs centrifugal and gravitational forces using a three-dimensional microfluidic disk platform to perform the simplified microbead FLISA protocol. This 3D microfluidic disk using multi-material features a hybrid structure comprising a 3D-printed chamber block and laser-cut PMMA layers. The 3D-printed block contains an incubation chamber with a tilted floor for controlling the mixing of reagents through the cyclic movement of microbeads. This microfluidic disk performs precise microfluidic control in the incubation chamber. Furthermore, the washing and detecting steps are achieved without requiring any manual intervention or additional processes. Therefore, the proposed method is expected to diminish the overall cost of manufacturing microfluidic disks.

2. Materials and Methods

2.1. Fabrication of 3D-Printed Block and Post Treatment

The disk layer is designed using CATIA V5 software for 3D modeling. The chamber model is converted to an STL file format required for the 3D printing software. The incubation chamber layer is fabricated via stereolithography apparatus. The 3D printing model (ProJet MJP 2500 Plus, 3D Systems, USA) has the ability to fabricate the plane surface with roughness average (R_a) of less than 0.4 µm, even at an angle of 45° from the printing surface [38]. The incubation chamber has a volume of 25 µL with a tilted floor. The 3D printing employing SLA enables the fabrication of micro-pillar structures with 15 µm surface roughness (see Figure S1 in the Supplementary Material). One side of the chamber is connected to the side wall of the block for liquid transport, and the other side of the chamber is connected to a ventilation hole in the top layer of the block; this configuration is demonstrated in Figure S2 of the Supplementary Material. After printing out the incubation chamber block, post-treatment is performed to remove the wax supports and to clean surfaces. First, the supporting wax in the chamber is roughly removed under a water vapor environment at atmospheric pressure for 30 min. Afterward, the hot oil is used to immerse the incubation chamber block for 2 h in order to entirely remove the remaining wax. After cooling the block at room temperature for 10 min, ultrasonic cleaning with detergents and water is used to remove the residual oil in the chamber; each of these processes are continued for 10 min. Lastly, the washed block is desiccated for 24 h in a convection oven to remove the moisture. After assembling 3D microfluidic disk, microfluidic components are washed to remove remaining debris by pipetting using phosphate-buffered saline (PBS, pH 7.4, Sigma-Aldrich, St. Louis, MO, USA) with 100 µL containing 0.5 wt% bovine serum albumin (BSA, Sigma-Aldrich) and once distilled water with 100 µL. Thereafter, the 3D microfluidic disk is desiccated in a convection oven for 24 h at room temperature.

2.2. Assembly of the 3D Microfluidic Disk

The 3D microfluidic disk is assembled in a layer-by-layer manner as shown in Figure 1a; the laser-cut PMMA disk layers are sequentially stacked along with the 3D-printed block. The washing chamber and microchannel are assembled by adopting pressure-sensitive adhesive (PSA) sheets between the laser-cut PMMA layers. A laser cutting machine (Nova24 Laser Engraver; Thunder Laser, China) was used to process the PSA film-attached PMMA sheets into the disk shape. The laser-cut pattern is aligned through the 5 mm holes on the disc layer to which disk layers are attached sequentially. Then, a press machine is used to apply 4.0 MPa pressure for the PSA layers between the 3D-printed block and the PMMA disk layers at room temperature for 10 min. The PSA sheet prevents the leakage of liquid from the clearance between the 3D-printed block and the PMMA layers. As shown in Figure 1b, the assembled 3D microfluidic disk features an incubation chamber (blue-dyed water) and a washing chamber (yellow-dyed water). Details of the microfluidic components are presented on the right side of Figure 1b. The microfluidic circuit comprises the following components: (1) incubation chamber, (2) washing chamber, (3) detection region for fluorescence signal, (4) vent hole of the incubation chamber, (5) inlet port of the incubation chamber, and (6) inlet port of the washing chamber. The microchannel bridge between the incubation and washing chambers has a rectangular cross-section with 300 ± 20 µm widths. A cross-sectional schematic view of the microfluidic chamber describes the dimensions as shown in Figure 1c. The microchannel has 500 µm height, which is also the thickness of the PMMA disk. The detection region at the end of the washing chamber has a sharp edge geometry. This edge geometry of the washing chamber is intended to aggregate fluorescence-labeled microbeads in order to intensify the emission light signal.

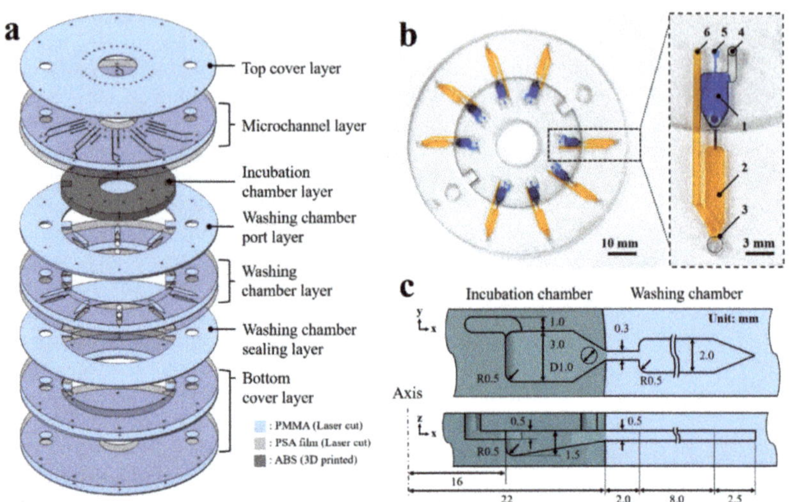

Figure 1. (**a**) Expanded view of the 3D microfluidic disk. (**b**) Image of the 3D microfluidic disk containing dyed water in the incubation (blue dye) and washing chambers (yellow dye). (**c**) Schematic of the cross-sectional illustrating the dimensions of the microfluidic chambers.

2.3. Preparation of VEGF Reagents

During the reagent preparation, antibodies of VEGF to capture (cAb, Human/Primate VEGF Antibody, R&D Systems, Minneapolis, MN, USA) and detect (dAb, Human VEGF 165 Antibody, R&D Systems) are bound to the surface of microbead and fluorescent dye, respectively. An antibody coupling kit is used for binding the cAb onto the 2.8 μm diameter epoxy magnetic bead surface (Dynabeads antibody coupling kit and M-270 Epoxy microbead, Thermo Fisher, Waltham, MA, USA). This binding procedure requires 24 h in accordance with the protocol manual. The cAb concentration on the microbead surfaces is 20 μg mg^{-1} in 1 mL of the solution with PBS containing 1 wt% BSA. Additionally, a fluorescence conjugation kit is used for binding the fluorescent dye with dAb with 1:1 volume ratio (DyLight 488 Conjugation kit, Abcam, Cambridge, UK); these are then left overnight at room temperature in dark condition. Consequently, the dAb-bound fluorescent dye is diluted in PBS to 1 μg mL^{-1} concentration. Serial ten-fold dilution of the standard VEGF antigen (Recombinant Human VEGF 165, R&D Systems) was performed using PBS to obtain from 1000 to 1 ng mL^{-1} concentration. Additionally, pure PBS without VEGF antigen is used for a control solution. A dextran (Mr 15,000~25,000, Sigma-Aldrich) is dissolved with 20 wt% in PBST (0.05 v/v% Triton-X100 contained PBS, Sigma-Aldrich) for wash buffer at 60 °C on a magnetic hot plate stirrer for 24 h.

2.4. Analysis and Detection of Fluorescence

A microscope was customized for fluorescence measurement at detection region in 3D microfluidic disk, which was placed over the spindle motor instantly after finishing rotation of disk. Fluorescence signal is measured in a darkroom for blocking outside light pollution. The 150 W halogen illuminator with optic lens and filters are used for the light source in the microscope. Optical filtration selectively passes only fluorescence wavelengths using excitation filter (average light transmission, T_{avg} over 93% in the 473–491 nm wavelength range), dichroic filter (T_{avg} > 93% in the 502–950 nm range, and T_{avg} < 7% in the 350–488 nm range), and emission filter (T_{avg} > 93% in the 506–534 nm range). Images from fluorescence microscopy are analyzed using image processing software (ImageJ 1.8.0). Inside the border of the washing chamber, the fluorescence excitation signal is amplified. Additionally, background noise is removed by high-pass filtration.

3. Results and Discussion

3.1. Validation of the Simplified Microbead FLISA Protocol

The immunological assay is one of the general analysis techniques using absorbent or fluorescent materials for the antibody-antigen interaction-based protein level quantification. The enzyme-linked immunosorbent assay (ELISA) process is performed on well-plate platforms; it employs a colorimetric reaction of enzymes that trigger color change according to the conjugation ratio with the substrate. In this technique, the light absorption ratio is varied according to the enzyme concentration. However, the light absorption mechanism of the ELISA protocol has limitations in applying to microfluidic systems. The optical spectroscopy light absorption characteristics have a relationship with the concentration of the reagent, which is expressed as the Lambert-Beer law following Equation (1):

$$I = I_0 \cdot 10^{-\varepsilon c l} \tag{1}$$

where I and I_0 are intensity of the transmittance and the incident light to the reagent. The ε, c, and l are the molar absorptivity coefficient (cm^2 mol^{-1}), the molar concentration (mol L^{-1}), and the optical path length (cm), respectively. In the microfluidic disk platform, the optical path length is geometrically restricted to spectroscopic analysis, which is the occurring degradation of the spectral resolution [39]. For this part, it is pointless to increase the thickness of the disk for improving the detection resolution in a microfluidic platform. Alternatively, the FLISA method is possible for the superposition of fluorescence signals, which improve the limit of detection for quantitative immunoassay even when using small sample volumes [40,41]. The microfluidic disk platforms employing microbead FLISA protocol are especially helpful in detecting the superposition of excitation signals from the aggregated fluorescence-linked microbeads. This "simplified" FLISA protocol facilitates the incubation step and diminishes the required time for the reaction between antibody and antigen. The reagents are loaded to the incubation chamber concurrently, where they are mixed and bound together.

In this study, the reagents are prepared as described in Figure 2. Prior to applying the microfluidic disk platform, the protocol is confirmed using a 96-well black polystyrene microplate. The variation of the incubation time and reagent volume condition is performed to verify the fluorescence signal linearity in the protocol. The cAb-bound microbeads, dAb-bound fluorescent dye, and human VEGF antigens are simultaneously loaded into the microwell with a volume ratio of 1:1:1. During the incubation process, shaking the incubator is used for gentle mixing of reagents with 120 rpm for 2 h at 37 °C. In the reagent loading and initial incubation steps, the cAb-bound microbead has a high probability opportunity to bind with the VEGF antigen due to the 20 times higher concentration compared with the dAbs in the reagents. After the VEGF antigens are bound to the microbead surfaces, other epitopes of the VEGF antigen are exposed to binding with the dAb-bound fluorescence dye in the reagents during the incubation step. In the washing step, unbound reagents are removed by pipetting for washing three times. The fluorescence microscope is used to confirm the amount of fluorescent dye-coupled microbeads, which affect the fluorescence excitation intensity with regard to VEGF concentration.

The schematic in Figure 3a describes the fluorescent dye-coupled microbead surface formation. The binding between the VEGF antigen-antibody is performed by the specificity of avidity-driven interactions. In Figure 3b, the microbeads are photographed with a long wave pass filter (>630 nm). The red image is based on the relatively high spectral reflectance characteristics of the microbead (ferrite oxide) in the red wavelength (>630 nm) to show the position of the microbeads. Figure 3c is a fluorescence microscopic image showing that the fluorescent dye is well bound to the surface of the microbead. The merged image represents that the green fluorescent dye is located in the same position as the microbeads, as shown in Figure 3d.

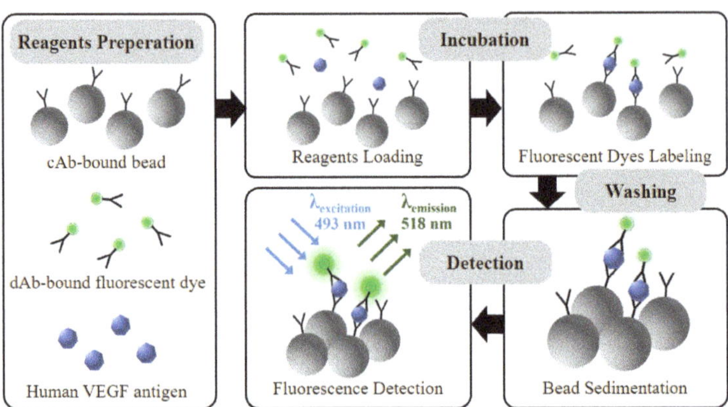

Figure 2. Schematic of the VEGF detection process via simplified microbead FLISA protocol. Biological assay between reagents is performed together in incubation step. The unbound dAb-fluorescent dye is eliminated in washing step. The fluorescence signal is measured for VEGF detection as much as the fluorescent dye attached to the microbead.

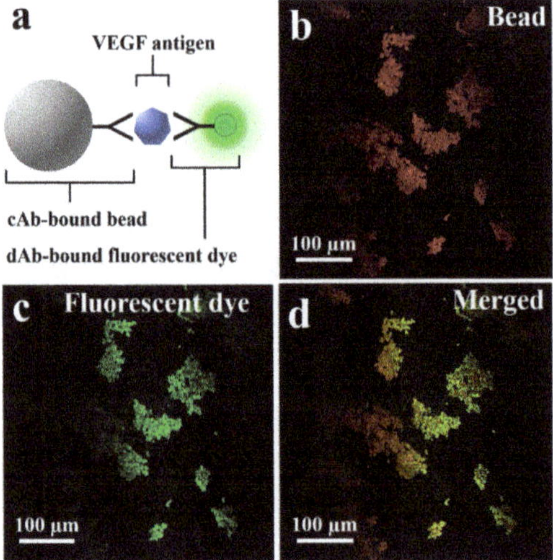

Figure 3. (**a**) Schematic of fluorescent dye-coupled VEGF on the microbeads. (**b**) The microbeads photographed with the longpass filter (>630 nm). (**c**) The dAb-bound fluorescent dye (green fluorescent protein) detected on the surface of the microbeads. (**d**) The merged image for the fluorescent dye and the microbeads.

The simplified microbead FLISA reduces the required reagent volumes and also the incubation time. The fluorescence signal was analyzed under variations in volume of the reagent and incubation time. The green fluorescence intensities are analyzed for VEGF concentrations of 1 μg mL^{-1} and 1 ng mL^{-1} as well as the pure PBS. In the simplified microbead FLISA protocol, 100 μL reagent volume (which is required volume for the traditional FLISA method) is reduced to 10 and 5 μL, while the incubation time of 2 h is reduced to 1 h.

The fluorescence intensities are presented under the concentration of VEGF antigen as shown in Figure 4. When using reagent volumes of 100, 10, and 5 μL for the concentration of 1 μg mL^{-1} VEGF antigen solution, the fluorescence intensities in the arbitrary unit are 48.08,

35.03, and 36.19, respectively. Moreover, for the 10 µL volume of VEGF antigen solution with 1 µg mL^{-1} concentration, the fluorescence excitation signal is reduced from 35.03 to 28.0 upon decreasing the incubation time from 2 to 1 h. Decreasing the reagent volume (from 100 to 10 µL) and the incubation time (from 2 to 1 h) causes reductions of fluorescent intensity of 34.8% and 20.5%, respectively. This shows that the fluorescence intensity is retained linearly, even when using small reagent volumes and reduced incubation periods. These results show possibilities, despite the reduction in reagent volume and incubation time, for analysis via a microfluidic system using image processing for high resolution through fluorescence signal amplification and the background noise elimination. Thus, the simplified microbead FLISA on the 3D microfluidic disk is performed employing a 1 h incubation time and 10 µL reagent volume. The microbead FLISA protocol shows the possibility of a 0.2~3.0 ng mL^{-1} range of VEGF level detecting, which can be a criterion for clinical signs for diagnosis and treatment [42,43].

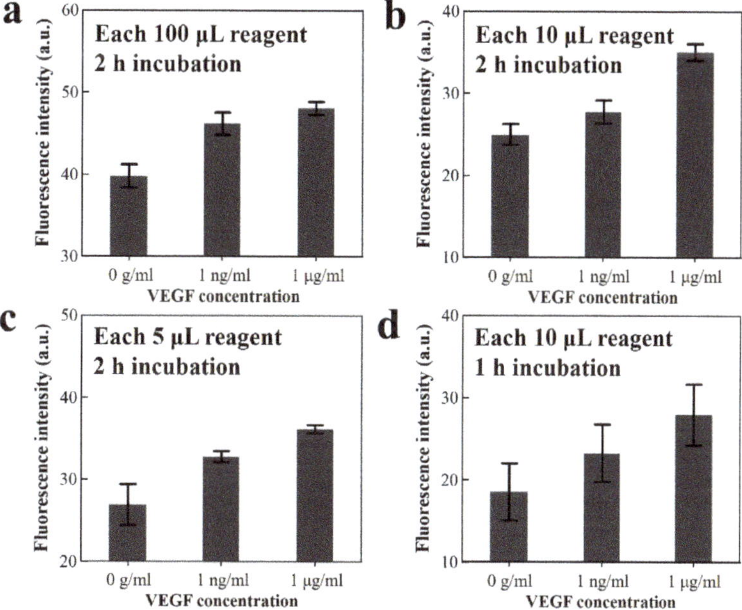

Figure 4. (**a–c**) Green fluorescence intensity of the fluorescent dye-coupled VEGF on the microbeads after 2 h incubation, with reagent volumes of 100, 10, and 5 µL, respectively, and (**d**) 10 µL reagent volume for 1 h incubation.

A flow chart in Figure 5 compares a traditional sandwich ELISA method [44,45] and the microbead assay method in a microfluidic system [46–48]. The protocol can diminish the volume of the antigen solution and the required time for the incubation step up to values of 1/10 and 1/5, respectively.

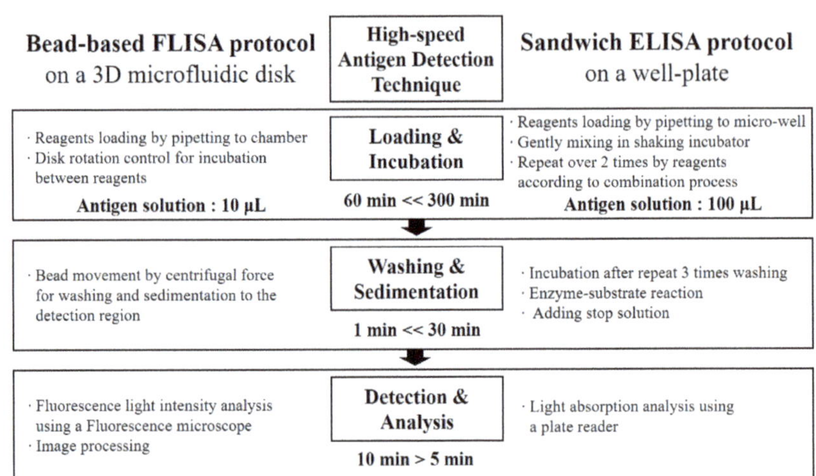

Figure 5. A flow chart comparison of antigen detection techniques between microbead FLISA and traditional sandwich ELISA protocols.

3.2. One-Step Simplified Microbead FLISA Using a 3D Microfluidic Disk

The 3D microfluidic disk using simplified microbead FLISA protocol is completed by a one-step process including incubation, washing, and detection steps in sequence. The schematic of Figure 6 describes the cross-section of the microfluidic circuit along the radial direction of the disk in order to highlight the proposed sequential process. First, the 30 µL wash buffer is loaded into the washing chamber through the inlet port. Then, the reagents with a volume of 10 µL in each of the cAb-bound microbeads, VEGF antigen diluted solution, and dAb-bound fluorescent dye are loaded into the incubation chamber using pipettes.

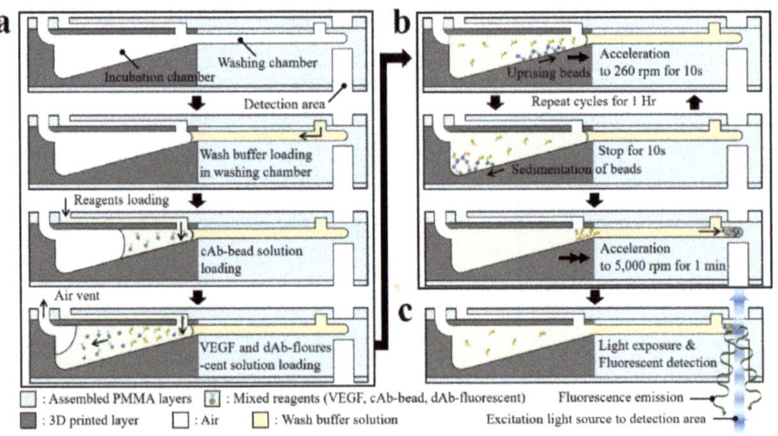

Figure 6. Schematic of sectional view of the microfluidic components highlighting the sequential protocol of the simplified microbead FLISA; the processes indicated are (**a**) loading the wash buffer and reagents, (**b**) incubation, and (**c**) fluorescence detection, respectively.

After loading the reagents, the 3D microfluidic disk is placed on a spindle motor, which is located in a dark room to avoid the photobleaching of the fluorescent dye during the incubation process. The inlet ports are blocked using commercial transparent tape to prevent reagent evaporation during the protocol. For the incubation step, disk rotation is regulated to ensure that the reagents mix sufficiently. In the incubation chamber with the tilted floor, microbeads move along the radial direction of the chamber surface, while accelerating at 260 rpm for 10 s. Afterward, when the disk is stopped for 10 s, the microbeads move back to the disk axis direction along the chamber bottom. The microbead is controlled by regulating rotation to enhance the mixing reagents. The angular velocity of the disk affects the bead movement during sedimentation, which can be expressed as Equation (2), referring to Stokes' law:

$$U_s = \frac{2}{9}\frac{(\rho_{bead} - \rho_f)}{\mu}aR^2 \tag{2}$$

where U_s is the sedimentation velocity; ρ_{bead} and ρ_f are the microbead and fluid densities, respectively; μ is the fluid viscosity; R is the radius of the microbead; and a is the acceleration. Here, for the microbeads in the incubation chamber with a tilted floor, the acceleration is changed to $a \cdot \cos\theta - g \cdot \sin\theta$, where θ and g are the slope of the incubation chamber and gravitational acceleration, respectively. The disk angular velocity with respect to time for the cycle is shown in Figure S3 of the Supplementary Material, including the microbead position after each step (reagent loading, reagent incubation, and washing). In the incubation chamber, when the disk rotates and then stops for 10 s, the microbeads move by 5.50 ± 2.86 μm along the radial direction at 260 rpm angular velocity and then move back by 3.22 μm, respectively, according to Stokes' law (Equation (2)). When the disk is rotating, the distance traversed by the microbead depends on its relative radial position to the central axis of the disk. For the stationary state of the disk, the microbeads on the tilted floor of the incubation chamber are only influenced by gravitational force and move toward the central axis, regardless of their position. The incubation step between reagents is achieved through gentle mixing for 1 h. At the end of the incubation process, the disk is accelerated for microbead sedimentation in the washing step to 5 krpm for 10 s and then continued for 1 min. The unbound dAb-fluorescent dye is separated due to the difference in density when the microbeads pass through the wash buffer. During this process, the fluorescent dye-coupled microbeads as well as the unlabeled microbeads settle together at the end of the washing chamber.

In Figure 7a, the images depict the microbeads with and without fluorescent dye-coupled VEGF at the end of the washing chamber. The dashed yellow line indicates the boundary of the washing chamber. In Figure 7b, the results of the fluorescence area ratio, A_f/A_b, is a parameter evaluated by the fluorescence area (A_f) and the microbead aggregation area (A_b). The value of A_f/A_b increases with the VEGF concentration between 1 μg mL^{-1} and 0 g mL^{-1}. The average coefficient of variation of five points is 5.50% in the calibration curve. To summarize the study, the one-step process is successfully performed employing the simplified microbead FLISA protocol. The proposed multi-material disk structure comprises laser-cut PMMA layers and a 3D-printed block. The microfluidic disk contains a washing chamber with microchannels and an incubation chamber. The pressurized PSA films grasp layers together while preventing leakage of reagents during the processes. In the proposed simplified microbead FLISA protocol, the excitation signal intensity of the fluorescence is linearly related to the VEGF concentration, regardless of adjustment of the incubation time and the volume of reagent. In the 3D microfluidic disk, the entire protocol of the simplified microbead FLISA can be terminated within 1 h without additional manual intervention to process. Furthermore, the fluorescence signals can be detected immediately on the disk after completing the rotation. The obtained green fluorescent images are analyzed through the image processing. The calibration curve shows clearly over 10 ng mL^{-1} to show quantitative detection for the VEGF concentration via the fluorescence area ratio, A_f/A_b. Through the fluorescence images of the fluorescent dye-coupled microbeads, the detection limit of the 1 ng mL^{-1} VEGF concentration is enough to distinguish with 0 g mL^{-1}. As a result, VEGF detection can be accomplished to a resolution with ng mL^{-1} through the fluorescence area ratio analysis (A_f/A_b) via a three-dimensional microfluidic system employing the one-step simplified microbead FLISA protocol. The detected VEGF level using a 3D microfluidic disk can be a diagnosis for clinical signs of retinal disorders. In detecting VEGF level with a 1 ng mL^{-1} resolution, difficulties of the quantitative analysis for VEGF concentration still remain in the 3D microfluidic system by using rapid antigen testing. Moreover, the reagents can be detected precisely by operating the disk angular velocity and the number of cycles with bi-direction rotation in order to enhance the resolution for target antigens.

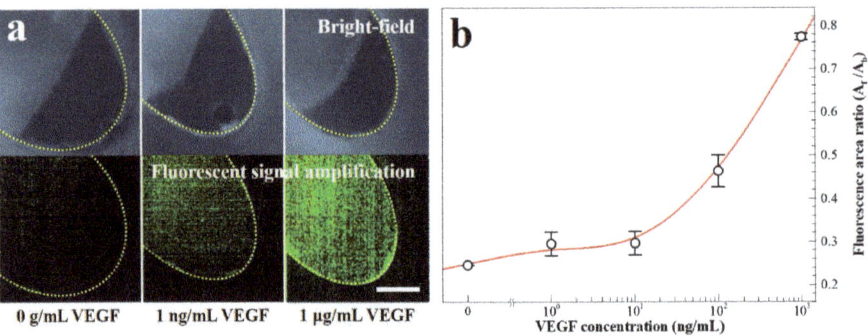

Figure 7. (**a**) Bright-field and fluorescence images for fluorescent dye-coupled microbeads, with VEGF concentrations of 0 g mL^{-1}, 1 ng mL^{-1}, and 1 µg mL^{-1} (scale bar = 100 µm). (**b**) Fluorescence intensity with varying VEGF concentrations.

4. Conclusions

A simplified microbead FLISA protocol using a multi-material-based 3D microfluidic disk is successfully implemented for the low-level VEGF detection. The 3D microfluidic disk consists of the PSA film-attached laser-cut PMMA layers and a 3D-printed block. In the proposed microfluidic disk, only the component for the incubation step requiring precise control is fabricated using a 3D printing method, whereas the remaining components are constituted of the laser-cut PMMA channel in order to reduce the manufacturing cost. the microbeads are utilized as the substrate for the immobilization of antibodies on their surface to apply the simplified microbead FLISA protocol. Even for a 30 µL volume of the reagents, the enlarged specific surface area of the microbeads provides support for the antigen–antibody interactions. Additionally, in the simplified microbead FLISA protocol, the reagent volumes are only required to be 1/10 compared with the commercial detection method. The linearity of green fluorescence signals is shown with respect to VEGF concentrations. The excited fluorescence signals observed during the detection step are superimposed from the aggregated microbeads at the edge of the washing chamber. The fluorescence area ratio, A_f/A_b, with respect to the VEGF concentration could be confirmed as characteristic with a ng mL^{-1} resolution. Therefore, the 3D microfluidic disk platform can be used to detect VEGF on the sequential benchtop process using only passive mixing in a simple clockwise rotation cycle within one hour. Regardless of the target antigen including VEGF, manipulating the conditions of disk acceleration and cycles with bi-directional rotation enhances the physical contact opportunity to an antigen–antibody interaction by microbead movement control. In the future, it can be applied as a low-cost, high-speed diagnostic system in developing countries for virus detection, such as for COVID-19. This is possible by providing a platform for detecting biochemical targets within an hour by utilizing a 3D microfluidic disk and a simple rotor.

Supplementary Materials: Supplementary Materials can be found at https://www.mdpi.com/article/10.3390/bios11080270/s1, Figure S1: SEM images of microstructure (designed diameter (D) = 500, 250, 200, and 100 μm) on 3D-printed surface fabricated in the (a–d) vertical and (e–h) lateral directions to show resolution of stereolithography apparatus, Figure S2: (a) Top view of 3D-printed block. (b) Isometric view and (c) sectional view of the 3D-printed block, Figure S3: Angular velocity of disk with respect to time for the mixing cycle during incubation and the bead washing process.

Author Contributions: Conceptualization: D.H.K., H.W.K. and S.-W.P.; methodology, H.W.K. and S.-W.P.; validation, N.K.K.; formal analysis, D.H.K. and N.K.K.; investigation, D.H.K. and S.-W.P.; resources, H.W.K. and S.-W.P.; writing—original draft preparation, D.H.K.; writing—review and editing, D.H.K., N.K.K., S.-W.P. and H.W.K.; supervision, S.-W.P. and H.W.K.; project administration, S.-W.P. and H.W.K.; funding acquisition, S.-W.P. and H.W.K. All authors have read and agreed to the published version of the manuscript.

Funding: This research was supported by a grant from the Korea Health Technology R&D Project through the Korea Health Industry Development Institute (KHIDI), funded by the Ministry of Health & Welfare, Republic of Korea (grant number: HI19C0642), the National Research Foundation of Korea (NRF), funded by the Ministry of Education (NRF-2019R1A6A3A13096916), and funded by a grant (CRI18017-1) from Chonnam National University Hospital Biomedical Research Institute.

Institutional Review Board Statement: Not applicable.

Informed Consent Statement: Not applicable.

Data Availability Statement: Not applicable.

Conflicts of Interest: The authors declare no conflict of interest.

Abbreviations

BSA	Bovine Serum Albumin
cAb	Capture Antibody
dAb	Detection Antibody
ELISA	Enzyme-Linked Immunosorbent Assay
FLISA	Fluorescence-Linked Immunosorbent Assay
PMMA	Poly(methyl methacrylate)
PBS	Phosphate Buffer Saline
PBST	Phosphate Buffer Saline with Tween Detergent
PSA	Pressure-Sensitive Adhesive
SLA	Stereo-Lithography Apparatus
VEGF	Vascular Endothelial Growth Factor

References

1. Sebastiani, P.; Thyagarajan, B.; Sun, F.; Schupf, N.; Newman, A.B.; Montano, M.; Perls, T.T. Biomarker signatures of aging. *Aging Cell* **2017**, *16*, 329–338. [CrossRef]
2. Longo, R.; Gasparini, G. Challenges for patient selection with VEGF inhibitors. *Cancer Chemother. Pharmacol.* **2007**, *60*, 151–170. [CrossRef]
3. Eljarrat-Binstock, E.; Pe'er, J.; Domb, A.J. New techniques for drug delivery to the posterior eye segment. *Pharm. Res.* **2010**, *27*, 530–543. [CrossRef]
4. Chang, J.H.; Garg, N.K.; Lunde, E.; Han, K.Y.; Jain, S.; Azar, D.T. Corneal neovascularization: An anti-VEGF therapy review. *Surv. Ophthalmol.* **2012**, *57*, 415–429. [CrossRef]
5. Sulaiman, R.S.; Basavarajappa, H.D.; Corson, T.W. Natural product inhibitors of ocular angiogenesis. *Exp. Eye Res.* **2014**, *129*, 161–171. [CrossRef] [PubMed]
6. Gupta, P.; Yadav, K.S. Applications of microneedles in delivering drugs for various ocular diseases. *Life Sci.* **2019**, *237*, 116907. [CrossRef] [PubMed]
7. Ng, E.W.; Shima, D.T.; Calias, P.; Cunningham, E.T.; Guyer, D.R.; Adamis, A.P. Pegaptanib, a targeted anti-VEGF aptamer for ocular vascular disease. *Nat. Rev. Drug Discov.* **2006**, *5*, 123–132. [CrossRef] [PubMed]
8. Witmer, A.N.; Vrensen, G.F.J.M.; Van Noorden, C.J.F.; Schlingemann, R.O. Vascular endothelial growth factors and angiogenesis in eye disease. *Prog. Retin. Eye Res.* **2003**, *22*, 1–29. [CrossRef]
9. Nowak, J.Z. Age-related macular degeneration (AMD): Pathogenesis and therapy. *Pharmacol. Rep.* **2006**, *58*, 353.

10. Lançon, A.; Frazzi, R.; Latruffe, N. Anti-oxidant, anti-inflammatory and anti-angiogenic properties of resveratrol in ocular diseases. *Molecules* **2016**, *21*, 304. [CrossRef]

11. Penn, J.S.; Madan, A.; Caldwell, R.B.; Bartoli, M.; Caldwell, R.W.; Hartnett, M.E. Vascular endothelial growth factor in eye disease. *Prog. Retin. Eye Res.* **2008**, *27*, 331–371. [CrossRef]

12. Schmidl, D.; Garhöfer, G.; Schmetterer, L. Nutritional supplements in age-related macular degeneration. *Acta Ophthalmol.* **2015**, *93*, 105–121. [CrossRef]

13. Gucciardo, E.; Loukovaara, S.; Salven, P.; Lehti, K. Lymphatic vascular structures: A new aspect in proliferative diabetic retinopathy. *Int. J. Mol. Sci.* **2018**, *19*, 4034. [CrossRef]

14. Masuda, T.; Shimazawa, M.; Hara, H. The kallikrein system in retinal damage/protection. *Eur. J. Pharmacol.* **2015**, *749*, 161–163. [CrossRef] [PubMed]

15. Wecker, T.; Ehlken, C.; Bühler, A.; Lange, C.; Agostini, H.; Böhringer, D.; Stahl, A. Five-year visual acuity outcomes and injection patterns in patients with pro-re-nata treatments for AMD, DME, RVO and myopic CNV. *Br. J. Ophthalmol.* **2017**, *101*, 353–359. [CrossRef]

16. Zhang, C.; Han, Y.; Lin, L.; Deng, N.; Chen, B.; Liu, Y. Development of quantum dots-labeled antibody fluorescence immunoassays for the detection of morphine. *J. Agric. Food Chem.* **2017**, *65*, 1290–1295. [CrossRef] [PubMed]

17. Lv, Y.; Wu, R.; Feng, K.; Li, J.; Mao, Q.; Yuan, H.; Shen, H.; Chai, X.; Li, L.S. Highly sensitive and accurate detection of C-reactive protein by CdSe/ZnS quantum dot-based fluorescence-linked immunosorbent assay. *J. Nanobiotechnol.* **2017**, *15*, 35. [CrossRef]

18. Xiong, S.; Zhou, Y.; Huang, X.; Yu, R.; Lai, W.; Xiong, Y. Ultrasensitive direct competitive FLISA using highly luminescent quantum dot beads for tuning affinity of competing antigens to antibodies. *Anal. Chim. Acta* **2017**, *972*, 94–101. [CrossRef] [PubMed]

19. Yao, J.; Xing, G.; Han, J.; Sun, Y.; Wang, F.; Deng, R.; Hu, X.; Zhang, G. Novel fluoroimmunoassays for detecting ochratoxin A using CdTe quantum dots. *J. Biophotonics* **2017**, *10*, 657–663. [CrossRef] [PubMed]

20. Walsh III, D.I.; Sommer, G.J.; Schaff, U.Y.; Hahn, P.S.; Jaffe, G.J.; Murthy, S.K. A centrifugal fluidic immunoassay for ocular diagnostics with an enzymatically hydrolyzed fluorogenic substrate. *Lab Chip* **2014**, *14*, 2673–2680. [CrossRef]

21. Hsu, M.Y.; Chen, S.J.; Chen, K.H.; Hung, Y.C.; Tsai, H.Y.; Cheng, C.M. Monitoring VEGF levels with low-volume sampling in major vision-threatening diseases: Age-related macular degeneration and diabetic retinopathy. *Lab Chip* **2015**, *15*, 2357–2363. [CrossRef]

22. Tabrizi, M.A.; Shamsipur, M.; Saber, R.; Sarkar, S.; Ebrahimi, V. A high sensitive visible light-driven photoelectrochemical aptasensor for shrimp allergen tropomyosin detection using graphitic carbon nitride-TiO2 nanocomposite. *Biosens. Bioelectron.* **2017**, *98*, 113–118. [CrossRef] [PubMed]

23. Jiang, S.H.; Li, J.; Dong, F.Y.; Yang, J.Y.; Liu, D.J.; Yang, X.M.; Wang, Y.H.; Yang, M.W.; Fu, X.L.; Zhang, X.X.; et al. Increased serotonin signaling contributes to the Warburg effect in pancreatic tumor cells under metabolic stress and promotes growth of pancreatic tumors in mice. *Gastroenterology* **2017**, *153*, 277–291. [CrossRef] [PubMed]

24. Wu, Y.; Yi, L.; Li, E.; Li, Y.; Lu, Y.; Wang, P.; Zhou, H.; Liu, J.; Hu, Y.; Wang, D. Optimization of Glycyrrhiza polysaccharide liposome by response surface methodology and its immune activities. *Int. J. Biol. Macromol.* **2017**, *102*, 68–75. [CrossRef] [PubMed]

25. Khang, H.; Cho, K.; Chong, S.; Lee, J.H. All-in-one dual-aptasensor capable of rapidly quantifying carcinoembryonic antigen. *Biosens. Bioelectron.* **2017**, *90*, 46–52. [CrossRef] [PubMed]

26. Weiss, M.; Frohnmayer, J.P.; Benk, L.T.; Haller, B.; Janiesch, J.W.; Heitkamp, T.; Börsch, M.; Lira, R.B.; Dimova, R.; Lipowsky, R.; et al. Sequential bottom-up assembly of mechanically stabilized synthetic cells by microfluidics. *Nat. Mater.* **2018**, *17*, 89–96. [CrossRef]

27. Barani, A.; Paktinat, H.; Janmaleki, M.; Mohammadi, A.; Mosaddegh, P.; Fadaei-Tehrani, A.; Sanati-Nezhad, A. Microfluidic integrated acoustic waving for manipulation of cells and molecules. *Biosens. Bioelectron.* **2016**, *85*, 714–725. [CrossRef] [PubMed]

28. Miller, H.; Zhou, Z.; Shepherd, J.; Wollman, A.J.; Leake, M.C. Single-molecule techniques in biophysics: A review of the progress in methods and applications. *Rep. Prog. Phys.* **2017**, *81*, 024601. [CrossRef]

29. Zhang, Y.N.; Zhao, Y.; Zhou, T.; Wu, Q. Applications and developments of on-chip biochemical sensors based on optofluidic photonic crystal cavities. *Lab Chip* **2018**, *18*, 57–74. [CrossRef]

30. Ye, D.; Li, L.; Li, Z.; Zhang, Y.; Li, M.; Shi, J.; Zuo, X. Molecular threading-dependent mass transport in paper origami for single-step electrochemical DNA sensors. *Nano Lett.* **2018**, *19*, 369–374. [CrossRef]

31. Lee, B.S.; Lee, J.N.; Park, J.M.; Lee, J.G.; Kim, S.; Cho, Y.K.; Ko, C. A fully automated immunoassay from whole blood on a disc. *Lab Chip* **2009**, *9*, 1548–1555. [CrossRef] [PubMed]

32. Kang, D.H.; Kim, N.K.; Park, S.W.; Lee, W.; Kang, H.W. A microfluidic circuit consisting of individualized components with a 3D slope valve for automation of sequential liquid control. *Lab Chip* **2020**, *20*, 4433–4441. [CrossRef] [PubMed]

33. Pourmand, A.; Shaegh, S.A.M.; Ghavifekr, H.B.; Aghdam, E.N.; Dokmeci, M.R.; Khademhosseini, A.; Zhang, Y.S. Fabrication of whole-thermoplastic normally closed microvalve, micro check valve, and micropump. *Sens. Actuators B-Chem.* **2018**, *262*, 625–636. [CrossRef]

34. Bressan, L.P.; Adamo, C.B.; Quero, R.F.; de Jesus, D.P.; da Silva, J.A. A simple procedure to produce FDM-based 3D-printed microfluidic devices with an integrated PMMA optical window. *Anal. Methods* **2019**, *11*, 1014–1020. [CrossRef]

35. Wondimu, S.F.; von der Ecken, S.; Ahrens, R.; Freude, W.; Guber, A.E.; Koos, C. Integration of digital microfluidics with whispering-gallery mode sensors for label-free detection of biomolecules. *Lab Chip* **2017**, *17*, 1740–1748. [CrossRef]
36. Matellan, C.; Armando, E. Cost-effective rapid prototyping and assembly of poly (methyl methacrylate) microfluidic devices. *Sci. Rep.* **2018**, *8*, 1–13. [CrossRef]
37. Park, J.M.; Cho, Y.K.; Lee, B.S.; Lee, J.G.; Ko, C. Multifunctional microvalves control by optical illumination on nanoheaters and its application in centrifugal microfluidic devices. *Lab Chip* **2007**, *7*, 557–564. [CrossRef]
38. Kang, K.; Oh, S.; Yi, H.; Han, S.; Hwang, Y. Fabrication of truly 3D microfluidic channel using 3D-printed soluble mold. *Biomicrofluidics* **2018**, *12*, 014105. [CrossRef]
39. Monaghan, T.; Harding, M.J.; Harris, R.A.; Friel, R.J.; Christie, S.D.R. Customisable 3D printed microfluidics for integrated analysis and optimization. *Lab Chip* **2016**, *16*, 3362–3373. [CrossRef]
40. Wu, W.; Liu, X.; Shen, M.; Shen, L.; Ke, X.; Cui, D.; Li, W. Multicolor quantum dot nanobeads based fluorescence-linked immunosorbent assay for highly sensitive multiplexed detection. *Sens. Actuator B Chem.* **2021**, *338*, 129827. [CrossRef]
41. Lee, L.G.; Nordman, E.S.; Johnson, M.D.; Oldham, M.F. A low-cost, high-performance system for fluorescence lateral flow assays. *Biosensors* **2013**, *3*, 360–373. [CrossRef]
42. Shimada, H.; Akaza, E.; Yuzawa, M.; Kawashima, M. Concentration gradient of vascular endothelial growth factor in the vitreous of eyes with diabetic macular edema. *Investig. Ophthalmol. Vis. Sci.* **2009**, *50*, 2953–2955. [CrossRef] [PubMed]
43. Aiello, L.P.; Avery, R.L.; Arrigg, P.G.; Keyt, B.A.; Jampel, H.D.; Shah, S.T.; Pasquale, L.R.; Thieme, H.; Iwamoto, M.A.; Park, J.E.; et al. Vascular endothelial growth factor in ocular fluid of patients with diabetic retinopathy and other retinal disorders. *N. Engl. J. Med.* **1994**, *331*, 1480–1487. [CrossRef]
44. Anthony, F.W.; Evans, P.W.; Wheeler, T.; Wood, P.J. Variation in detection of VEGF in maternal serum by immunoassay and the possible influence of binding proteins. *Ann. Clin. Biochem.* **1997**, *34*, 276–280. [CrossRef]
45. Teng, J.; Huang, L.; Zhang, L.; Li, J.; Bai, H.; Li, Y.; Ding, S.; Zhang, Y.; Cheng, W. High-sensitive immunosensing of protein biomarker based on interfacial recognition-induced homogeneous exponential transcription. *Anal. Chim. Acta* **2019**, *1067*, 107–114. [CrossRef]
46. Shan, S.; He, Z.; Mao, S.; Jie, M.; Yi, L.; Lin, J.M. Quantitative determination of VEGF165 in cell culture medium by aptamer sandwich based chemiluminescence assay. *Talanta* **2017**, *171*, 197–203. [CrossRef] [PubMed]
47. Henares, T.G.; Mizutani, F.; Hisamoto, H. Current development in microfluidic immunosensing chip. *Anal. Chim. Acta* **2008**, *611*, 17–30. [CrossRef]
48. Zhao, Z.; Al-Ameen, M.A.; Duan, K.; Ghosh, G.; Lo, J.F. On-chip porous microgel generation for microfluidic enhanced VEGF detection. *Biosens. Bioelectron.* **2015**, *74*, 305–312. [CrossRef] [PubMed]

Article

Stochastic Time Response and Ultimate Noise Performance of Adsorption-Based Microfluidic Biosensors

Ivana Jokić [1,*], Zoran Djurić [2,3], Katarina Radulović [1], Miloš Frantlović [1], Gradimir V. Milovanović [3,4] and Predrag M. Krstajić [1]

[1] Institute of Chemistry, Technology and Metallurgy, National Institute of the Republic of Serbia, University of Belgrade, Njegoševa 12, 11000 Belgrade, Serbia; kacar@nanosys.ihtm.bg.ac.rs (K.R.); frant@nanosys.ihtm.bg.ac.rs (M.F.); pkrstajic@nanosys.ihtm.bg.ac.rs (P.M.K.)
[2] Institute of Technical Sciences of SASA, Knez Mihailova 35, 11000 Belgrade, Serbia; zdjuric@itn.sanu.ac.rs
[3] Serbian Academy of Sciences and Arts, Knez Mihailova 35, 11000 Belgrade, Serbia; gvm@mi.sanu.ac.rs
[4] Mathematical Institute of SASA, Knez Mihailova 36, 11000 Belgrade, Serbia
* Correspondence: ijokic@nanosys.ihtm.bg.ac.rs; Tel.: +38-16-4116-1683

Abstract: In order to improve the interpretation of measurement results and to achieve the optimal performance of microfluidic biosensors, advanced mathematical models of their time response and noise are needed. The random nature of adsorption–desorption and mass transfer (MT) processes that generate the sensor response makes the sensor output signal inherently stochastic and necessitates the use of a stochastic approach in sensor response analysis. We present a stochastic model of the sensor time response, which takes into account the coupling of adsorption–desorption and MT processes. It is used for the analysis of response kinetics and ultimate noise performance of protein biosensors. We show that slow MT not only decelerates the response kinetics, but also increases the noise and decreases the sensor's maximal achievable signal-to-noise ratio, thus degrading the ultimate sensor performance, including the minimal detectable/quantifiable analyte concentration. The results illustrate the significance of the presented model for the correct interpretation of measurement data, for the estimation of sensors' noise performance metrics important for reliable analyte detection/quantification, as well as for sensor optimization in terms of the lower detection/quantification limit. They are also incentives for the further investigation of the MT influence in nanoscale sensors, as a possible cause of false-negative results in analyte detection experiments.

Keywords: microfluidic adsorption-based sensor; stochastic model; adsorption; mass transfer; ultimate noise performance; detection limit; quantification limit

Citation: Jokić, I.; Djurić, Z.; Radulović, K.; Frantlović, M.; Milovanović, G.V.; Krstajić, P.M. Stochastic Time Response and Ultimate Noise Performance of Adsorption-Based Microfluidic Biosensors. *Biosensors* **2021**, *11*, 194. https://doi.org/10.3390/bios11060194

Received: 19 April 2021
Accepted: 4 June 2021
Published: 12 June 2021

Publisher's Note: MDPI stays neutral with regard to jurisdictional claims in published maps and institutional affiliations.

1. Introduction

Microfluidic sensors are promising tools for chemical and biological detection [1–4]. The operation of a large class of such devices, known as adsorption-based sensors, relies on the adsorption–desorption (AD) process of a target substance on the surface of a sensing element. These include SPR (Surface Plasmon Resonance), CNT (Carbon NanoTube) or NWFET (NanoWire Field Effect Transistor), resistive graphene-based, potentiometric, SAW (Surface Acoustic Wave), FBAR (thin Film Bulk Acoustic wave Resonator), microcantilever sensors, etc. [5–13]. The sensing element of microfluidic sensors is typically located in a flow-through reaction chamber, where the sample to be analyzed is introduced (Figure 1). The AD process is coupled with mass transfer (MT) processes of target particles in the microfluidic chamber. Via MT processes (convection and diffusion), particles are transported to specific sites on a sensing surface where adsorption occurs, and away from the adsorption sites after desorption. A coupled effect of AD and MT processes determines the temporal change in the number of particles adsorbed on the sensing surface, $N(t)$, which causes a change in a measurable sensor parameter, yielding the sensor response. Hence, the sensor time response can be considered as determined by the time evolution of the number of adsorbed target particles, $N(t)$. As it contains information on the target substance presence and its concentration in the analyzed sample, the time response analysis enables both detection and quantification of chemical substances or biological specimens, thus having important applications in environmental protection, public and personal healthcare and security, the food industry, agriculture, and defense. Analysis of response kinetics can provide information as early as in the transient regime, i.e., before the binding process of target particles reaches the steady state, which significantly shortens the time needed to obtain the data. The response kinetics also contains information on the parameters that characterize the interaction process of target particles and surface binding sites [14], thus enabling the characterization of bimolecular binding reactions, important for the fundamental understanding of vital biochemical processes and pharmacology.

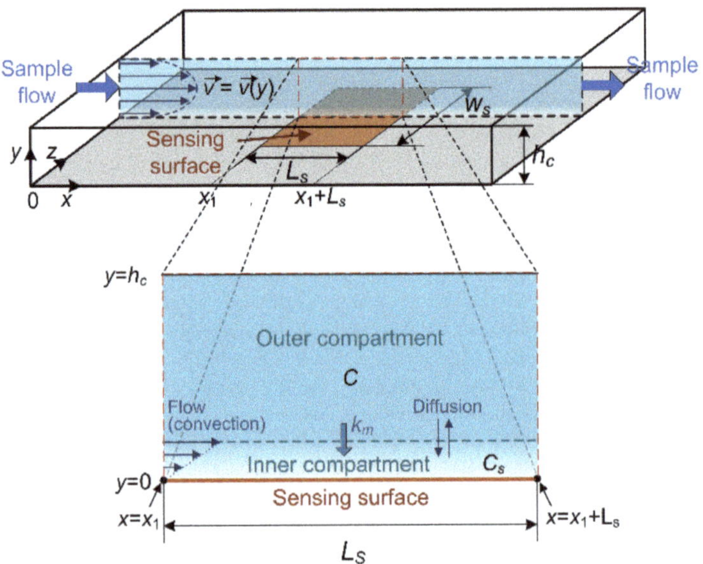

Figure 1. Adsorption-based microfluidic sensor: schematic representation of the typical system geometry with designations of dimensions and coordinate axes. The magnified partial cross-section of the microfluidic reaction chamber in the sensing surface zone is given as an illustration of the two-compartment model approximation of the spatially and time-dependent target substance concentration, affected by coupled adsorption–desorption and mass transfer processes of target analyte particles.

The random nature of the AD process coupled with MT causes fluctuations in the number of adsorbed particles, which result in sensor signal stochastic fluctuations known as AD noise, binding/unbinding noise, biological or chemical noise [15–19]. The total fluctuations in the sensor signal also depend on other kinds of noise originating from the sensor transduction mechanism and the read-out circuitry, but the unavoidable AD noise determines the sensor's ultimate noise performance and poses fundamental detection and quantification limits inherent to all adsorption-based sensors. The contribution of AD noise to the total sensor noise can even be dominant [16,17,20,21]. Thus, the analysis of AD noise and related parameters of stochastic sensor response becomes an important tool for the optimization of adsorption-based chemical and biological microfluidic sensors in terms of reliable analyte detection and quantification, and also in terms of improved sensing performance (i.e., higher signal-to-noise ratio and lower minimal detectable and quantifiable concentrations). This is especially true because miniaturization is a general trend in the field of chemical and biological sensors, focusing on adsorption-based micro-

and nanodevices, where achieving a sufficiently high signal-to-noise ratio (SNR) can be a challenge [20]. As AD noise also contains information about the quantity of the target substance in the sample, about binding process parameters, and about substance parameters useful for its recognition [15,22–25], mathematical modeling and more profound knowledge of AD fluctuations characteristics, both in the transient regime and in the steady state of sensor response, can enable the development of new measurement methods based on stochastic (i.e., noise) analysis in micro/nanosensors, as an addition to the existing conventional methods.

Stochastic mathematical models of sensor response consider the time-dependent number of adsorbed particles as a random process, $N(t)$, whose expected value reveals the binding (i.e., the sensor response) kinetics, and the variance is a measure of the sensor's AD noise. As stochastic models take into account the influence of individual events of particle binding and unbinding to the surface adsorption sites on the sensor response, as well as the inherent random nature of these events, they are more accurate in describing the binding kinetics than deterministic models. They describe response fluctuations, which are always present. For the analysis of stochastic sensor response, stochastic simulations are often used. However, analytical approximations of stochastic models are very useful, because they offer a good insight into the dependences of the response statistics on various sensor system parameters, while being more efficient than simulations in which high accuracy requires large computing resources and a long computation time. Stochastic models for the analysis of sensor response should take into account MT effects, as the randomness of the number of adsorbed particles originates from the coupling of the inherently stochastic AD process and MT.

Approximate mathematical models that enable the analysis of statistical parameters (expected value, standard deviation, and variance) of the stochastic time response of adsorption-based (both chemical and biological) sensors and their noise performance metrics (AD noise power spectral density and signal-to-noise ratio) have been developed for some practically significant cases [15,16,18,20,21,24–30]. Based on these references, it can be concluded that the closed-form solutions for the mentioned quantities are only devised for simplified cases, during the transient regime of the binding process on the sensing surface, or after the steady state of the binding process is reached. For example, in the analysis given in [26], analytical solutions were used for the time-varying expected value and relative fluctuations in the response of an adsorption-based plasmonic sensor, which fluctuate only due to the stochastic nature of the AD process. Analyte transport processes to and from adsorption sites, and the depletion of analyte particles from the sample during adsorption were not taken into account in the closed-form expressions. That corresponds to the idealized situation when MT is fast enough compared to the AD process, and when the number of analyte particles available for adsorption in the sensor's reaction chamber is much greater than the number of adsorbed particles at any given time, so the particle concentration in the chamber is considered as constant in time and uniform in space. The authors of [20] considered the scaling effects of biosensor systems through the stochastic analysis that takes into account the probabilistic capturing (i.e., adsorption) process. The MT effects were neglected in the derived mathematical model. The time evolutions of the expected value and standard deviation of the number of adsorbed particles were numerically calculated for the regime of constant analyte concentration in the reaction chamber volume, and for the regime of analyte depletion. In the former case, the sensor signal-to-noise ratio (SNR) for one fixed moment in time was analyzed, considering the effects of the sensing area reduction. However, it is well known that MT can significantly alter the sensor response kinetics [31–33], so it is important to consider this effect in the analysis of stochastic response. It was also experimentally shown that a suppression of MT influence leads to a great improvement of the biosensor limit of detection [34], which implies that MT affects detection limits. Therefore, fluctuations and noise models used for the estimation of the ultimate sensing performance should also include the MT influence.

In reference [27], a stochastic model of analyte diffusion within the biosensor chamber was presented, which incorporates the probabilistic model for the specific binding of analyte particles to immobilized probes at a sensing surface, as one of the boundary conditions. The model was used for the analysis of two idealized situations: (1) when there is an infinite adsorption capacity of the sensing surface, and (2) when the number of probes is finite, but a very small fraction of analyte particles present in the system is captured by the probes (i.e., no sample depletion by the binding events; thus, the number of free analyte particles in the chamber, available for adsorption, is considered constant). The closed-form solutions for the statistical parameters of biosensor response and for noise figures of merit were derived only for the biochemical equilibrium for these two cases. By using the stochastic modeling of the analyte capturing, considering the binding kinetics and the mass transfer by diffusion, the expressions for the equilibrium statistical response parameters and settling time approximation were obtained in [16]. In references [15,24,25,28], a theory was presented with closed-form expressions, as well as an analysis of the sensor AD noise power spectral density in the steady state, when the fluctuations are caused by coupled stochastic AD process and MT.

In [18,29,30], a stochastic simulation was used for the analysis of the change in the expected value and variance of the number of adsorbed particles, and the sensor signal-to-noise ratio in time, considering the transport of analyte particles by diffusion. The emphasis was on the influence of the target substance concentration and probe density on the mentioned time dependences, while MT influence was not analyzed in particular.

None of the mentioned works provided a stochastic model of sensor time response that takes into account both the diffusion and convection of analyte particles as processes that constitute MT in microfluidic sensors. In addition, none of them analyzed the MT influence neither on the stochastic temporal response, including its time-dependent expected value and variance (i.e., AD noise), nor on the SNR that determines the ultimate detection and quantification limit.

In this paper, we aim to investigate the temporal change in the statistical parameters of the biosensor stochastic response from the beginning of the adsorption process on the sensing surface until the steady state is reached, taking into account the mass transfer of analyte particles by both convection and diffusion, which corresponds to the realistic operating conditions in microfluidic biosensors. We first present the theoretical model for the expected value and variance of the number of adsorbed particles. The model is devised by applying the approach based on the master equation for the random processes known as birth–death processes in probability theory, to which the considered random process N belongs, and by introducing the effective probabilities of the increase and decrease in the number of adsorbed particles. The effective probabilities combine the influences of the inherently random AD and MT processes on the change in the number of adsorbed particles. By using the obtained analytical model, we investigate the response kinetics and AD noise of a protein biosensor, through the analysis of the expected value and the variance of the number of adsorbed particles, both in the transient regime and in the steady state of the binding process, for practically relevant analyte concentrations, mass transfer coefficients, and sensing surface areas. We also analyze the sensor signal-to-noise ratio, which sets the fundamental detection and quantification limits. One of the goals of our analysis is to investigate both the qualitative and the quantitative influence of MT on the kinetics of sensor stochastic response, on sensor AD noise, and on the maximal achievable SNR value, affecting both the reliable analyte detection and determination of analyte concentration. Although such an analysis can reveal new guidelines for the optimization of sensor design and operating conditions, to the best of the authors' knowledge, it does not exist in the published literature. Another goal is to determine the applicability boundaries of the simple stochastic model of sensor response (i.e., the one that neglects the mass transfer influence), and thus the conditions under which it becomes necessary to use the more comprehensive stochastic model (that takes into account the AD process coupled with the MT of analyte particles) in order to improve the interpretation of the measurement results and the estimation of sensor performance metrics such as noise, SNR, and analyte detection and quantification limits.

2. Method—Mathematical Modeling of Biosensor Stochastic Time Response

The change in the number of target analyte particles adsorbed on a sensing surface in unit time, assuming reversible adsorption, is determined by the difference between the instantaneous rates of the increase and decrease in $N(t)$, denoted by $a(t)$ and $d(t)$, respectively. These rates include the combined effects of adsorption–desorption and mass transfer processes. Namely, both AD and MT (convection and diffusion) processes change the target analyte concentration in a microfluidic sensor chamber, affecting the temporal change in N. The use of the two-compartment model (TCM) for the approximation of the spatial- and time-dependent analyte concentration in a flow-through chamber, when modeling the sensor response kinetics influenced by mass transfer, is experimentally verified for various adsorption-based biosensors [31–33,35]. It covers the case of surface reaction (i.e., AD process) coupled with convection and diffusion, when a thin zone (the inner compartment), depleted of analyte particles, is formed adjacent to the sensing surface, while the remaining part of the sensor chamber (the outer compartment) approximately retains the spatially uniform and time-constant analyte concentration C, equal to the analyte concentration in the sample introduced in the sensor chamber [31–33,35], as illustrated in Figure 1. By assuming the 1:1 binding of analyte molecules to the surface binding sites, the uniformity of all binding sites, and no interaction between analyte molecules, TCM yields Equation (1), which defines the temporal change in N [15,31,36]:

$$\frac{dN}{dt} = k_a C_S (N_m - N) - k_d N = k_a \frac{C + \frac{k_d}{k_m A} N}{1 + \frac{k_a}{k_m A}(N_m - N)}(N_m - N) - k_d N = a(N) - d(N), \quad (1)$$

where k_a and k_d are the adsorption and desorption rate constants, respectively, C_S is the analyte concentration in the immediate vicinity of the binding sites on the sensing surface of area A, N_m is the number of binding sites on the surface, and k_m is the mass transfer coefficient, introduced in TCM as $k_m = 1.467(D^2 v_m/(L_s h_c))^{1/3}$ [31] in order to characterize the transport of analyte particles by both convection and diffusion between the bulk solution and the surface binding sites (D is the diffusion coefficient of analyte particles, v_m is the mean convection velocity, L_s is the adsorption zone length, and h_c is the sensor chamber height). According to TCM, all quantities are averaged across the sensing surface. The effective rates of the increase and decrease in the number of adsorbed particles, $a(t)$ and $d(t)$, respectively, do not depend explicitly on t, but on the instantaneous value of N, thus $a(N)$ and $d(N)$ in Equation (1).

Equation (1) is derived for the diffusion-limited regime. The equation for the case of the adsorption-limited regime (i.e., the "rapid mixing" regime, when the analyte concentration in the whole flow-through reaction chamber is considered uniform in space and constant in time, due to fast MT compared to adsorption) neglects MT effects, and it is given as:

$$\frac{dN}{dt} = k_a C (N_m - N) - k_d N = a_{RM}(N) - d_{RM}(N), \quad (2)$$

where $a_{RM}(N)$ and $d_{RM}(N)$ are actual adsorption and desorption rates.

The time evolution of the number of adsorbed particles $N(t)$ is obtained from the deterministic kinetic Equations (1) or (2) (with or without taking MT into account, respectively) for given initial conditions. It enables the analysis of the deterministic time response of a sensor, as a function of the number of adsorbed particles, which is assumed to be a deterministic quantity.

The number of adsorbed particles at any given time is a result of a sequence of random bindings and unbindings of target particles to and from the surface adsorption sites. Therefore, the number of adsorbed particles on the sensor's active surface, N, is actually a stochastic quantity, determined by the stochastic nature of the AD process coupled with transport processes of analyte particles. Hence, N randomly fluctuates around its expected value <N>, and consequently, the sensor response is also a fluctuating quantity.

Observed in time, the stochastic number of adsorbed analyte particles at the sensor's surface, $N(t)$, is a random birth–death process [37]. By assuming that in a time interval $dt \rightarrow 0$, N can be either changed by one, or unchanged (i.e., there can be an adsorption of one particle, a desorption of one particle, or a lack of AD events), the probability distribution of the random variable N in an arbitrary moment of time t ($t \geq 0$), $P_N(n, t)$, for the given initial state $N(0) = n_0$ (here, $n_0 = 0$ as $t = 0$ is assumed as the moment when the AD process starts on the sensing surface), is given by the master equation:

$$\frac{d}{dt} P_N(n,t) = P_N(n-1,t) \cdot A(n-1) + P_N(n+1,t) \cdot D(n+1) - P_N(n,t) \cdot (A(n) + D(n)) \tag{3}$$

where n is the actual value of the random variable N at the given moment, and it denotes the state of the process ($n \in \{0,1,2...N_m\}$, where N_m is the total number of adsorption sites on the sensing surface), while $A(n)dt$ and $D(n)dt$ are the probabilities of transition from the state n to the state $n + 1$, and from the state n to the state $n - 1$ during the time interval $dt \rightarrow 0$, respectively. $A(n)$ and $D(n)$ are the probability of the increase in the number of adsorbed particles by 1 in unit time, and the probability of the decrease in N by 1 in unit time, respectively, when the current state is $N = n$. Equation (3) is valid for $n = 0$ if we define $P_N(-1,t) = 0$ and $A(-1) = 0$, and it is also valid for $n = N_m$ assuming $P_N(N_m + 1,t) = 0$ and $D(N_m + 1) = 0$ ($D(0) = 0$ due to the nature of the desorption process, and $A(N_m) = 0$ because the sensing surface adsorption capacity is limited to N_m).

As mentioned in the introduction, we are interested in the expected value of N, which reveals the response kinetics, and in the variance of N, as it is a measure of the AD noise. Starting from the definitions for the first and the second moment of the random variable:

$$\langle N \rangle = \sum_{n=0}^{N_m} n P_N(n,t), \; \sigma^2 = \langle (\Delta N)^2 \rangle = \sum_{n=0}^{N_m} (n - \langle N \rangle)^2 P_N(n,t) \tag{4}$$

and using the master equation (Equation (3)), the exact equations for the expected value, <N>, and the variance, σ^2, of the random number of adsorbed particles are obtained [37]:

$$\frac{d}{dt}\langle N \rangle = \langle N A(N) - D(N) \rangle \tag{5}$$

$$\frac{d\sigma^2}{dt} = \langle A(N) + D(N) \rangle + 2\langle (N - \langle N \rangle)[A(N) - D(N)] \rangle \tag{6}$$

The transition rate $A(N)$ depends on the adsorption rate constant, on the fraction of adsorption sites available for adsorption, and on the amount of particles that are available to participate in adsorption, when the number of adsorbed particles is $N = n$, while $D(N)$ depends on the desorption rate constant and on the fraction of occupied sites on the surface $N = n$. Thus, when the combined effect of AD and MT processes determines the probabilities of the change in the random variable N, the use of TCM for the approximation of the amount of particles that are available to participate in adsorption (those located in the immediate vicinity of the surface binding sites, as explained for Equation (1)) yields $A(n) = k_a C_s(N_m - n) = k_a(C + k_d n/(k_m A))(N_m - n)/(1 + k_a(N_m - n)/(k_m A))$ and $D(n) = k_d n$. In this way, the expressions are obtained for the effective probabilities of the increase and decrease in the number of adsorbed particles per unit time, which combine the influences of the AD and MT processes. $A(n)$ and $D(n)$ depend on the current state $N = n$ (which is a feature of birth–death processes). After representing the nonlinear transition rate as a Taylor series centered at the expected value, Equations (5) and (6) take the approximate form, which includes the first and the second moments:

$$\frac{d\langle N \rangle}{dt} = A(\langle N \rangle) - D(\langle N \rangle) + (A'' - D'') \cdot \frac{\sigma^2}{2}, \tag{7}$$

$$\frac{d\sigma^2}{dt} = A(\langle N\rangle) + D(\langle N\rangle) + \left[2(A' - D') + \frac{1}{2}(A'' + D'')\right]\cdot\sigma^2, \tag{8}$$

(A', D', A'', and D'' are the first and second derivatives of A and D with respect to n, calculated for $n = <N>$) [38]. After substituting the functions A and D and their derivatives in Equations (7) and (8), a system of equations is obtained:

$$\frac{d\langle N\rangle}{dt} = \frac{k_a C(N_m - \langle N\rangle) - k_d\langle N\rangle}{1 + \frac{k_a}{k_m A}(N_m - \langle N\rangle)} - \frac{k_a}{k_m A}\frac{\frac{k_a k_d}{k_m A}N_m + k_a C + k_d}{\left[1 + \frac{k_a}{k_m A}(N_m - \langle N\rangle)\right]^3}\cdot\sigma^2, \tag{9}$$

$$\frac{d\sigma^2}{dt} = \frac{k_a(C + 2\frac{k_d}{k_m A}\langle N\rangle)(N_m - \langle N\rangle) + k_d\langle N\rangle}{1 + \frac{k_a}{k_m A}(N_m - \langle N\rangle)} - \frac{\frac{k_a k_d}{k_m A}N_m + k_a C + k_d}{\left[1 + \frac{k_a}{k_m A}(N_m - \langle N\rangle)\right]^3}\left\{2\left[1 + \frac{k_a}{k_m A}(N_m - \langle N\rangle)\right] + \frac{k_a}{k_m A}\right\}\cdot\sigma^2, \tag{10}$$

which is solved for $\langle N\rangle$ and σ^2 (with the conditions $\langle N\rangle = 0$ and $\sigma^2 = 0$ at the moment $t = 0$).

The time-dependent SNR is defined as:

$$SNR(t) = \frac{\langle N\rangle}{\sigma}. \tag{11}$$

As σ is a measure of fluctuations resulting from the stochastic nature of the processes (AD coupled with MT) upon which the sensor operation is based, these fluctuations constitute the fundamental, i.e., unavoidable noise. Thus, the SNR defined in this way is the best possible SNR (also known as the quantum-limited SNR in the literature [16,20]) for a given adsorption-based sensor design and parameter set.

The steady-state expected value and variance of the number of adsorbed particles according to the presented model are obtained from Equations (9) and (10), respectively, for d$\langle N\rangle$/dt = 0 and dσ^2/dt = 0:

$$\langle N\rangle_e = \frac{N_m k_a C}{k_a C + k_d\left(1 + \frac{k_a}{k_m A}\right)}, \tag{12}$$

$$\sigma_e^2 = k_d\langle N\rangle_e\frac{\left[1 + \frac{k_a}{k_m A}(N_m - \langle N\rangle_e)\right]^2}{k_a C + k_d + \frac{k_a k_d N_m}{k_m A}}, \tag{13}$$

$$\sigma_e^2 = k_d k_a C N_m \cdot \frac{\left[k_a C + k_d\left(1 + \frac{k_a}{k_m A}\right)\left(1 + \frac{k_a}{k_m A}N_m\right)\right]^2}{\left[k_a C + k_d\left(1 + \frac{k_a}{k_m A}\right)\right]^3\left[k_a C + k_d\left(1 + \frac{k_a}{k_m A}N_m\right)\right]}, \tag{14}$$

and the steady-state SNR is:

$$SNR_e = \frac{\langle N\rangle_e}{\sigma_e}. \tag{15}$$

For the rapid mixing regime (i.e., adsorption-limited binding), the mass transfer influence is neglected, and the transition probabilities per unit time are $A_{RM}(n) = k_a C(N_m - n)$ and $D_{RM}(n) = k_d n$. In that case, Equations (5) and (6) yield the system of exact equations:

$$\frac{d\langle N\rangle}{dt} = A_{RM}(\langle N\rangle) - D_{RM}(\langle N\rangle), \tag{16}$$

$$\frac{d\sigma^2}{dt} = A_{RM}(\langle N\rangle) + D_{RM}(\langle N\rangle) - 2(k_d + k_a C)\cdot\sigma^2, \tag{17}$$

well-known in the literature [39], and whose solutions for the time evolution of both the expected value and the variance of N are:

$$\langle N \rangle_{RM} = \langle N \rangle_{RM,e}(1 - e^{-t/\tau_{RM}}), \tag{18}$$

$$\sigma^2_{RM} = \sigma^2_{RM,e}(1 - e^{-t/\tau_{RM}})\left(1 + \frac{k_aC}{k_d}e^{-t/\tau_{RM}}\right). \tag{19}$$

Here:

$$\langle N \rangle_{RM,e} = \frac{k_aCN_m}{k_d + k_aC}, \tag{20}$$

$$\sigma^2_{RM,e} = k_d\tau_{RM}\langle N \rangle_{RM,e} = \frac{k_dk_aCN_m}{(k_d + k_aC)^2} \tag{21}$$

are the expected value and the variance in the steady state, respectively, and $\tau_{RM} = 1/(k_d + k_aC)$. The time-dependent SNR and its steady-state value in the case of neglected MT influence are obtained from Equations (11) and (15), respectively, by using $<N>_{RM}$, σ_{RM}, $<N>_{RM,e}$, and $\sigma_{RM,e}$ instead of the corresponding parameters of the model that includes the MT effect.

Equations (9)–(15) constitute a theoretical model that enables the investigation of the microfluidic sensor stochastic response, and also of the sensor AD noise and SNR, in the case of adsorption–desorption coupled with analyte convection and diffusion. The same quantities, but in the case of neglected MT influence, can be investigated by using the theoretical model given by Equations (11), (15), (18)–(21).

3. Results and Discussion

The theoretical models presented in Section 2 are used here for the investigation of statistical parameters of the stochastic response and noise performance of a biosensor for the detection of proteins in a liquid sample (the model is applicable to various receptor–ligand pairs, i.e., various biological analytes (not only proteins), whose binding to the adsorption sites can be described by Equation (1); the parameter values used in our analysis are very close to those in [31], which are within the ranges found in BIACORE experiments with proteins). As the time response of adsorption-based sensors is a function (preferably linear) of the number of adsorbed target particles, we perform the stochastic analysis of the random process $N(t)$. That enabled us to obtain some general conclusions that are valid for the various types of adsorption sensors, regardless of their measurement parameter (optical, electrical, or mechanical, such as the refractive index, conductance, or mechanical resonance frequency), whose adsorption-induced time change constitutes the sensor response.

We first analyze and discuss the temporal change in both the expected value and the variance of the number of adsorbed particles, and of the sensor maximal achievable SNR (Section 3.1). Subsequently, we present the analysis of the same quantities after the established steady state of all the influencing transient processes (Section 3.2). Various practically relevant analyte concentrations, mass transfer coefficients, and sensing surface areas are considered.

3.1. Analysis of Time Evolution of the Expected Value and Variance of the Number of Adsorbed Particles and Sensor Signal-to-Noise Ratio, Considering MT Influence

Figure 2a,b show the time-dependent expected value of the number of adsorbed particles, $<N>$, for different concentrations of the target protein in the sample (ranging from $6{\cdot}10^{16}$ to $6{\cdot}10^{18}$ m^{-3}). The curves shown as solid lines in the presented diagrams are obtained by using the stochastic model given by Equations (9) and (10), which is numerically solved. The diagrams enable the investigation of the kinetics of the stochastic sensor response, considering MT effects. The AD process parameters are $k_a = 1.33{\cdot}10^{-19}$ m^{3}s^{-1} and $k_d = 0.08$ s^{-1}, there are $N_m = 3{\cdot}10^6$ adsorption sites on the sensing surface of area $A = 10^{-9}$ m^2, and the mass transfer coefficients are $k_{m1} = 2{\cdot}10^{-2}$ ms^{-1} for Figure 2a and

$k_{m2} = 2 \cdot 10^{-5}$ ms^{-1} for Figure 2b. All the parameter values are very close to those used in [31], for which the TCM applicability has been demonstrated in the same work.

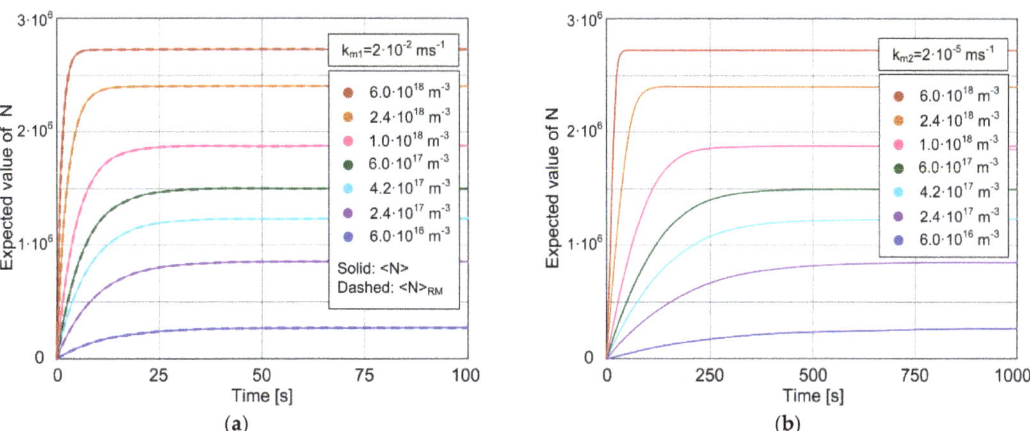

Figure 2. The expected value of the number of adsorbed particles in time, reflecting the kinetics of a biosensor stochastic response for different concentrations of the target protein in the analyzed sample: (**a**) The case of negligible MT influence— the curves <N> (shown by solid lines) according to the model that considers MT for $k_{m1} = 2 \cdot 10^{-2}$ ms^{-1} show the overlapping with the curves predicted by the model that neglects MT, <N>$_{RM}$ (dashed lines, entirely covered by solid lines). (**b**) The case of MT influenced kinetics ($k_{m2} = 2 \cdot 10^{-5}$ ms^{-1}).

As it can be seen in Figure 2a,b, the expected response is a monotonically increasing function of time for all concentrations, for both values of k_m. The slower mass transfer process (low k_m) prolongs the transient period of the time response at all concentrations, while the influence on the equilibrium expected value is not noticeable for the given set of parameter values. These conclusions are in accordance with those corresponding to the response kinetics described by the deterministic model (Equation (1)) [28,31].

The curves corresponding to the values $k_m > k_{m1}$ match those shown in Figure 2a, which means that for $k_m = k_{m1}$, mass transfer is already sufficiently fast to be of a negligible influence on the response kinetics. This explains the matching of curves (solid lines in Figure 2a), obtained by using the stochastic model that takes into account the coupling of stochastic AD and MT processes characterized by the coefficient k_{m1}, with those obtained by using the stochastic model that neglects the influence of MT (dashed lines in Figure 2a, entirely covered by solid lines). Indeed, the expressions given in Section 2 show that, for a sufficiently high k_m, transition rates according to the model that takes into account MT, $A(n)$ and $D(n)$, become equal to $k_a C(N_m - n)$ and $k_d n$, respectively, which are the well-known expressions for transition rates $A_{RM}(n)$ and $D_{RM}(n)$, respectively, valid when MT effects are negligible. This means that, for a sufficiently high k_m, the derived stochastic model, given by Equations (9) and (10), reduces to the model presented by Equations (18) and (19), i.e., the former model is a superset of the latter. Therefore, the model that takes into account the coupling of AD and MT processes covers the cases of both the pronounced and negligible MT effects on the stochastic response, and it is in this example applied for the research of microfluidic sensor kinetics both in the case when the MT influence is significant, i.e., in the mass transfer-limited regime (as shown in Figure 2b), and in the case when the MT influence is negligible, i.e., in the regime of rapid mixing or the adsorption-limited kinetics (as shown in Figure 2a).

The variance, which is a measure of AD noise, is another statistical parameter of the sensor stochastic response that we analyze. Figure 3a,b show the time-dependent variance of the protein biosensor for which the expected value of the number of adsorbed particles is shown in Figure 2a,b for the same seven concentrations. The diagram in Figure 3a is obtained for $k_{m1} = 2 \cdot 10^{-2}$ ms^{-1}, and it corresponds to the expected value given in Figure 2a, while the diagram in Figure 3b is for $k_{m2} = 2 \cdot 10^{-5}$ ms^{-1}, and it constitutes a pair with the diagram in Figure 2b. The solid line curves in Figure 3a,b are obtained by using the theoretical model that takes into account AD and MT processes (Equations (9) and (10)), while the dashed lines in Figure 3a represent the variances determined according to the model that neglects the MT influence (Equations (18) and (19)).

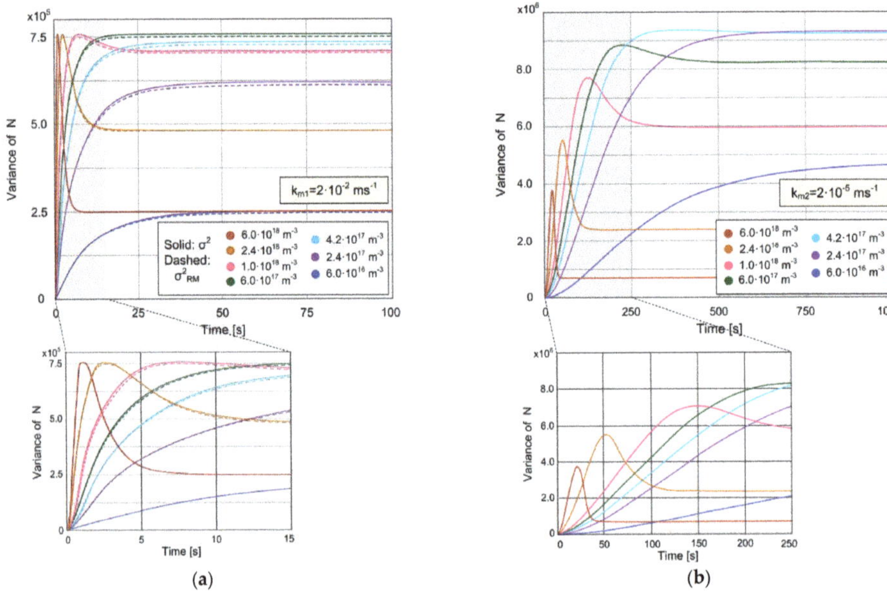

Figure 3. Variance of the number of adsorbed particles, revealing the behavior of the sensor response variance, i.e., sensor's AD noise during time, for different MT coefficient values: (**a**) $k_{m1} = 2 \cdot 10^{-2}$ ms^{-1} (solid lines correspond to the stochastic model that takes into account the coupling of AD and MT processes, dashed lines correspond to the model that neglects the MT influence); (**b**) $k_{m2} = 2 \cdot 10^{-5}$ ms^{-1} (according to the model that considers the combined effects of AD and MT).

The analysis of the solid line curves in Figure 3a shows that at the concentration of $6 \cdot 10^{17}$ m^{-3} and lower, $\sigma^2(t)$ is a monotonically increasing function. As the concentration increases to $6 \cdot 10^{17}$ m^{-3}, the transient regime duration decreases, and the equilibrium variance value increases. With the further increase in the concentration, the dependence $\sigma^2(t)$ has an increasingly prominent peak, the duration of the transient regime continues to decrease, while the steady-state variance value decreases.

Figure 3b shows that slower MT causes the appearance of a peak in the dependence $\sigma^2(t)$ at lower concentrations (noticeable even at $C = 4.2 \cdot 10^{17}$ m^{-3}). As the concentration increases, the transient regime duration decreases, and the steady-state variance value first increases and then decreases with the concentration, in the same way as in the case of high k_m (Figure 3a). The peak becomes increasingly pronounced with the concentration beyond $4.2 \cdot 10^{17}$ m^{-3}, and the maximal variance (corresponding to the peak) noticeably decreases. These conclusions stemming from our model, regarding the time-dependent variance when the MT influence is pronounced, are in accordance with the results of the stochastic computer simulation, which is based on the model that takes into account coupled AD and diffusion (called "coupled hybridization-diffusion process" in the mentioned reference) in nanowire biosensors and presented in [18].

By comparing the diagrams shown in Figure 3a,b, it can be concluded that slow MT causes the increase in the variance at all the concentrations from the analyzed range. It also prolongs the time needed for the variance to reach the equilibrium value.

In Figure 3a a small difference is noticeable between the solid and dashed curves for the same concentration value. That means that k_{m1} is very close to the specific value above which the MT influence on the variance becomes negligible. For every k_m value greater than that specific value, the model that takes into account MT yields the same curve $\sigma^2(t)$ as $\sigma_{RM}^2(t)$ for a given C. When the MT influence is negligible, the analysis of Equation (19) shows that the function $\sigma_{RM}^2(t)$ has a maximum (peak) at concentrations $C > k_d/k_a = 6 \cdot 10^{17}$ m^{-3}, and that this maximum does not change as the concentration increases further (the peak height is independent of C and equal to $N_m/4$), as can be seen in Figure 3a. At the moment when the variance is at its peak value, the expected value $<N>_{RM}$ equals $N_m/2$ (which is obtained from Equation (18)). $\sigma_{RM}^2(t)$ is a monotonically increasing function for $C < k_d/k_a$ (then $<N>_{RM} < N_m/2$), and $\sigma_{RM}^2(t) \approx <N>_{RM}(t)$ is valid for $C << k_d/k_a$. If the measurement of the signal fluctuation is used as an analytical tool in biosensing, the position and the value of the variance maximum can provide information in addition to those obtained by the noise analysis in the steady state. For example, in the case of negligible mass transfer influence, the value of the variance maximum can be used for the estimation of the number of surface binding sites N_m, which is a parameter important for the estimation and optimization of the sensor performance. In addition, the existence of the variance overshoot indicates that $C > k_d/k_a$.

Figure 4a,b show the time dependence of the best possible SNR (as defined in Equation (11)) of the biosensor in the case of the nearly negligible mass transfer influence (k_{m1}), and in the case when the mass transfer influence is pronounced (k_{m2}), respectively, for different analyte concentrations (the values of all parameters are given at the beginning of Section 3.1, and they are the same as for Figure 2a,b and Figure 3a,b). The curves corresponding to the cases when the MT influence is negligible (obtained according to the model that neglects MT, given by Equations (18) and (19)) are so close to those shown in Figure 4a (for $k_{m1} = 2 \cdot 10^{-2}$ ms^{-1}) that the difference between them is not noticeable in the diagram of that scale. The diagrams show that the SNR decreases with the decrease in C, for every t, both for rapid and slow mass transfer. Mass transfer increases the time needed for SNR to achieve its maximum value (corresponding to the steady state) at a given concentration. In addition, slow MT decreases the SNR for the given analyte concentration. Therefore, it depends on the value of k_m whether or not it is possible to reach the required SNR for reliable detection and quantification of an analyte by using a given sensor with a given set of parameter values.

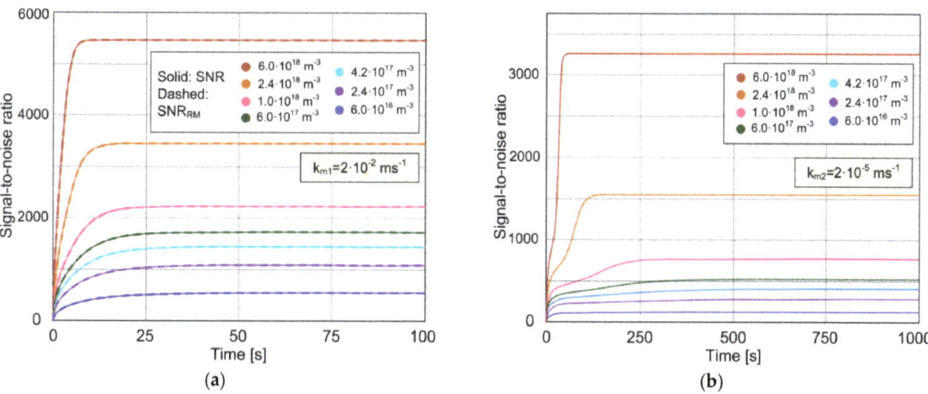

Figure 4. Time dependence of the sensor signal-to-noise ratio according to the model that considers the combined AD and MT effects (solid lines), for two different values of the MT coefficient: (**a**) $k_{m1} = 2 \cdot 10^{-2}$ ms^{-1}; (**b**) $k_{m2} = 2 \cdot 10^{-5}$ ms^{-1}. The curves obtained according to the model that neglects MT (dashed lines) match those predicted by the model that takes into account the coupling of AD and MT processes for k_{m1} (solid lines), as shown in (**a**).

3.2. Analysis of MT Influence on the Sensor Stochastic Response and Noise Performance in Steady State

Here, we show the analysis of the expected value and variance of the number of adsorbed particles after the steady state of all relevant transient processes has been reached. We also present the steady-state analysis of the maximal achievable SNR. In the following diagrams, denoted with solid lines are the curves obtained by using the model (given by Equations (12) and (13)) that considers fluctuations in the number of adsorbed particles as a consequence of the coupling of stochastic AD process and the mass transfer of analyte particles. The curves corresponding to the stochastic model that neglects MT (Equations (20) and (21)) are denoted with dashed lines. When MT is sufficiently fast, its effects become negligible, as shown by the matching of the corresponding expressions for the steady-state statistical parameters determined by using the two mentioned stochastic models, for the high enough values of k_m. The difference between the corresponding quantities determined by using the two models can thus be used as a measure of MT influence. The MT with the coefficient $k_m = 2 \cdot 10^{-5}$ ms^{-1} is assumed, the adsorption sites surface density is $n_m = N_m/A = 3 \cdot 10^{15}$ m^{-2} for all analyzed sensors, and the remaining parameter values are those given at the beginning of Section 3.1, unless otherwise noted.

Figure 5a shows the expected value of the number of adsorbed particles in the steady state, as a function of the sensing surface area ranging from 10^{-12} to 10^{-9} m^2, for different concentrations of the target protein. The curves obtained according to the two stochastic models match, which means that the influence of MT on the steady-state expected value is negligible for the given set of parameter values. The expressions for the steady-state expected value according to the two models, given by Equations (12) and (20), yield the ratio:

$$\frac{\langle N \rangle_{RM,e}}{\langle N \rangle_e} = 1 + \frac{k_a/(k_m A)}{1 + k_a C/k_d} \tag{22}$$

and, thus, the condition at which MT does not influence the expected number of adsorbed particles (when $<N>_{RM,e}/<N>_e \approx 1$ is valid):

$$k_m \gg \frac{k_a/A}{1 + k_a C/k_d} = k_{m,ev}. \tag{23}$$

This is also the condition for the applicability of the simpler stochastic model that does not take MT into account. The most stringent condition for a given k_m corresponds to the case when $k_{m,ev}$ has the highest value, i.e., at the lowest analyte concentration, and in sensors of the smallest sensing surface area from the considered range. For the sets of parameter values used in our analysis (Figure 5a), the maximal $k_{m,ev}$ equals $1.2 \cdot 10^{-7}$ ms^{-1}, so the most stringent condition for the expected values obtained according to the two models to be approximately equal is $k_m \gg 1.2 \cdot 10^{-7}$ ms^{-1}, which is fulfilled for $k_m = 2 \cdot 10^{-5}$ ms^{-1}. This explains the matching of the curves obtained by using the two models (Figure 5a). The condition (23) is satisfied in a wide range of parameter values. However, in sensors with extremely small sensing surfaces (such as nanowire or carbon nanotube mechanical or electrical sensors), as well as in detection of particles present in ultra-low concentrations, the value of $k_{m,ev}$ can be such that the condition given by Equation (23) is not satisfied, which implies that MT could influence the expected value of the sensor stochastic response by decreasing it. This conclusion is in accordance with the result of a computer simulation performed for a nanowire biosensor in [29], which showed the decrease in the expected value of the number of adsorbed particles in the case of binding influenced by diffusion. This deserves further investigation by using a model particularly considering nanoscale sensors.

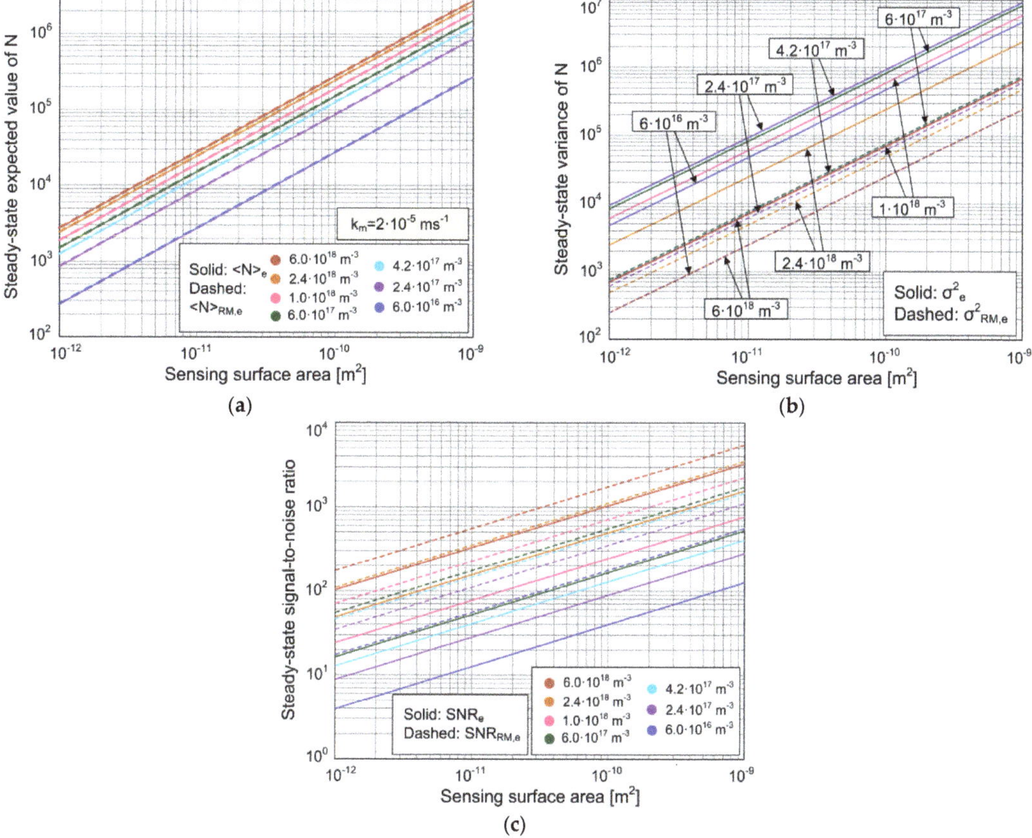

Figure 5. Expected value (**a**) and variance (**b**) of the number of adsorbed particles in the steady state, and the steady-state signal-to-noise ratio (**c**), as a function of the sensing surface area, for different concentrations of the target protein, according to the stochastic model that takes into account the combined effect of AD and mass transfer processes (solid lines), and according to the stochastic model that neglects mass transfer (dashed lines).

The dependence of the steady-state variance of the number of adsorbed particles on the sensing surface area, determined according to the two stochastic models, is shown in Figure 5b. The target protein concentration is used as the parameter for the shown curves.

Although the curves in Figure 5a for the same sets of parameter values show a negligible difference between the steady-state expected values obtained by using the two models, the corresponding steady-state variances differ significantly, which can be seen in Figure 5b. The diagram shows that MT with the coefficient $k_m = 2 \cdot 10^{-5}$ ms^{-1} causes a significant increase in the steady-state variance at a given concentration and sensing surface area. The condition for MT to be of negligible influence on the variance is obtained from $\sigma^2_e \approx \sigma^2_{RM,e}$. It can be formulated through the ratio:

$$\frac{\sigma^2_e}{\sigma^2_{RM,e}} = \frac{(k_a C + k_d)^2 \left[k_a C + k_d \left(1 + \frac{k_a}{k_m A}\right)\left(1 + \frac{k_a n_m}{k_m}\right)\right]^2}{\left[k_a C + k_d \left(1 + \frac{k_a}{k_m A}\right)\right]^3 \left[k_a C + k_d \left(1 + \frac{k_a n_m}{k_m}\right)\right]} \approx 1, \tag{24}$$

which is obtained from Equations (12), (13), (20) and (21). The condition (24) is more complex than the condition for approximately equal expected values (Equation (23)). By analyzing the ratio of the steady-state variances, the following ultimate condition for k_m is obtained, under which the ratio approximately equals 1:

$$\frac{\frac{k_a}{k_m A} + \frac{k_a n_m}{k_m} + \frac{k_a^2 n_m}{k_m^2 A}}{1 + \frac{k_a C}{k_d}} \ll 1. \tag{25}$$

When the condition given by Equation (25) is not satisfied, it is necessary to use the stochastic model that takes into account MT in order to perform the steady-state variance (i.e., AD noise) analysis. None of the combinations of C and A values from the considered range can satisfy the condition (25).

Figure 5b shows that the variance increases with a greater sensing surface area in cases when the MT influence is significant (solid lines), as well as in cases when MT is sufficiently fast and is thus of negligible influence (dashed lines). It can be seen from the diagram that solid and dashed lines that correspond to the same C value are parallel, which means that the ratio of variances determined according to the two models at the same concentration is independent of A in the whole considered range of A. In other words, the amount of MT influence on the change in steady-state variance (expressed as $\sigma^2_e/\sigma^2_{RM,e}$) does not depend on A. Indeed, the analysis of the expression $\sigma^2_e/\sigma^2_{RM,e}$ shows that for $A \gg k_a/k_m = 6.65 \cdot 10^{-15}$ m^2 (satisfied for all A values within the considered range), the following is valid:

$$\frac{\sigma_e^2}{\sigma_{RM,e}^2} \approx 1 + \frac{k_a n_m / k_m}{1 + k_a C / k_d}, \tag{26}$$

yielding the condition for the value of k_m at which the MT influence on the variance is insignificant, i.e., the applicability condition for the simpler stochastic model:

$$k_m \gg \frac{k_a n_m}{1 + k_a C / k_d} = k_{m,var}. \tag{27}$$

The condition given by Equation (27) can be easiest to satisfy at the greatest C value from the considered range, i.e., $C = 6 \cdot 10^{18}$ m^{-3}, but even then, $k_{m,var} \approx 3.6 \cdot 10^{-5}$ ms^{-1}. Due to that, the variances determined according to the two models differ significantly when the MT coefficient equals $2 \cdot 10^{-5}$ ms^{-1}, for every C value from the considered range.

Figure 5c shows the biosensor steady-state SNR obtained by using the two stochastic models, as a function of the sensing surface area, where the target analyze concentration is a parameter. The SNR decreases as the sensing surface area decreases (the standard deviation also decreases, but the decrease in the expected value is more pronounced, as it can be seen in Figure 5a,b), and also when the target protein concentration is lower, according to both models. A significant difference between the SNRs determined by the two models can be seen in the diagram. The decrease in the steady-state SNR, caused by MT, exists at all considered values of C and A. The condition under which the MT influence on the SNR is negligible is given by Equation (27) for the cases analyzed here (as $A \gg k_a/k_m = 6.65 \cdot 10^{-15}$ m^2 is valid). By using the model that takes into account the MT influence, and for k_m that satisfies the condition (27), the curves are obtained that match those shown as dashed lines in Figure 5c.

The value of k_m influences the maximal achievable SNR value of a sensor with a given sensing surface area. The diagram shown in Figure 5c can be used to determine whether or not it is possible to achieve an SNR value required to reliably detect and quantify the target substance concentration using a sensor of a given surface area. It can be seen that the steady-state SNR of a sensor with the sensing surface area of 10^{-12} m^2, protein concentration of $6 \cdot 10^{16}$ m^{-3}, and MT coefficient of $2 \cdot 10^{-5}$ ms^{-1} approximately equals 3, which is the minimal value needed for protein detection [40,41]. The same sensor in the case of negligible MT influence has an SNR of almost 20, so it satisfies the more stringent condition (SNR $\geq$ 10 [40]) for reliable quantification of the concentration. A sensor with a sensing surface area of 10^{-11} m^2 in the case of $k_m = 2 \cdot 10^{-5}$ ms^{-1} has an SNR higher than 10, so it satisfies the conditions for both analyte detection and quantification of the concentration even when the MT influence is pronounced. The same diagram enables the determination of the minimal detectable and measurable concentrations (i.e., the fundamental detection and quantification limits, as they are determined by the fundamental noise) of the given sensor, for the required SNR value (e.g., in [42], SNR = 1 was used for the estimation of the minimum detectable change in the measured quantity in a graphene ISFET, as the theoretical limit of performance determined by intrinsic noise; in [43], SNR = 1 was also used for the determination of the detection threshold in silicon nanowire sensors). These results show that the fundamental detection and quantification limits depend on the MT rate.

A diagram that enables the steady-state analysis of the dependences of both the sensor's time response variance and SNR on the target substance concentration, in the cases of pronounced and negligible MT influence, is given in Figure 6a. Figure 6b shows the ratio of variances when the MT influence exists, and when it is negligible, and the ratio of SNRs for the same two cases. The curves in Figure 6a,b are for a sensor with $A = 10^{-9}$ m^2, and they correspond to the steady-state values of time-dependent variances and SNRs shown in Figures 3 and 4, respectively. For the remaining surface areas considered in Section 3.2 (from 10^{-12} to 10^{-10} m^2), the conclusions will be the same as those obtained based on Figure 6b about the MT-influenced change in the variance and SNR (expressed through the ratios $\sigma^2_e/\sigma^2_{RM,e}$ and $SNR_e/SNR_{RM,e}$) in the considered concentration range. This is because the analysis given in the comment for Figure 5b,c showed that the magnitude of the MT influence on these two quantities does not depend on the active surface area when $A \gg k_a/k_m = 6.65 \cdot 10^{-15}$ m^2 (then, the ratios $\sigma^2_e/\sigma^2_{RM,e}$ (Equation (26)) and $SNR_e/SNR_{RM,e}$ do not depend on A). In addition, as $\sigma^2_{RM,e}$ is proportional to A, and it can be shown that for surface areas $A \gg k_a/k_m = 6.65 \cdot 10^{-15}$ m^2, σ^2_e is also proportional to A, all conclusions about the dependences of σ^2_e and $\sigma^2_{RM,e}$ on C, based on Figure 6a, will be valid for all sensing surface areas in the range from 10^{-12} to 10^{-9} m^2. For a similar reason (SNR_e and $SNR_{RM,e}$ are proportional to $A^{1/2}$), conclusions based on Figure 6a about the influence of both MT and target substance concentration on the change in SNR of a sensor with the sensing surface area of 10^{-9} m^2 are valid for other sensors of different sensing surface areas from the mentioned range.

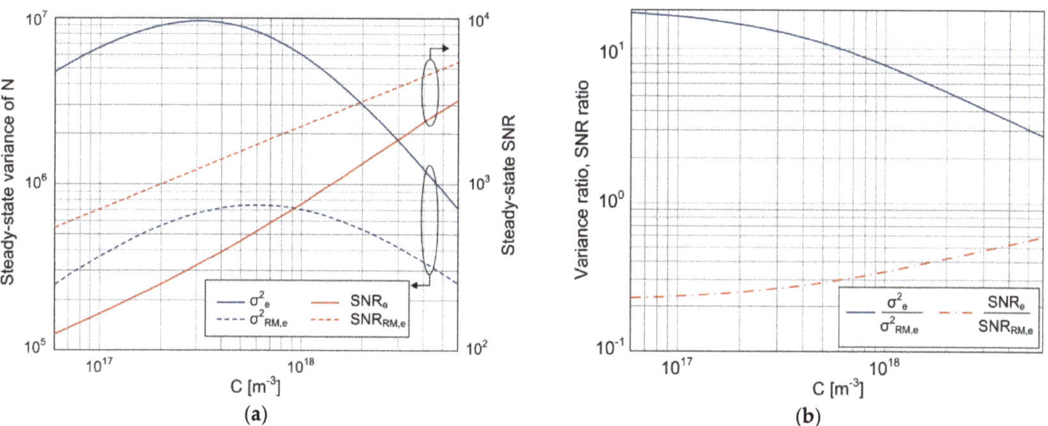

Figure 6. (a) Dependence of the variance of the number of adsorbed particles and sensor's SNR ($A = 10^{-9}$ m^2) in the steady state on the target substance concentration, according to the stochastic model that takes into account MT (solid lines), and the model that neglects it (dashed lines). (b) Ratios of steady-state variances ($\sigma^2_e/\sigma^2_{RM,e}$) and SNRs ($SNR_e/SNR_{RM,e}$) according to the two models, depending on the concentration.

Figure 6a shows that the steady-state variance obtained by using the model that takes into account the AD process, diffusion, and convection has a maximum at a certain concentration. This result for the MT-influenced binding of target particles is in accordance with the computer simulation results obtained for the variance of silicon nanowire field-effect biosensors whose response fluctuates due to the coupling of the AD process and mass transfer by diffusion [29,30].

In Figure 6a, it can be seen that MT influences the concentration value at which AD noise (i.e., variance) has its maximum. When MT is negligible (dashed line in Figure 6a), the variance reaches its maximum at $C_{max,RM} = k_d/k_a = 6 \cdot 10^{17}$ m^{-3}. Starting from the expression for σ^2_e (Equation (13)), which is simplified under the condition $A \gg k_a/k_m$, it can be analytically shown that the variance influenced by MT has the maximum at the concentration $C_{max} \geq C_{max,RM}$. Thus, due to the influence of MT, the AD noise maximum moves toward lower concentrations of the target substance.

Figure 6b shows that, for a given sensor, the influence of MT on the increase in variance becomes more pronounced at lower analyte concentrations. As the concentration decreases, the ratio $\sigma^2_e/\sigma^2_{RM,e}$, given by Equation (26), asymptotically approaches the maximum value $1 + k_a n_m/k_m \approx 21$, while, when C increases, the ratio of variances approaches 1 (i.e., for given k_m, the variances according to the two models are approximately equal at a sufficiently high concentration).

The dependences $SNR_e(C)$ and $SNR_{RM,e}(C)$, shown in Figure 6a, increase monotonically as the concentration increases. The influence of MT on the decrease in the sensor's steady-state SNR is concentration-dependent. This can be quantitatively analyzed based on the diagram shown in Figure 6b. MT causes the greatest decrease in the sensor's SNR at low concentrations. With the decrease in concentration, the ratio $SNR_e/SNR_{RM,e}$ asymptotically approaches its minimum.

The dependence $SNR_e(C)$ obtained by using the derived analytical expression and shown in Figure 6a is in accordance with that obtained in [30] by computer simulation for the case of diffusion-influenced binding. Diagrams of this kind (Figure 6a) enable the determination of the concentration detection and quantification limits for a given sensor and given experimental conditions, as the values of C at which the SNR has the minimal required values for reliable analyte detection and quantification, respectively.

4. Conclusions

A theoretical model was presented that enables efficient analysis of the stochastic time response and ultimate noise performance of adsorption-based microfluidic chemical and biological sensors, taking into account the influence of mass transfer (MT) of the analyte particles. It was shown that for sufficiently fast MT, the model we devised match the commonly used model that neglects mass transfer, so it is applicable in a wider parameter range, covering the cases of both pronounced and negligible MT influence.

Two models (one that neglects, and the other that takes into account MT effects) were used for the analysis of statistical parameters of the protein biosensor stochastic response, for various analyte concentrations, mass transfer coefficients, and sensing surface areas. The sensor signal-to-noise ratio (SNR), which sets the fundamental detection and quantification limit, was also investigated. The comparison of results obtained according to the two stochastic models has led to the conclusions about the qualitative and quantitative influence of MT on the sensor response kinetics and noise performance metrics, both in the transient regime and in steady state.

The analysis showed that MT can significantly alter the time dependence of the expected value and variance of the number of adsorbed particles, and thus the response kinetics, adsorption–desorption (AD) noise, and SNR of the sensor.

Slow mass transfer decelerates the response kinetics. The analysis indicated that MT can also influence the steady-state response value by decreasing it, in sensors of extremely small sensing surface areas, or when particles present in ultra-low concentrations need to be detected. This secondary effect of MT, which degrades the sensor's sensitivity and may cause false negative/positive results when interpreting measurement data, should be further investigated by using a model considering nanoscale sensors.

According to the analysis results, MT prolongs the time needed for the variance to achieve the steady-state value. Both stochastic models predict a peak in the time dependence of the variance, at concentrations greater than a certain critical value. However, MT influences the appearance of the peak in the variance transient regime at lower analyte concentrations. In the case of pronounced MT, the height of the peak decreases with the concentration. When MT is negligible, the peak height remains constant for different concentrations. The condition was formulated under which MT influences the variance in the steady state. MT can significantly increase the steady-state variance, i.e., the sensor AD noise. This effect is not limited to extremely small sensing surface areas, and it is especially pronounced at low analyte concentrations. MT also shifts the maximum of the steady-state AD noise toward the lower analyte concentrations.

An important conclusion of the analysis is that MT influences the sensor's maximal achievable SNR value. Slow MT decreases the SNR, and therefore, for the given sensor and target substance, it depends on MT parameters whether or not it it possible to reach the required SNR value for reliable detection and quantification of an analyte in the concentration range of interest. Another effect of MT is that the maximum SNR value (i.e., the steady-state value) is reached more slowly at slower MT. The MT influence on the SNR is more pronounced at lower analyte concentrations. The results have shown that even when mass transfer does not influence the expected value, it can significantly influence both the variance of the response (which is a measure of the inevitable adsorption–desorption noise) and the signal-to-noise ratio (which sets the fundamental detection and quantification limits of adsorption-based sensors). Therefore, mass transfer can exhibit a significant influence on the ultimate sensor performance, including the minimal detectable and quantifiable analyte concentrations.

The steady-state analysis of the difference between the corresponding quantities determined by using the two stochastic models also enables the obtaining of the applicability conditions for the simpler stochastic model that does not take into account MT. At the same time, these conditions reveal the criteria that can be used to establish when it becomes necessary to apply the model that takes into account the coupling between the AD process and mass transfer. It is shown that when the sensor's noise performance analysis is intended, the application of the stochastic model that takes MT into account may be necessary for the set of parameter values at which the simpler stochastic model is applicable for the analysis of the response kinetics.

To the best of the authors' knowledge, these are the first results regarding the influence of MT on the sensor stochastic response and its noise characteristics. The results have illustrated the significance of the presented theoretical model for the correct interpretation of measurement results, for the estimation of sensors' noise performance metrics important for reliable analyte detection and quantification, as well as for sensor optimization in terms of the lower detection and quantification limits. The presented model is also a useful tool for the development of new methods for the detection and quantification of substances, based on the analysis of sensor signal fluctuations. In general, the use of the stochastic model that takes into account MT becomes necessary as the sensing surface area and analyte concentration decrease. Due to the ongoing efforts toward the miniaturization of sensing devices, and the lower detectable concentrations in the latest sensor generation, it is expected for this theoretical model to be increasingly useful.

Author Contributions: Conceptualization, I.J.; methodology, I.J. and Z.D.; formal analysis, I.J., Z.D. and M.F.; software, K.R. and G.V.M.; investigation, I.J., Z.D. and M.F.; visualization, I.J., M.F. and P.M.K.; writing—original draft preparation, I.J. and M.F.; validation, Z.D., K.R. and P.M.K. All authors have read and agreed to the published version of the manuscript.

Funding: This research was funded by the Ministry of Education, Science and Technological Development of the Republic of Serbia, grant number 451-03-68/2020-14/200026, and the Serbian Academy of Sciences and Arts, grant number F-150 and F-96.

Institutional Review Board Statement: Not applicable.

Informed Consent Statement: Not applicable.

Data Availability Statement: The data presented in this study are available on request from the corresponding author.

Conflicts of Interest: The authors declare no conflict of interest.

References

1. Sackmann, E.K.; Fulton, A.L.; Beebe, D.J. The present and future role of microfluidics in biomedical research. *Nature* **2014**, *507*, 181–189. [CrossRef]
2. Luka, G.; Ahmadi, A.; Najjaran, H.; Alocilja, E.; DeRosa, M.; Wolthers, K.; Malki, A.; Aziz, H.; Althani, A.; Hoorfar, M. Microfluidics integrated biosensors: A leading technology towards lab-on-a-chip and sensing applications. *Sensors* **2015**, *15*, 30011–30031. [CrossRef]
3. Liu, K.-K.; Wu, R.-G.; Chuang, Y.-J.; Khoo, H.S.; Huang, S.-H.; Tseng, F.-G. Microfluidic systems for biosensing. *Sensors* **2010**, *10*, 6623–6661. [CrossRef]
4. Bhalla, N.; Pan, Y.; Yang, Z.; Payam, A.F. Opportunities and Challenges for Biosensors and Nanoscale Analytical Tools for Pandemics: COVID-19. *ACS Nano* **2020**, *14*, 7783–7807. [CrossRef]
5. Singh, P. SPR Biosensors: Historical Perspectives and Current Challenges. *Sens. Actuators B Chem.* **2016**, *229*, 110–130. [CrossRef]
6. Wang, D.S.; Fan, S.K. Microfluidic Surface Plasmon Resonance Sensors: From Principles to Point-of-Care Applications. *Sensors* **2016**, *16*, 1175. [CrossRef]
7. Zhang, Y.; Luo, J.; Flewitt, A.J.; Cai, Z.; Zhao, X. Film bulk acoustic resonators (FBARs) as biosensors: A review. *Biosens. Bioelectron.* **2018**, *116*, 1–15. [CrossRef]
8. Peña-Bahamonde, J.; Nguyen, H.N.; Fanourakis, S.K.; Rodrigues, D.F. Recent advances in graphene-based biosensor technology with applications in life sciences. *J. Nanobiotechnol.* **2018**, *16*, 75. [CrossRef]
9. Ambhorkar, P.; Wang, Z.; Ko, H.; Lee, S.; Koo, K.; Kim, K.; Cho, D. Nanowire-Based Biosensors: From Growth to Applications. *Micromachines* **2018**, *9*, 679. [CrossRef]
10. Liu, S.; Guo, X. Carbon nanomaterials field-effect-transistor-based biosensors. *NPG Asia Mater.* **2012**, *4*, e23. [CrossRef]
11. Voiculescu, I.; Nordin, A.N. Acoustic wave based MEMS devices for biosensing applications. *Biosens. Bioelectron.* **2012**, *33*, 1–9. [CrossRef]
12. Arlett, J.L.; Myers, E.B.; Roukes, M.L. Comparative advantages of mechanical biosensors. *Nat. Nanotechnol.* **2011**, *6*, 203–215. [CrossRef]
13. Zheng, F.; Wang, P.; Du, Q.; Chen, Y.; Liu, N. Simultaneous and Ultrasensitive Detection of Foodborne Bacteria by Gold Nanoparticles-Amplified Microcantilever Array Biosensor. *Front. Chem.* **2019**, *7*, 232. [CrossRef]
14. Xu, S.; Zhan, J.; Man, B.; Jiang, S.; Yue, W.; Gao, S.; Guo, C.; Liu, H.; Li, Z.; Wang, J.; et al. Real-time reliable determination of binding kinetics of DNA hybridization using a multi-channel graphene biosensor. *Nat. Commun.* **2017**, *8*, 14902. [CrossRef]
15. Jokić, I.; Djurić, Z.; Frantlović, M.; Radulović, K.; Krstajić, P.; Jokić, Z. Fluctuations of the number of adsorbed molecules in biosensors due to stochastic adsorption-desorption processes coupled with mass transfer. *Sens. Actuators B Chem.* **2012**, *166–167*, 535–543. [CrossRef]
16. Hassibi, A.; Zahedi, S.; Navid, R.; Dutton, R.W.; Lee, T.H. Biological shot-noise and quantum-limited signal-to-noise ratio in affinity-based biosensors. *J. Appl. Phys.* **2005**, *97*, 084701. [CrossRef]
17. Bellando, F.; Mele, L.J.; Palestri, P.; Zhang, J.; Ionescu, A.M.; Selmi, L. Sensitivity, Noise and Resolution in a BEOL-Modified Foundry-Made ISFET with Miniaturized Reference Electrode for Wearable Point-of-Care Applications. *Sensors* **2021**, *21*, 1779. [CrossRef]
18. Tulzer, G.; Heitzinger, C. Noise and fluctuations in nanowire biosensors. *IFAC Pap.* **2015**, *48*, 761–765. [CrossRef]
19. Georgakopoulou, K.; Birbas, A.; Spathis, C. Modeling of fluctuation processes on the biochemically sensorial surface of silicon nanowire field-effect transistors. *J. Appl. Phys.* **2015**, *117*, 104505. [CrossRef]
20. Das, S.; Vikalo, H.; Hassibi, A. On scaling laws of biosensors: A stochastic approach. *J. Appl. Phys.* **2009**, *105*, 102021. [CrossRef]
21. Mele, L.J.; Palestri, P.; Selmi, L. General model and equivalent circuit for the chemical noise spectrum associated to surface charge fluctuation in potentiometric sensors. *IEEE Sens. J.* **2020**, *20*, 6258–6269. [CrossRef]
22. Aćimović, S.S.; Šípová-Jungová, H.; Emilsson, G.; Shao, L.; Dahlin, A.B.; Käll, M.; Antosiewicz, T.J. Antibody—Antigen interaction dynamics revealed by analysis of single-molecule equilibrium fluctuations on individual plasmonic nanoparticle biosensors. *ACS Nano* **2018**, *12*, 9958–9965. [CrossRef]
23. Lüthgens, E.; Janshoff, A. Equilibrium coverage fluctuations: A new approach to quantify reversible adsorption of proteins. *ChemPhysChem* **2005**, *6*, 444–448. [CrossRef]
24. Jokić, I.; Frantlović, M.; Djurić, Z.; Radulović, K.; Jokić, Z. Adsorption-desorption noise in microfluidic biosensors operating in multianalyte environments. *Microel. Eng.* **2015**, *144*, 32–36. [CrossRef]

25. Djurić, Z.; Jokić, I.; Peleš, A. Fluctuations of the number of adsorbed molecules due to adsorption-desorption processes coupled with mass transfer and surface diffusion in bio/chemical MEMS sensors. *Microel. Eng.* **2014**, *124*, 81–85. [CrossRef]
26. Jakšić, O.; Jakšić, Z.; Čupić, Ž.; Randjelović, D.; Kolar-Anić, L.Z. Fluctuations in transient response of adsorption-based plasmonic sensors. *Sens. Actuators B Chem.* **2014**, *190*, 419–428. [CrossRef]
27. Hassibi, A.; Vikalo, H.; Hajimiri, A. On noise processes and limits of performance in biosensors. *J. Appl. Phys.* **2007**, *102*, 014909. [CrossRef]
28. Frantlović, M.; Jokić, I.; Djurić, Z.; Radulović, K. Analysis of the competitive adsorption and mass transfer influence on equilibrium mass fluctuations in affinity-based biosensors. *Sens. Actuators B Chem.* **2013**, *189*, 71–79. [CrossRef]
29. Tulzer, G.; Heitzinger, C. Fluctuations due to association and dissociation processes at nanowire-biosensor surfaces and their optimal design. *Nanotechnology* **2015**, *26*, 025502. [CrossRef]
30. Tulzer, G.; Heitzinger, C. Brownian-motion based simulation of stochastic reaction-diffusion systems for affinity based sensors. *Nanotechnology* **2016**, *27*, 165501. [CrossRef]
31. Myszka, D.G.; He, X.; Dembo, M.; Morton, T.A.; Goldstein, B. Extending the range of rate constants available from BIACORE: Interpreting mass transport-influenced binding data. *Biophys. J.* **1998**, *75*, 583–594. [CrossRef]
32. Schuck, P.; Zhao, H. The role of mass transport limitation and surface heterogeneity in the biophysical characterization of macromolecular binding processes by SPR biosensing. *Methods Mol. Biol.* **2010**, *627*, 15–54. [CrossRef]
33. Kusnezow, W.; Syagailo, Y.V.; Rüffer, S.; Klenin, K.; Sebald, W.; Hoheisel, J.D.; Gauer, C.; Goychuk, I. Kinetics of antigen binding to antibody microspots: Strong limitation by mass transport to the surface. *Proteomics* **2006**, *6*, 794–803. [CrossRef]
34. Soleymani, L.; Li, F. Mechanistic Challenges and Advantages of Biosensor Miniaturization into the Nanoscale. *ACS Sens.* **2017**, *2*, 458–467. [CrossRef] [PubMed]
35. Anderson, H.; Wingqvist, G.; Weissbach, T.; Wallinder, D.; Katardjiev, I.; Ingemarsson, B. Systematic investigation of biomolecular interactions using combined frequency and motional resistance measurements. *Sens. Actuators B Chem.* **2011**, *153*, 135–144. [CrossRef]
36. Djurić, Z.; Jokić, I.; Milovanović, G. Signal-to-noise ratio in adsorption-based microfluidic bio/chemical sensors. *Procedia Eng.* **2016**, *168*, 642–645. [CrossRef]
37. Schuster, P. *Stochasticity in Processes: Fundamentals and Applications to Chemistry and Biology*; Springer: Cham, Switzerland, 2016. [CrossRef]
38. Lee, C.H. A Moment closure method for stochastic chemical reaction networks with general kinetics. *MATCH Commun. Math. Comput. Chem.* **2013**, *70*, 785–800.
39. Van Kampen, N.G. *Stochastic Processes in Physics and Chemistry*; North-Holland Publishing Company: Amsterdam, The Netherlands; New York, NY, USA; Oxford, UK, 1981. [CrossRef]
40. Shrivastava, A.; Gupta, V.B. Methods for the determination of limit of detection and limit of quantification of the analytical methods. *Chron. Young Sci.* **2011**, *2*, 21–25. [CrossRef]
41. Wang, T.; Huang, D.; Yang, Z.; Xu, S.; He, G.; Li, X.; Hu, N.; Yin, G.; He, D.; Zhang, L. A Review on Graphene-Based Gas/Vapor Sensors with Unique Properties and Potential Applications. *Nano-Micro Lett.* **2016**, *8*, 95–119. [CrossRef]
42. Fakih, I.; Mahvash, F.; Siaj, M.; Szkopek, T. Sensitive Precise pH Measurement with Large-Area Graphene Field-Effect Transistors at the Quantum-Capacitance Limit. *Phys. Rev. Appl.* **2017**, *8*, 044022. [CrossRef]
43. Lee, J.W.; Jang, D.; Kim, G.T.; Mouis, M.; Ghibaudo, G. Analysis of charge sensitivity and low frequency noise limitation in silicon nanowire sensors. *J. Appl. Phys.* **2010**, *107*, 044501. [CrossRef]

 *biosensors*

Article

A Multichannel Microfluidic Sensing Cartridge for Bioanalytical Applications of Monolithic Quartz Crystal Microbalance

María Calero [1], Román Fernández [1,2], Pablo García [2], José Vicente García [2], María García [2], Esther Gamero-Sandemetrio [2,3], Ilya Reviakine [2,4,5,*], Antonio Arnau [1] and Yolanda Jiménez [1,*]

[1] Centro de Investigación e Innovación en Bioingeniería, Universitat Politècnica de València, 46022 Valencia, Spain; macaal3@teleco.upv.es (M.C.); rfernandez@awsensors.com (R.F.); aarnau@eln.upv.es (A.A.)

[2] Advanced Wave Sensors S.L. Paterna, 46988 Valencia, Spain; pgarcia@awsensors.com (P.G.); jvgarcia@awsensors.com (J.V.G.); mgarcia@awsensors.com (M.G.); egamero@florida-uni.es (E.G.-S.)

[3] Unidad de Educación, Florida Universitaria, 46470 Valencia, Spain

[4] IMBB-FORTH and Department of Biology, University of Crete, Heraklion, 70013 Crete, Greece

[5] Department of Bioengineering, University of Washington, Seattle, WA 98150, USA

* Correspondence: ilyareviakine@awsensors.com (I.R.); yojiji@eln.upv.es (Y.J.)

Received: 20 October 2020; Accepted: 20 November 2020; Published: 24 November 2020

Abstract: Integrating acoustic wave sensors into lab-on-a-chip (LoC) devices is a well-known challenge. We address this challenge by designing a microfluidic device housing a monolithic array of 24 high-fundamental frequency quartz crystal microbalance with dissipation (HFF-QCMD) sensors. The device features six 6-µL channels of four sensors each for low-volume parallel measurements, a sealing mechanism that provides appropriate pressure control while assuring liquid confinement and maintaining good stability, and provides a mechanical, electrical, and thermal interface with the characterization electronics. We validate the device by measuring the response of the HFF-QCMD sensors to the air-to-liquid transition, for which the robust Kanazawa–Gordon–Mason theory exists, and then by studying the adsorption of model bioanalytes (neutravidin and biotinylated albumin). With these experiments, we show how the effects of the protein–surface interactions propagate within adsorbed protein multilayers, offering essentially new insight into the design of affinity-based bioanalytical sensors.

Keywords: HFF-QCM (high fundamental frequency quartz crystal microbalance); mass transport; flow cell; biosensor; food safety; PoC (point of care); MQCM (monolithic quartz crystal microbalance)

---

1. Introduction

Inexpensive, fast, parallel, small-volume sensors for detecting minute amounts of analytes in complex samples are required for wide-spread application of lab-on-a-chip (LoC) devices in research-, industrial-, and health-related areas [1,2]. Equipped with such sensors, LoC devices can overcome the drawbacks of the conventional bulk analytical approaches that rely on expensive centralized facilities and play their singular role in the individualization of health care promised by personalized medicine [3].

Quartz crystal microbalance with dissipation, or QCMD, offers the technology needed for developing and mass-producing sensors for LoC applications. QCMD is known for its thin film monitoring [4] and biomolecular interaction measurement capabilities [5]. Digital, label-free, and very simple—its basic principle is based on electrically measuring resonance properties (frequency and dissipation) of a quartz crystal resonator [6]—QCMD can be easily automated. Key challenges for integrating QCMD technology into LoC-based applications are low throughput, high cost per sensor and per assay (related to sample/reagents volume), and lack of versatile and effective microfluidic/mechanical interfaces that can assure reliable field operation. Further improvements in sensitivity would also be beneficial for rapidly detecting small amounts of analytes in dilute samples.

To address these challenges, we introduce a microfluidic cartridge housing a monolithic high fundamental frequency (HFF) QCMD sensor array. We picked HFF-QCMD sensors because they are more sensitive than their low fundamental frequency counterparts, and can be used with low sample volumes (a few µL), and because they offer the possibility of integration into monolithic arrays [7–9]. Higher sensitivity is due to the smaller thickness of the HFF resonators [4,10], which in turn leads to a reduced surface area and therefore, to smaller sample volume [11]. Integration overcomes two major impediments for developing QCMD-based sensors for LoC applications: low throughput and relatively high individual sensor cost. QCMD and HFF-QCMD sensors have already been used to develop immunoassays [12–17] and DNA hybridization assays with the limit of detection (LoD) in the tens of femtomoles [18] and in complex samples (blood, saliva) [19,20], but currently, advanced commercial systems offer measurements with at most four sensors at a time, each of which has to be operated individually. This negates other advantages of QCMD technology in high-volume LoC applications. To this end, we have recently reported our design of monolithic 150 MHz HFF-QCMD sensor arrays and characterized their performance [7]. Here, we present the design of an integrated microfluidic cartridge for housing the arrays (Figure 1) and evaluate its performance for measuring biomolecular interactions in a biotin/neutravidin based system. Our work advances the state-of-the-art by integrating sensing and microfluidic analyte delivery compatible with LoC applications.

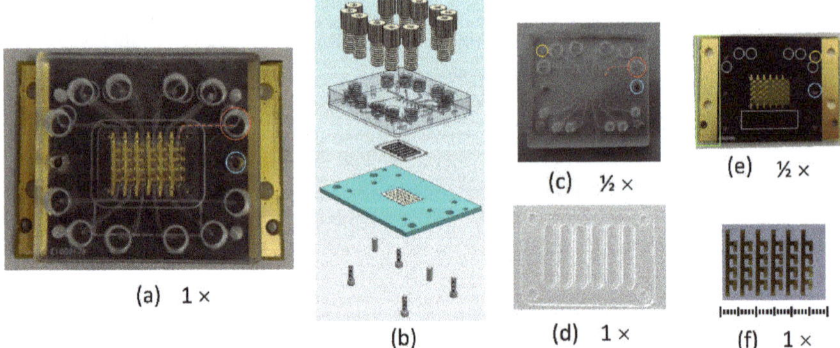

Figure 1. Multichannel Microfluidic Sensor Cartridge. Assembled cartridge (**a**). Schematic of cartridge assembly (**b**) from components (from top to bottom): microfluidic connectors, Poly(methyl methacrylate) (PMMA) cell, Poly(dimethylsiloxane) (PDMS) gasket, Printed Circuit Board (PCB) with the array, assembly screws and alignment pins. Photos of its components: PMMA cell (**c**), PDMS gasket with six channels (**d**), PCB (**e**), and the 24-element HFF-QCMD array (**f**, see Figure S1 in the Supporting Information for an enlarged view). Note the difference in scale: (**c**,**e**) are shown at half the size relative to (**a**,**d**,**f**), which are drawn to scale. Array dimensions are 14.25 mm × 9.05 mm.

The development of robust and simple HFF-QCMD detection systems for LoC applications is very much in its infancy [2]. Integration of the fluidic circuitry with the array of quartz sensors faces a number of challenges [21,22], the key of which is insuring independent operation of the individual sensor elements that is free of the interaction between them. Interactions between sensors can be mediated electrically, mechanically, or via the fluidic path affecting analyte transport to the surface. Therefore, our evaluation of the cartridge and array performance focused as much on the reproducible measurements of frequency and dissipation shifts in response to the introduction of bioanalytes as on the independent operation of the individual array elements, as evidenced by the absence of artifacts originating from the interaction between them through any of the three pathways (electrical, mechanical, fluidic). Our experimental results were consistent with the literature and with the results of numerical calculations of molecular transport rates based on the channel and array geometry, molecular diffusion coefficients, and fluid flow rates.

2. Materials and Methods

2.1. Materials

Nanopure water used in this study was either analytical grade water (Panreac Química SLU, Barcelona, Spain), or produced with a Smart2Pure UVUF water purification system (Thermo Fisher Scientific, Barcelona, Spain). Nitrogen was from Air Liquide España S.A. (Valencia, Spain). Phosphate buffered saline (PBS) tablets for preparing 0.01 M phosphate buffer containing 0.0027 M potassium chloride and 0.137 M sodium chloride, pH 7.4, at 25 °C, and Bovine Serum Albumin (BSA) were purchased from Sigma Aldrich Química, S.L.U. (Madrid, Spain). NeutrAvidin (Nav), biotinylated BSA (bBSA), and Sodium Dodecyl Sulfate (SDS) 20% solution were purchased from Fisher Scientific S.L. (Madrid, Spain). COBAS Cleaner was purchased from Sanilabo S.L. (Valencia, Spain).

Poly(methyl methacrylate), PMMA, was from Monje Hermanos S.L. (Valencia, Spain); poly(dimethylsiloxane), PDMS, was from Ellsworth Adhesives Iberica, (Madrid, Spain). Single-component conductive epoxy AA-DUCT 900 was from Atom Adhesives (Providence, RI, USA). DuPont™ Teflon® FEP film with a thickness of 76 μm was from Dupont (Wilmington, DE, USA).

2.2. Multichannel Microfluidic Sensing Cartridge Design

The multichannel sensor cartridge consists of an array, a custom-made Printed Circuit Board (PCB) for mounting the array, and a microfluidic cell consisting of a gasket and the body. These components are shown in Figure 1 while their design and manufacturing are described below.

2.3. Arrays and PCBs

Monolithic HFF-QCMD sensor arrays (AWS-Array2-24-150.0M, Advanced Wave Sensors (AWSensors) S. L., Valencia, Spain) were manufactured as described previously [7]. They were mounted on rectangular 52.02 mm × 36.02 mm × 1.55 mm PCBs custom-designed using the ALTIUM Designer 18 software package (Altium, San Diego, CA, USA) following high-frequency signal routing and crosstalk prevention considerations and manufactured from FR4-type material by Eurocircuits (Mechelen, Belgium). The mounting was done with a single-component conductive epoxy that was deposited by stencil printing technology (eC-stencil-mate, Eurocircuits). The stencil was configured to leave a tiny quantity (≈0.16 mg) of epoxy on each contact area, sufficient for making electrical contact, but not in excess in order not to affect the resonant behavior of the sensors. Before the deposition of the epoxy, the PCBs were degreased with acetone and dried at 50 °C for 5 min. Once the conducting epoxy was deposited, the arrays were placed on the PCBs manually with the assistance of a USB Microscope (Shenzhen Andonstar Tech Co. Ltd., Shenzhen, China) and a manual pick and place machine (eC-placer, Eurocircuits). The epoxy was cured at 150 °C for 1 h in an oven (eC-reflow-mate v4, Eurocircuits, Mechelen, Belgium). During the curing process, a 70 g weight was placed over the array to ensure a good contact between the array and the PCB. The array surface was protected with the transparent Fluorinated Ethylene Propylene (FEP) film.

2.4. Microfluidic Cell

The microfluidic cell consisted of two parts: a flexible gasket containing six independent flow channels, one for each column of the array, and the body of the cell for housing the gasket that contained inlet and outlet fluidic channels, the opening for the mounting screws, and the alignment pins (Figure 1b).

The gaskets were fabricated from PDMS using a mold, 9 gaskets at a time. To this end, the required volume of the PDMS curing agent and PDMS monomer was mixed at the 1:10 ratio. Vacuum was applied to the mixture for 45 min to remove air. The mold was dried at 60 °C and the freshly degassed PDMS mixture was cast onto the mold. Curing took place in an oven at 60 °C for 80 min. The height of the channels is determined by the thickness of the gasket being 480 μm while the width is 1.46 mm. Consequently, the total volume of each fluidic channel was less than 6 μL. (See Section S3 of the Supporting Information).

The body of the microfluidic cell consisted of two parts: the top that contained the openings for standard microfluidic connectors (VacuTight Ferrule Tefzel Red P-840 (ETFE) 1/16, Fitting P-844, IDEX Health & Science, Rohnert Park, CA, USA) for connecting the cell to the external flow control device, and the bottom, that contained inlet and outlet microchannels, 12 in total, connecting those openings to the fluidic channels in the gasket.

The mold for manufacturing PDMS gaskets, and the two parts of the microfluidic cell, were manufactured from PMMA in the vertical CNC (Computer Numerical Control) machining center (Chevalier 1418VMC-Plus, Falcon Machine Tools Co. Ltd., Chang Hua, Taiwan). The two pieces of the microfluidic cell were joined by first softening their surfaces with ethanol, and then pressing them against each other over a period of 1 min at 70 °C, as described in ref. [23]. In this manner, a one-piece microfluidic cell was obtained. The PDMS gasket was mounted on the cell, and the cell with the gasket was placed on top of the PCB-mounted array. The assembly was guided by the alignment pins and held together with screws, as shown in Figure 1b.

2.5. Evaluation of Cartridge Performance

The performance of the array-based sensor cartridge was evaluated by monitoring resonance frequency and dissipation in situ and in real time as the arrays were exposed to liquid flow and as proteins were adsorbed to the surfaces of the arrays (in separate experiments). The results of the protein adsorption experiments were further compared with those obtained with the commercially available, individual 50 MHz HFF-QCMD sensors (AWsensors S.L., Valencia, Spain).

Resonance frequency and bandwidth (or dissipation) of the arrays and individual HFF sensors were measured as a function of time with the AWSensors X24 and A20+ platforms, respectively, running the AWSuite software package. Individual sensors were mounted in the standard, commercially available QuickLock® measurement cells (AWSensors S.L., Valencia, Spain). Fluid flow and sample injection were controlled with the AWSensors multichannel F20+ fluidics controller connected to the inlets of the sensing cartridge (for the arrays) or the inlets of the QuickLock® measurement cells.

For the measurements of the air-to-liquid transition, baseline frequency and bandwidth signals were acquired in air for ~15 min, followed by flowing PBS buffer, at a flow rate of 20 μL/min.

For the protein adsorption experiments, baseline frequency and dissipation signals were acquired in buffer flowing at a rate of 14.5–20 μL/min, depending on the experiment, for ~5–10 min, followed by the injection of the proteins, Nav (at a concentration of 100–200 μg/mL in PBS) or bBSA (at a concentration of 100 μg/mL). Each injection was always repeated twice, 100 μL each time, to ensure that adsorption reached saturation.

2.6. Array, Cell and Sensor Cleaning and Preparation

Prior to the experiments, PCB-mounted arrays and individual sensors were immersed in freshly filtered 2% SDS solution for 30 min. They were rinsed with Nanopure water and dried under a stream of nitrogen directed at the array and sensor surfaces through a 0.45 μm pore diameter filter housed in hand-held filter gun assembly (Skan AG, Allschwil, Switzerland) and then cleaned for 30 min in a UV/Ozone Cleaner (BioForce Nanosciences, Salt Lake City, UT, USA) that was preheated for 30 min prior to use. After cleaning, array and sensor surfaces were rinsed with ethanol to reduce the oxide that forms as a result of the UV-Ozone treatment [24] and once again dried with a stream of filtered nitrogen.

The PDMS gasket and the PMMA cell were immersed in COBAS cleaner for 30 min, rinsed with, and then sonicated in Nanopure water for 15 min to remove traces of the detergent, and dried with a stream of filtered nitrogen.

Individual sensors were mounted in the QuickLock™ measurement cells cleaned with the COBAS cleaning solution for 30 min, followed by repeated rinsing with water and ethanol, and drying with a stream of filtered nitrogen.

3. Results and Discussion

3.1. Multichannel Microfluidic Sensing Cartridge Design and Assembly

A microfluidic cartridge holding a 24 HFF-QCM sensor array operating at 150 MHz [7] was designed and implemented (Figure 1a). It consists of a microfluidic cell that delivers fluid to the sensing elements arranged in six channels of four elements each (Figure 1c), a soft PDMS gasket forming the six microfluidic channels (Figure 1d) that seals the PMMA fluid cell to the array, and the array mounted on the PCB that served as the bottom of the cell (Figure 1e,f). PMMA is a low-cost, transparent material with good thermal insulating capacity that is suitable for disposable cartridges and is used routinely in biomedical applications (e.g., bone cement [25,26]). It has been selected because of its adequate rigidity to confer robustness to the microfluidic assembly design and its mass fabrication potential [23]. The PDMS gasket is fixed in the PMMA cell with the help of four guides (Figure 1d). The PMMA cell contains sockets for the fluidic connectors on the top (outlined in red in Figure 1a,c), as well as sockets for the guiding pins (blue in Figure 1a,c,e) used to align the array, and mounting screws (yellow in Figure 1c,e) used to fix it. The assembly process is very simple and results in a rectangular chamber design which avoids non-uniform sample distribution [27,28]. The PCB offers mechanical protection, ease of handling, improves the thermal stability of the array during measurements, and forms the contacts between the array and the acquisition electronics. To this end, we integrated into the PCB design two gold sidebands (highlighted in a green rectangle in Figure 1e) that are connected to the Peltier elements of the temperature control system, providing a direct thermal path. Our design allows simultaneous measurements with six different solutions. Furthermore, within each of these six measurements, it is possible to have four replicas with identically treated array element surfaces. Alternatively, each array element could be used to immobilize a different biomolecule, allowing four independent measurements with each of the six solutions at the expense of statistics.

3.2. Cartridge Performance

The performance of the cartridge was evaluated in two separated sets of experiments: the responses of the sensors to changing the media from air to liquid were measured in a first set of experiments, and sequential adsorption of proteins neutravidin and biotinylated BSA (bBSA) on the sensor surface was followed in the second set. Cartridge and array performance were evaluated by comparing our experimental results with theoretical predictions and with other experimental data, either obtained in this study with individual HFF-QCMD sensors, or reported in the literature with the classical low-fundamental frequency QCMD sensors. The distribution of sensor responses was analyzed as a function of sensor position in the array to assure independent operation of the individual sensors. Finally, biomolecular transport in the cartridge has also been studied to evaluate possible limitations of the microfluidic cartridge design.

3.2.1. Air-to-Liquid Transition

Changes in the frequency and bandwidth of a quartz resonator upon exposure to a Newtonian liquid are well-understood as a result of the classical works by Mason [29] and Kanazawa [30]. When discussing the air-liquid shifts, it is more convenient to use bandwidth than dissipation, because the frequency and bandwidth changes are expected to be equal and opposite: $\Delta f = -\Delta \Gamma = -f_{res}^{3/2}\left(\eta_l \rho_l / \pi \rho_q \mu_q\right)^{1/2}$; this is the so-called Kanazawa–Gordon–Mason equation, where f_{res} is the fundamental resonance frequency of the resonator, η_l and ρ_l are the liquid viscosity and density, respectively, ρ_q is the density of quartz, and μ_q is its shear elastic modulus. Experimental results are shown in Figure 2, where the shifts in both signals upon the transition from air to buffer are plotted as a function of the array element (Supplementary Figure S1). We find no systematic trends in the shifts of either of the two signals as a function of the array element position, as is expected for the normal functioning of the array, the flow system, and the electronics. It is interesting to note the increase in Δf and $-\Delta\Gamma$ in sensors 20 to 24. Although it seems a systematic trend, this effect was not observed in other measurements, confirming its coincidental nature.

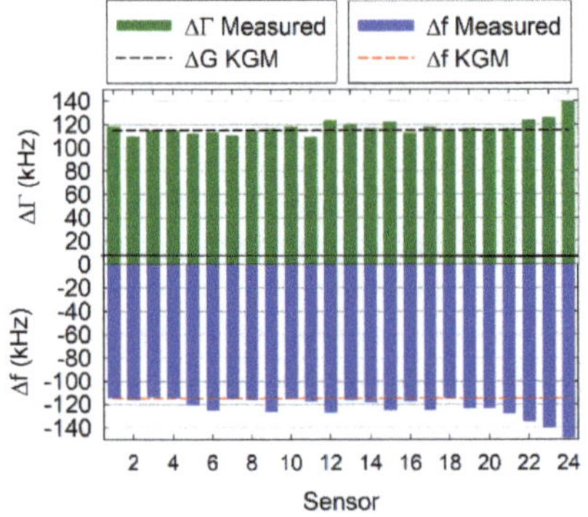

Figure 2. Air-buffer measurements. A plot of the typical set of air-to-buffer shifts for all the elements of one array (bars) with the expected values calculated based on the Kanazawa-Gordon-Mason equation (dashed lines).

On average, we observed shifts of -120 ± 15 kHz for Δf and 115 ± 9 kHz for $\Delta\Gamma$, averaged over 110 individual air-to-buffer measurements. The theoretical values for the arrays used in these measurements are $\pm (115 \pm 2)$ kHz, where the negative and the positive signs refer to the frequency and the dissipation shifts, respectively, and the error arises from the variation of the elements' resonance frequencies [7]. The larger experimentally observed value of Δf is expected, because finite roughness typically masquerades as an additional Sauerbrey shift that does not affect the bandwidth [31,32]; a 5 kHz difference in frequency would correspond to a Sauerbrey film of ~9.9 Å with a density of 1 g/cm^3, which is reasonable given the roughness values of ~1 nm we reported previously for the surfaces of these arrays [7].

3.2.2. Protein Adsorption and Interaction Studies

The functionality of the multichannel cartridges for sensing biomolecular interactions was evaluated using neutravidin (Nav) and biotinylated BSA (bBSA) as a model system that is widely used in biotechnology and bioanalytics [33,34].

Proteins were allowed to adsorb in sequence, starting with Nav in one set of experiments, and starting with bBSA in another set. Each sequence consisted of three sets of (Nav + bBSA) pairs, with non-biotinylated BSA injected between each step to test for non-specific adsorption. The purpose of the two different sequences was to establish a protocol for subsequent applications of these arrays in bioanalytical sensing. One such sequence is shown in Figure 3a.

Average results from multiple arrays and individual HFF sensors are shown in Figure 3b (for the Nav–BSA sequence) and in Figure 3c (for the bBSA–Nav) sequence. The adsorbed layer masses, calculated from the frequency shifts using the Sauerbrey relationship [4], are summarized in Table S1 in the Supporting Information. The masses expected on the basis of the sizes of the molecules and the corresponding literature values are presented in Supplementary Table S2. The adsorption processes are depicted schematically in Figure 4.

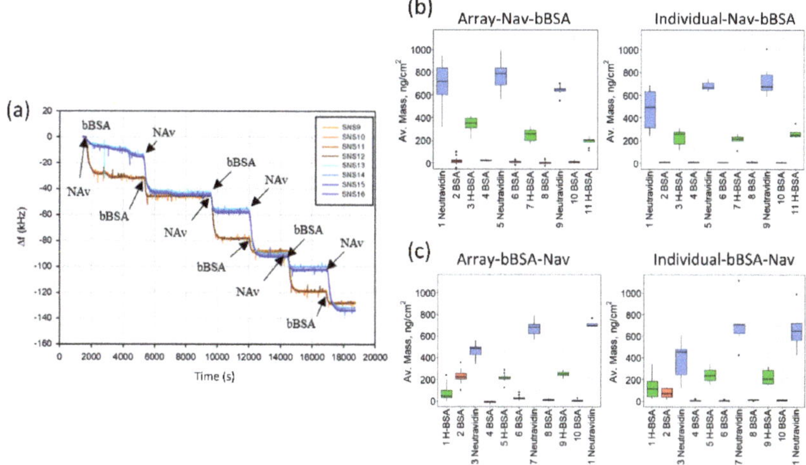

Figure 3. Protein adsorption measurements. (a) Frequency changes due to the sequential adsorption of proteins for two columns of the array relative to a baseline in buffer. In one of the two columns (SNS13 to SNS16), proteins were adsorbed in sequence bBSA-Nav- bBSA-Nav- bBSA-Nav, in the other (SNS9 to SNS12) Nav-bBSA-Nav-bBSA- Nav-bBSA. SNSi with $i = 9$ to 16 refers to the element position on the array according to Figure S1 in the Supporting Information. (b,c) Mass changes observed in the experiments performed with arrays (**left**) and individual 50 MHz resonators at the 3rd overtone (**right**) for the two adsorption sequences, respectively. Error bars are standard deviations.

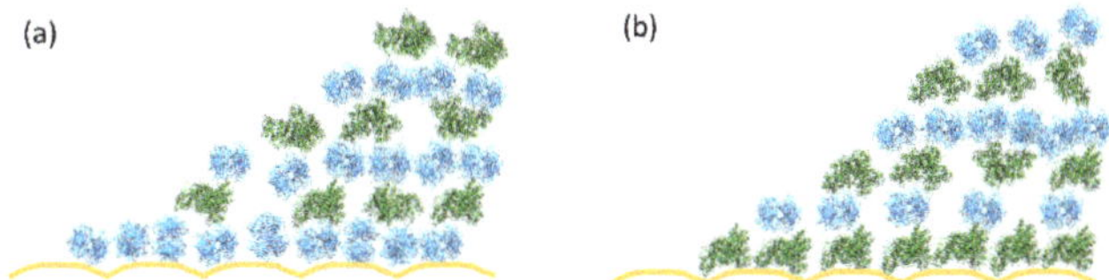

Figure 4. Schematic representation of the surface architecture achieved in the adsorption experiments. (a) Shows the adsorption sequence starting with Nav, and (b)—starting with bBSA. The gold is shown in yellow, Nav in light blue, and BSA in green. No distinction is made between the biotinylated and non-biotinylated BSA. The vertical dimension of the roughness of the gold is to scale with the sizes of the proteins; similarly, the protein dimensions are also drawn to scale relative to each other, but Nav aggregation is ignored for simplicity. On average, there is ~ one bBSA molecule for every two Nav molecules adsorbed on gold. Away from the surface, there are ~ three Nav molecules per every bBSA molecule. See Section S2 in the Supporting Information for further discussion. This figure was prepared using USCF Chimera version 1.14 [35].

Several trends can be noted in the experimental data. First and foremost, there is a good agreement between the results obtained with the arrays and with the individual HFF resonators (Figure 3, Supplementary Table S1). Second, the amounts of protein adsorbed observed with both systems (arrays and the individual resonators) are in good agreement with the literature. Focusing on the adsorption of Nav on gold (Figure 3b), we find 700 ± 180 ng/cm^2 for the array and 590 ± 200 ng/cm^2 for the individual sensors (Supplementary Table S1). The limiting value of the adsorbed Nav mass is also ~700 ng/cm^2 ("limiting" here refers to the apparent saturation of the values of the adsorbed mass as a function of the adsorption step in Figure 3b,c). This is in good agreement with the literature values for Nav adsorption on gold that range between ~700 and ~1300 ng/cm^2 (Supplementary Table S2). The variability here is due to the tendency of this protein to aggregate and corresponds to the age and treatment of the neutravidin solution. The aggregation tendency also explains the difference between the observed masses and their expected values based on the protein dimensions and between Nav and a very similar protein, streptavidin (Sav) (Supplementary Table S2) [36].

The amount of Nav adsorbed onto gold pre-coated with bBSA is significantly smaller, than directly onto gold (~400 ng/cm^2, nearly identical for the arrays and the individual sensors; Figure 3c and Supplementary Table S1). Also, the variation in the mass of Nav adsorbed, expressed as a standard error, is 15 ng/cm^2 when it is adsorbed on the bBSA layer, but 38 ng/cm^2 when it is adsorbed onto gold directly, indicating that orienting Nav on a biotinylated substrate results in a more homogeneous layer. The value of the adsorbed mass of Nav on bBSA is in good agreement with that adsorbed on the biotinylated SLBs [36] (566 ng/cm^2, Supplementary Table S2). One of the factors that contribute to the observed difference in the adsorbed mass is the difference in protein orientation: it is quasi-random when Nav is adsorbed on gold but fixed by the underlying bBSA layer when Nav adsorbs on bBSA. Such an effect has already been reported for Sav (Supplementary Table S2): a larger amount of this protein adsorbs on the mercaptoundecanoic acid (MUA) SAM or on gold than on a layer of biotinylated molecules (lipids or alkane thiols).

With Nav, there is the additional effect of aggregation of this protein, an effect that is absent in the case of Sav. Indeed, the difference between Nav adsorbed on gold and Nav adsorbed on bBSA or bSLB is much greater, than between streptavidin adsorbed on gold vs. bSLB or bSAM (Supplementary Table S2).

Notably, the tendency of neutravidin to aggregate appears to be reduced, when it is adsorbed in a fixed orientation on the biotinylated substrates (bBSA, Table S1, or bSLB, Supplementary Table S2), compared to when it is adsorbed on gold. This could also be due to the increased stability of this protein in the biotin-bound conformation [37], or to the steric limitations arising from the number of accessible biotins presented by the biotinylated surfaces. This effect is gradually lost in the subsequent adsorption steps (steps 2 and 3 in Supplementary Table S1 and Figure 4) since the limiting values of the adsorbed amounts of Nav are independent of the direction of the adsorption.

The value of the limiting mass of bBSA (~230 ng/cm^2, steps 2 and 3 in Supplementary Table S1) nearly identical for the arrays and the individual sensors, is independent of the direction of adsorption, and is consistent with what has been reported by others for bBSA on Nav on gold or on Nav on SLBs (Table S2). It is worth noting that there is a significant difference in the behavior of the biotinylated and non-biotinylated versions of BSA on bare gold (c.f. our results quoted above with Table S2 in the supporting information). This has been noted before by Kim et al. when adsorbing BSA/bBSA mixtures on gold [38]. The effect has not been investigated further, but probably originates from the effect of biotinylation on the adsorbed protein orientation, conformation, or both.

We would like to underscore the importance of the comparison between the array and the individual sensors: since we [11–16] and others [17,39–41] have previously demonstrated the robustness and reproducibility of HFF resonators, the observed correspondence shows (1) that integration into an array does not affect their performance, and (2), demonstrates the functionality of the microfluidic cartridge. The comparison with the published results obtained with the same bioanalytical systems using classical, low-frequency individual sensors confirms that our array-based cartridge is capable quantitatively and accurately assaying bioanalytical systems based on specific binding.

3.2.3. Biomolecular Transport in the Microfluidic Cartridge

One concern with an integrated array of sensing elements and the corresponding fluidics is the presence of systematic artifacts arising from transport limitations in the fluidic channels. Here, we first analyze the variation in the adsorbed protein layer masses detected by the different array elements (Figure 5) and then compare transport kinetics of the analytes in the fluidic cartridge with theoretical predictions based on molecular diffusion coefficients and the geometry of the fluidic channels and the arrays (Figure 6). We demonstrate lack of such systematic errors.

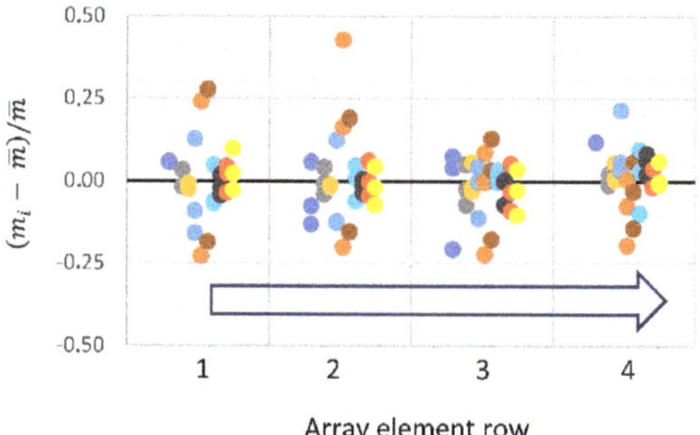

Figure 5. Variation in the adsorbed mass as a function of the sensor position. The absolute deviation of the adsorbed mass, normalized by the mean, is plotted against the row of the array element. m_i is the adsorbed mass detected by the i-th element of the array (see Figure S1). $\bar{m}$ is the average mass in a given column. Different colors represent different experiments. Results of each individual experiment are offset from each other for visibility. Blue open arrow indicates the direction of the fluidic flow up the column.

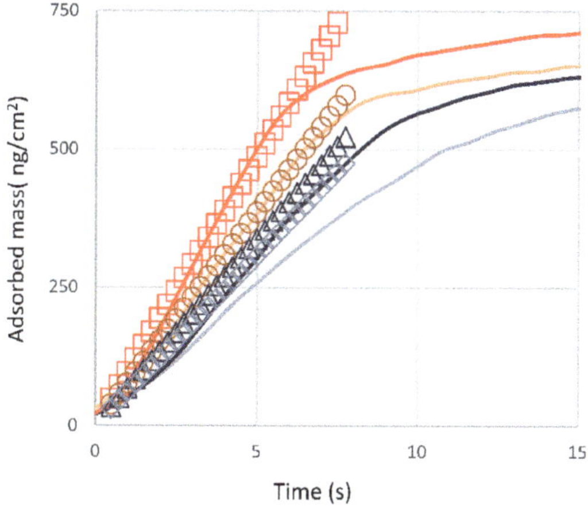

Figure 6. Comparison of the experimental and theoretical molecular transport rates at the array elements. Solid lines: experimental results for one array column obtained with Nav adsorbing on gold. Open symbols: calculated $\Gamma(x, t)$, where x specifies the distance between the inlet and the array element.

Indeed, it can be seen in Figure 5 that the distribution of the adsorbed masses of Nav as a function of the array element location is random. For most of the experiments, the variation is <15%, in terms of the absolute deviation. The data shown in Figure 5 also illustrate the robustness of the cartridge and array system to repeated cleaning and assembly cycles; arrays survive up to 30 such cycles without loss of performance.

Finally, we show that the experimentally observed transport rates can be compared semi-quantitatively with the predictions of simple calculations based on the channel geometry and molecular diffusion coefficient.

In the case of a flow through a rectangular channel with the width 2W and height 2R at a flow rate Q, the evolution of the surface coverage as a function of time and position along the channel, x, is described by $\Gamma(x, t) = 0.49\, D^{\frac{2}{3}} \dfrac{Q^{\frac{1}{3}}}{R^{\frac{2}{3}} W^{\frac{1}{3}}} \dfrac{1}{x^{\frac{1}{3}}} C_b\, t$, where $\Gamma(x, t)$ is the adsorbed mass at location x and time t, D is the diffusion coefficient (60 μm²/s), and C_b is the bulk analyte concentration [42,43]. This equation describes transport to a surface by a combination of flow and diffusion. The results are plotted in Figure 6 for our channel dimensions, Q = 4 μL/min, C_b = 0.195 mg/mL and three different locations x, such that the first sensor is 1.8 mm from the inlet and the distance between the sensors is 1.8 mm. In the example shown in Figure 6, only the last of the four sensor elements displays significant deviation. The leveling off of the adsorption curves at longer times is due to surface saturation that is not taken into account by this equation, which only considers the transport of the molecules to the surface.

This equation applies to situations where surfaces act as "perfect sinks" (adsorption kinetics is transport-limited, or, equivalently, the rate, at which the molecule binds to the surface >> than the transport rate), and when the flow-mediated transport is much faster, than the diffusion-mediated transport so that the bulk concentration profile in the x-direction is uniform in time. The first of the two conditions is met in cases of quasi-irreversible adsorption of protein molecules at bare surfaces in the initial stages of the adsorption [44,45]. The second needs to be evaluated by comparing transport and diffusive timescales in our geometry. This can be formally done in terms of a pair of Peclet numbers, one comparing transport and diffusive timescales along the channel normal ($Pe_H = \frac{Q}{2DW}$.), and the other—along the channel length ($Pe_S = 6\lambda^2 Pe_H$, where $\lambda = L_s/2R$ is the ratio of sensor length to the channel height, and other variables are as defined above) [46]. For our geometry, for all relevant flow rates, both $Pe_H \gg 1$ and $Pe_S \gg 1$, indicating that indeed, the flow-mediated transport along the channel is much faster, than the diffusion across the channel, and substantiating the validity of the above equation for $\Gamma(x,t)$. A more detailed analysis of the Peclet numbers for different analyte species can be found in Supplementary Section S3.

The differences between experiment and calculation, such as the one visible in Figure 6 for the last sensor in a column, must originate from the deviations of the channel geometry from that of a straight rectangular channel (inlet and outlet at 90° to the channel axis) and variation of the surface properties (advantageous contamination) between the sensor elements.

One further important assumption underpins the above analysis: that the material of the fluid cell is essentially non-adsorbing. This is because the well, where the analyte is placed (Figure 1, red circle), is located some 11 mm from the first sensor. However, it is impossible to fit the data for all four sensors if x for the first sensor is set to 11 mm, because the differences between sensor locations (1.8 mm apart) become irrelevant in comparison with the long inlet tube. The observed differences in the adsorption rates for the four sensors shown in Figure 6 therefore indicate that there is a minimal loss of the analyte on the inlet tube itself.

4. Conclusions

A microfluidic cartridge designed to host an array of 24 HFF-QCM resonators in bioanalytical applications was implemented. It works as a mechanical, electrical and thermal interface between the QCMD sensing array, the characterization instrument, and the fluid analyte. Currently, the assembly is aimed at research laboratories that reuse the array and allows fast and easy assembly and disassembly of the cartridge. Re-use of 30+ times with repeated surface cleaning is demonstrated. Use of low cost of the materials (PMMA, PDMS, custom PCB) makes the design appropriate for disposable applications. The functionality of the array + cartridge combination is evaluated using air-liquid shifts and adsorption of biomacromolecules (biotinylated albumin and neutravidin). Experimental results on the adsorbed masses and adsorption rates demonstrate quantitative agreement with the literature and theoretical considerations (protein geometry, transport conditions). Systematic errors arising from flow geometry are ruled out. The array + cartridge combination is used to show that the first adsorption step is critical in defining the molecular properties of the sensing interface: adsorbing neutravidin onto biotinylated BSA pre-adsorbed on the gold results in a more controlled layer of biotin sites for the subsequent sensing application than adsorbing neutravidin directly on the gold. In summary, the device is shown to be robust and function reliably in complex biosensing applications. In the future, it will be used to test human DNA samples for single base mutations in the colorectal cancer liquid biopsy samples and extended to immunoassays for pesticide and antibiotic detection in honey samples.

Supplementary Materials: The following are available online at http://www.mdpi.com/2079-6374/10/12/189/s1, Element positions on the Monolithic 150 MHz HFF QCMD sensor array. Quantification of adsorbed protein masses with the array and the individual sensors. Literature values for the dimensions and adsorbed layer masses for BSA, bBSA, and Nav. Peclet numbers and depletion thickness layer calculation. Figure S1: Monolithic 150 MHz HFF-QCMD sensor arrays. Numbers indicate the i-th element of the array, where i = r +4(c − 1) with r = 1–4 and c = 1–6 row and column indexes, respectively. Table S1: Quantification of adsorbed protein masses. Table S2: Literature values for the dimensions and adsorbed layer masses for BSA, bBSA, and Nav. Table S3: Pe_H calculated for different flow rates and biomolecules. Table S4: $⟦Pe⟧\_S$ calculated for different gasket deformation.

Author Contributions: Investigation: M.C., M.G., E.G.-S., I.R., and P.G.; formal analysis: M.C., M.G., I.R., and R.F.; data curation: M.C., M.G., and E.G.-S.; visualization: M.C., M.G., I.R., and R.F.; writing—original draft: R.F., M.C., Y.J.; writing—review & editing: I.R., Y.J., M.C., and R.F.; methodology: R.F., I.R., and J.V.G.; conceptualization: A.A., Y.J., R.F., I.R., and J.V.G.; funding acquisition: Y.J.; supervision: R.F. and I.R.; resources: P.G. and J.V.G.; software: R.F.; project administration: Y.J. All authors have read and agreed to the published version of the manuscript.

Funding: This work was supported in part by Ministerio de Economía, Industria y Competitividad de España—Agencia Estatal de Investigación with FEDER (Fondo Europeo de Desarrollo Regional) funds under Project AGL2016-77702-R and in part by the European Commission Horizon 2020 Programme, Capturing non-Amplified Tumor Circulating DA with Ultrasound Hidrodynamics, under Grant H2020-FETOPEN-2016-2017/737212-CATCH-U-DNA. M. Calero is the recipient of the doctoral fellowship BES-2017-080246 from the Ministerio de Economía, Industria y Competitividad de España.

Conflicts of Interest: AWSensors is a company that designs, develops, builds, and markets research-grade and custom-made QCMD systems, sensors, as well as the associated fluidics, software, and electronics founded by A.A. (founder and CEO) and Y.J. (co-founder). R.F., P.G., J.V.G., M.G., E.G.-S. and I.R. were employed by AWSensors at the time of writing. M.C. declares no conflict of interest.

References

1. Soper, S.A.; Brown, K.; Ellington, A.; Frazier, B.; Garcia-Manero, G.; Gau, V.; Gutman, S.I.; Hayes, D.F.; Korte, B.; Landers, J.L.; et al. Point-of-care biosensor systems for cancer diagnostics/prognostics. *Biosens. Bioelectron.* **2006**, *21*, 1932–1942. [CrossRef] [PubMed]

2. Lafleur, J.P.; Jönsson, A.; Senkbeil, S.; Kutter, J.P. Recent advances in lab-on-a-chip for biosensing applications. *Biosens. Bioelectron.* **2016**, *76*, 213–233. [CrossRef] [PubMed]

3. Nasseri, B.; Soleimani, N.; Rabiee, N.; Kalbasi, A.; Karimi, M.; Hamblin, M.R. Point-of-care microfluidic devices for pathogen detection. *Biosens. Bioelectron.* **2018**, *117*, 112–128. [CrossRef] [PubMed]

4. Sauerbrey, G. Verwendung von Schwingquarzen zur Wägung dünner Schichten und zur Mikrowägung. *Z. Phys.* **1959**, *155*, 206–222. (In German) [CrossRef]

5. Reviakine, I.; Johannsmann, D.; Richter, R.P. Hearing What You Cannot See and Visualizing What You Hear: Interpreting Quartz Crystal Microbalance Data from Solvated Interfaces. *Anal. Chem.* **2011**, *83*, 8838–8848. [CrossRef] [PubMed]

6. Johannsmann, D. *The Quartz Crystal Microbalance in Soft Matter Research*; Soft and Biological Matter; Springer International Publishing: Cham, Switzerland, 2015; ISBN 978-3-319-07835-9.

7. Fernandez, R.; Calero, M.; Reiviakine, I.; Garcia, J.V.; Rocha-Gaso, M.I.; Arnau, A.; Jimenez, Y. High Fundamental Frequency (HFF) Monolithic Resonator Arrays for Biosensing Applications: Design, Simulations, Experimental Characterization. *IEEE Sens. J.* **2020**, 1. [CrossRef]

8. Tuantranont, A.; Wisitsora-At, A.; Sritongkham, P.; Jaruwongrungsee, K. A review of monolithic multichannel quartz crystal microbalance: A review. *Anal. Chim. Acta* **2011**, *687*, 114–128. [CrossRef]

9. Kao, P.; Allara, D.; Tadigadapa, S. Fabrication and performance characteristics of high-frequency micromachined bulk acoustic wave quartz resonator arrays. *Meas. Sci. Technol.* **2009**, *20*, 124007. [CrossRef]

10. Zimmermann, B.; Lucklum, R.; Hauptmann, P.; Rabe, J.; Büttgenbach, S. Electrical characterisation of high-frequency thickness-shear-mode resonators by impedance analysis. *Sens. Actuators B Chem.* **2001**, *76*, 47–57. [CrossRef]

11. Fernández, R.; García, P.; García, M.; García, J.; Jiménez, Y.; Arnau, A. Design and Validation of a 150 MHz HFFQCM Sensor for Bio-Sensing Applications. *Sensors* **2017**, *17*, 2057. [CrossRef]

12. March, C.; García, J.V.; Sánchez, Á.; Arnau, A.; Jiménez, Y.; García, P.; Manclús, J.J.; Montoya, Á. High-frequency phase shift measurement greatly enhances the sensitivity of QCM immunosensors. *Biosens. Bioelectron.* **2015**, *65*, 1–8. [CrossRef] [PubMed]

13. Cervera-Chiner, L.; Juan-Borrás, M.; March, C.; Arnau, A.; Escriche, I.; Montoya, Á.; Jiménez, Y. High Fundamental Frequency Quartz Crystal Microbalance (HFF-QCM) immunosensor for pesticide detection in honey. *Food Control* **2018**, *92*, 1–6. [CrossRef]

14. Cervera-Chiner, L.; March, C.; Arnau, A.; Jiménez, Y.; Montoya, Á. Detection of DDT and carbaryl pesticides in honey by means of immunosensors based on high fundamental frequency quartz crystal microbalance (HFF-QCM). *J. Sci. Food Agric.* **2020**, *100*, 2468–2472. [CrossRef] [PubMed]

15. Cervera-Chiner, L.; Jiménez, Y.; Montoya, Á.; Juan-Borrás, M.; Pascual, N.; Arnau, A.; Escriche, I. High Fundamental Frequency Quartz Crystal Microbalance (HFF-QCMD) Immunosensor for detection of sulfathiazole in honey. *Food Control* **2020**, *115*, 107296. [CrossRef]

16. Montoya, A.; March, C.; Montagut, Y.; Moreno, M.J.; Manclús, J.J.; Arnau, A.; Jimenez, Y.; Jaramillo, M.; Marin, P.; Torres, R.A. A High Fundamental Frequency (HFF)-based QCM Immunosensor for Tuberculosis Detection. *Curr. Top. Med. Chem.* **2017**, *17*, 1623–1630. [CrossRef]

17. Grammoustianou, A.; Gizeli, E. Acoustic Wave–Based Immunoassays. In *Handbook of Immunoassay Technologies*; Sandeep, K.V., John, H.T.L., Eds.; Academic Press: London, UK, 2018; pp. 203–239. ISBN 9780128117620.

18. Milioni, D.; Mateos-Gil, P.; Papadakis, G.; Tsortos, A.; Sarlidou, O.; Gizeli, E. Acoustic Methodology for Selecting Highly Dissipative Probes for Ultrasensitive DNA Detection. *Anal. Chem.* **2020**, *92*, 8186–8193. [CrossRef]

19. Papadakis, G.; Palladino, P.; Chronaki, D.; Tsortos, A.; Gizeli, E. Sample-to-answer acoustic detection of DNA in complex samples. *Chem. Commun.* **2017**, *53*, 8058–8061. [CrossRef]

20. Papadakis, G.; Murasova, P.; Hamiot, A.; Tsougeni, K.; Kaprou, G.; Eck, M.; Rabus, D.; Bilkova, Z.; Dupuy, B.; Jobst, G.; et al. Micro-nano-bio acoustic system for the detection of foodborne pathogens in real samples. *Biosens. Bioelectron.* **2018**, *111*, 52–58. [CrossRef]

21. El Fissi, L.; Fernández, R.; García, P.; Calero, M.; García, J.V.; Jiménez, Y.; Arnau, A.; Francis, L.A. OSTEMER polymer as a rapid packaging of electronics and microfluidic system on PCB. *Sens. Actuators A Phys.* **2019**, *285*, 511–518. [CrossRef]

22. Papadakis, G.; Friedt, J.M.; Eck, M.; Rabus, D.; Jobst, G.; Gizeli, E. Optimized acoustic biochip integrated with microfluidics for biomarkers detection in molecular diagnostics. *Biomed. Microdevices* **2017**, *19*, 1–11. [CrossRef]

23. Wan, A.M.D.; Moore, T.A.; Young, E.W.K. Solvent bonding for fabrication of PMMA and COP microfluidic devices. *J. Vis. Exp.* **2017**, *2017*. [CrossRef] [PubMed]

24. Ron, H.; Matlis, S.; Rubinstein, I. Self-assembled monolayers on oxidized metals. 2. Gold surface oxidative pretreatment, monolayer properties, and depression formation. *Langmuir* **1998**, *14*, 1116–1121. [CrossRef]

25. Charnley, J. The Classic: The Bonding of Prostheses to Bone by Cement. *Clin. Orthop. Relat. Res.* **2010**, *468*, 3149–3159. [CrossRef] [PubMed]

26. Jaeblon, T. Polymethylmethacrylate: Properties and Contemporary Uses in Orthopaedics. *Am. Acad. Orthop. Surg.* **2010**, *18*, 297–305. [CrossRef]

27. Jönsson, M.; Anderson, H.; Lindberg, U.; Aastrup, T. Quartz crystal microbalance biosensor design II. Simulation of sample transport. *Sens. Actuators B Chem.* **2007**, *123*, 21–26. [CrossRef]

28. Jaruwongrungsee, K.; Maturos, T. Analysis of Quartz Crystal Microbalance Sensor Array with Circular Flow Chamber. *J. Appl.* **2009**, 50–54.

29. Mason, W.P. *Piezoelectric Crystals and Their Application to Ultrasonics*; Van Nostrand: New York, NY, USA, 1950.

30. Kanazawa, K.K.; Gordon, J.G. Frequency of a quartz microbalance in contact with liquid. *Anal. Chem.* **1985**, *57*, 1770–1771. [CrossRef]

31. Daikhin, L. Michael Urbakh Influence of surface roughness on the quartz crystal microbalance response in a solution New configuration for QCM studies. *Faraday Discuss.* **1997**, *107*, 27–38. [CrossRef]

32. Martin, S.J.; Granstaff, V.E.; Frye, G.C. Characterization of a quartz crystal microbalance with simultaneous mass and liquid loading. *Anal. Chem.* **1991**, *63*, 2272–2281. [CrossRef]

33. Wilchek, M.; Bayer, E.A. The Avidin-Biotin Complex in Bioanalytical Applications. *Anal. Biochem.* **1988**, *1*, 1–32. [CrossRef]

34. Diamandis, E.P.; Christopoulos, T.K. The Biotin-(Strept)Avidin System: Principles and Applications in Biotechnology. *Clin. Chem.* **1991**, *37*, 625–636. [CrossRef] [PubMed]

35. Pettersen, E.F.; Goddard, T.D.; Huang, C.C.; Couch, G.S.; Greenblatt, D.M.; Meng, E.C.; Ferrin, T.E. UCSF Chimera—A visualization system for exploratory research and analysis. *J. Comput. Chem.* **2004**, *25*, 1605–1612. [CrossRef] [PubMed]

36. Wolny, P.M.; Spatz, J.P.; Richter, R.P. On the Adsorption Behavior of Biotin-Binding Proteins on Gold and Silica. *Langmuir* **2010**, *26*, 1029–1034. [CrossRef] [PubMed]

37. Marttila, A.T.; Laitinen, O.H.; Airenne, K.J.; Kulik, T.; Bayer, E.A.; Wilchek, M.; Kulomaa, M.S. Recombinant NeutraLite Avidin: A non-glycosylated, acidic mutant of chicken avidin that exhibits high affinity for biotin and low non-specific binding properties. *FEBS Lett.* **2000**, *467*, 31–36. [CrossRef]

38. Kim, N.H.; Baek, T.J.; Park, H.G.; Seong, G.H. Highly sensitive biomolecule detection on a quartz crystal microbalance using gold nanoparticles as signal amplification probes. *Anal. Sci.* **2007**, *23*, 177–181. [CrossRef] [PubMed]

39. Ogi, H.; Naga, H.; Fukunishi, Y.; Hirao, M.; Nishiyama, M. 170-MHz Electrodeless Quartz Crystal Microbalance Biosensor: Capability and Limitation of Higher Frequency Measurement. *Anal. Chem.* **2009**, *81*, 8068–8073. [CrossRef]

40. Sagmeister, B.P.; Graz, I.M.; Schwödiauer, R.; Gruber, H.; Bauer, S. User-friendly, miniature biosensor flow cell for fragile high fundamental frequency quartz crystal resonators. *Biosens. Bioelectron.* **2009**, *24*, 2643–2648. [CrossRef]

41. Uttenthaler, E.; Schräml, M.; Mandel, J.; Drost, S. Ultrasensitive quartz crystal microbalance sensors for detection of M13-Phages in liquids. *Biosens. Bioelectron.* **2001**, *16*, 735–743. [CrossRef]

42. Hermens, W.T.; Beneš, M.; Richter, R.; Speijer, H. Effects of flow on solute exchange between fluids and supported biosurfaces. *Biotechnol. Appl. Biochem.* **2004**, *39*, 277–284.

43. Dahlin, A.B. *Plasmonic Biosensors. An integrated view of Refractometric Detection*; Ios Press: Amsterdam, The Netherlands, 2012; Volume 4, ISBN 978-1-60750-966-0.

44. Green, N.M. AVIDIN. 1. The use of [14C]BIOTIN for kinetic studies and for assay. *Biochem. J.* **1963**, *89*, 585–591. [CrossRef]

45. Wayment, J.R.; Harris, J.M. Biotin–Avidin Binding Kinetics Measured by Single-Molecule Imaging. *Anal. Chem.* **2009**, *81*, 336–342. [CrossRef] [PubMed]

46. Squires, T.M.; Messinger, R.J.; Manalis, S.R. Making it stick: Convection, reaction and diffusion in surface-based biosensors. *Nat. Biotechnol.* **2008**, *26*, 417–426. [CrossRef] [PubMed]

Publisher's Note: MDPI stays neutral with regard to jurisdictional claims in published maps and institutional affiliations.

 *biosensors* 

Article

Microfluidic-Based Electrochemical Immunosensing of Ferritin

Mayank Garg [1,2,3], **Martin Gedsted Christensen** [3], **Alexander Iles** [3], **Amit L. Sharma** [1,2], **Suman Singh** [1,2,*] and **Nicole Pamme** [3,*]

[1]   CSIR-Central Scientific Instruments Organisation, Sector 30-C, Chandigarh 160030, India; mayankgarg93@gmail.com (M.G.); amitsharma_csio@yahoo.co.in (A.L.S.)

[2]   Academy of Scientific and Innovative Research (AcSIR), Ghaziabad 201002, India

[3]   Department of Chemistry and Biochemistry, University of Hull, Cottingham Road, Hull HU6 7RX, UK; martingedstedchristensen@gmail.com (M.G.C.); a.iles@hull.ac.uk (A.I.)

*   Correspondence: ssingh@csio.res.in (S.S.); N.Pamme@hull.ac.uk (N.P.)

Received: 9 July 2020; Accepted: 2 August 2020; Published: 5 August 2020

Abstract: Ferritin is a clinically important biomarker which reflects the state of iron in the body and is directly involved with anemia. Current methods available for ferritin estimation are generally not portable or they do not provide a fast response. To combat these issues, an attempt was made for lab-on-a-chip-based electrochemical detection of ferritin, developed with an integrated electrochemically active screen-printed electrode (SPE), combining nanotechnology, microfluidics, and electrochemistry. The SPE surface was modified with amine-functionalized graphene oxide to facilitate the binding of ferritin antibodies on the electrode surface. The functionalized SPE was embedded in the microfluidic flow cell with a simple magnetic clamping mechanism to allow continuous electrochemical detection of ferritin. Ferritin detection was accomplished via cyclic voltammetry with a dynamic linear range from 7.81 to 500 ng·mL^{-1} and an LOD of 0.413 ng·mL^{-1}. The sensor performance was verified with spiked human serum samples. Furthermore, the sensor was validated by comparing its response with the response of the conventional ELISA method. The current method of microfluidic flow cell-based electrochemical ferritin detection demonstrated promising sensitivity and selectivity. This confirmed the plausibility of using the reported technique in point-of-care testing applications at a much faster rate than conventional techniques.

Keywords: amine functionalization; graphene oxide; immunosensor; electrochemistry; ferritin; microfluidics

---

1. Introduction

Anemia is a key healthcare challenge in India, especially for women, and is the top cause for maternal deaths in India (~50%) [1]. Other than India, countries like Bangladesh, Nepal, Bhutan, Afghanistan, many African, and some Central American counties face severe anemia prevalence in pregnant women (WHO Global Database on Anemia, World Health Organization). Furthermore, the highest prevalence of anemia is seen in Africa followed by Southeast Asia (according to the World Health Organization). Ferritin is one of the key biomarkers for anemia and it is an iron storing and transport protein [2]. It helps in preventing iron overload and associated toxicity effects [3], and is very critical in pregnancy and thus requires regular monitoring [4]. Apart from being an active biomarker for anemia, it also indicates level of oxidative stress. According to WHO guidelines, the normal level for ferritin should be between 15 and 200 ng·mL^{-1} for males and 15 and 150 ng·mL^{-1} for females, respectively. Imbalance in its concentration is associated with a range of fatal diseases, including cardiovascular disease [5], chronic kidney disease [6], Still's disease [7], and hemophagocytic syndrome [8]. In cancer also, ferritin is known to be upregulated and has been reported to be a marker for the diagnosis of this deadly disease [9]. Low levels of iron in pregnancy are readily treatable with tablets or diet recommendation to help fetal development. In a country like India, people often have to be assisted to attend health care facilities for check-ups. This results in loss of daily wages not only of the patient but also the person accompanying them, and thus causes a significant financial burden. The vision for the research conducted here is to develop a biosensor to be distributed by NGOs or other health monitoring agencies with access to the rural population.

Like many clinical markers, ferritin is conventionally detected using an Enzyme-Linked Immunosorbent Assay (ELISA). However, it requires not only skilled personnel but also involves use of specialized reagents such as enzyme linked antibodies and chromogens which make the method costly and time consuming [10]. The ELISA plate used for measuring an analyte has to be treated first to encourage binding of the biomolecules to the well surface. Following treatment, it then needs to be pre-coated with the primary antibodies. Once an antigen binds to the primary antibodies, secondary labeled antibodies and a chromogen are required to generate a color. Radioimmunoassays (RIA) on the other hand require usage of radiolabeled antibodies which are hazardous for daily handling. Though these conventionally used techniques are sensitive and selective to a large extent, generally they are not portable or do not provide a fast response. There is therefore a need for a dynamic, portable, and sensitive system for ferritin detection. Electrochemical methods are often attractive in clinical diagnostics settings, due to their ease of operation, minimal sample preparation, and easy data interpretation [11]. Recently, ferritin detection using electrochemical techniques has gained attraction [12,13], employing static systems [12–18]. Researchers of this manuscript have also recently reported the use of biosurfactant functionalized tungsten disulfide quantum dots for the electrochemical detection of ferritin using screen printed electrodes as a platform [19]. However, the availability of biosurfactant for synthesis of quantum dots is a limitation as the yield of biosurfactant is low from the microbe in use.

Continuous flow systems for in-line measurements in real-time allow running of patient samples one after another in a clinical setup. Thus, here, we investigated the electrochemical analysis of ferritin in continuous flow using a lab-on-a-chip-based approach with a microfluidic flow cell with an integrated, yet, readily exchangeable electrode. Lab-on-a-chip-based analysis platforms have been developed previously for electrochemical sensing of hydrogen ions, metals, nitrate and nitrite ions, phenolic compounds, pesticides and herbicides, and bacteria [20], including electrochemical detection of cancer biomarkers (carcinoembryonic antigen, prostate specific antigen, and cancer antigen 15-3 (CA15-3) [21]. Ko et al. reported ferritin immunosensing using a polymeric microfluidic device with gold electrodes [22]. However, the process of electrode modification and chip design are very complex. In our work, the lab-on-a-chip platform was integrated with electrochemically active carbon-coated screen-printed electrode (SPCE). The screen-printed electrodes provide a cost-effective platform for electrochemical measurements, as they are mass produced with minimum batch to batch variation, can be purchased in bulk, and can be read out with portable potentiostats, enabling point-of-care or even in-the-field application [23,24]. Owing to these advantages, researchers have combined screen-printed electrodes with flow cells for the electrochemical detection of various analytes [25–29]. Most of these flow cells allow screen-printed electrodes to be integrated in such a fashion that the flow cell cannot be reused once the measurements are made. This means that both the electrode and flow cell have to be disposed of. Furthermore, the flow cell designs are very intricate and do not provide a very controlled flow over the electrode surface. Thus, there is a need for a simple flow cell with integrated screen-printed electrodes for analysis of biomarkers such as ferritin, which allows ready exchange of electrodes and also provide a controlled flow over the electrode surface to enable precise control of time for binding and washing steps.

Therefore, we employed a simple flow cell and a magnetic clamping mechanism, previously developed in our group for glucose analysis [30] to study the electrochemical detection of ferritin. The different concentrations of the ferritin antigen were run one after the other on the electrode system with buffer washings in between them as is required for a clinical device. To the best of our knowledge, the use of a graphene-based 2D layered material as a platform for the continuous electrochemical sensing of ferritin in a simple flow cell has not yet been reported. Here, layered materials refer to those structures wherein layers of atoms are stacked on top of one another. Based on the available literature for the continuous electrochemical detection of ferritin, it can be concluded that a simple flow cell is required which can house the functionalized electrodes, thus increasing its flexibility of use. The study therefore aims to design and fabricate a flow cell with an integrated commercially available electrode which is modified with amine-functionalized graphene oxide, onto which anti-ferritin antibodies were immobilized for the continuous flow-based detection of ferritin, which is a novelty.

2. Experimental Section

2.1. Materials

Graphite powder was purchased from Alfa Aeser (Mumbai, India). Potassium permanganate, amino-terephthalic acid (NH_2-BDC), and sodium nitrate were sourced from Sigma-Aldrich (Bengaluru, India). Sulfuric acid and hydrogen peroxide were obtained from Merck (Mumbai, India). Ferritin antigen, anti-ferritin antibody, myoglobin, hemoglobin, bovine serum albumin (BSA), human serum (USA origin, sterile-filtered), potassium ferrocyanide, potassium ferricyanide, ethylene diamine tetra acetic acid (EDTA), and phosphate buffer saline tablets (PBS) (pH 7.4) were bought from Sigma-Aldrich (Dorset, UK). A ferritin ELISA kit was used for the validation studies from Orgentec (Mainz, Germany). DropSens carbon-coated screen-printed electrodes (SPCE, DS110) were procured from Metrohm (Runcorn, UK). A Milli-Q reverse osmosis system was used to obtain deionized water and was used for performing all experiments (Millipore, UK). Details of NH_2-GO synthesis and its characterization are given in Supplementary Information ESI1.

2.2. Fabrication of Microfluidics Flow Cell and Surface Modification of Electrode

The flow cell was designed on AutoDesk Inventor 2019 (San Rafael, CA, USA) and was fabricated from two layers of poly(methyl methacrylate) (PMMA) via a CNC milling (Datron M7, Datron, Germany) with 0.5, 1, 3, and 6 mm carbide drill bits (Figure 1). The outer dimensions of the flow cell were 35 mm × 30 mm × 4 mm. The top layer featured a 29 mm long, 10 mm wide, and 1 mm deep groove to house the DropSens SPE electrode. This layer also featured a spiral channel for transporting liquid over the surface of the working electrode. The spiral had a 3.8 mm diameter and the channel was 0.6 mm wide and 140 µm deep, with a total length of 19.45 mm, capable of handling an internal volume of 3.30 µL. At either end, there was an inlet and outlet hole (1 mm diameter). Both plates also featured wells of 6.2 mm diameter and 3.5 mm depth to hold NdFeB magnets (6 mm diameter, 3 mm height). The NdFeB magnets were obtained from Magnet Sales (Wiltshire, UK) and were used for clamping the plates together with the electrode. An O-ring (Size BS012, Nitrile, Simrit Service Centre Cramlington Ltd., UK) was used to ensure leak-free operation. Tubing (0.3 mm i.d., 1.58 mm o.d., Supelco) and capillary (100 µm i.d. and 363 µm o.d., Polymicro Technologies) were glued into the inlet and outlet holes with Araldite glue. At the inlet site, the tubing was interfaced to a syringe (1 mL), using an adapter (Luer (Female) to 1/4"-28 flat bottom (Female)) and flangeless fitting (1/4"-28 flat bottom). At the outlet site, tubing was inserted into a collection vessel. The liquid was pumped through the microfluidic chip via a syringe pump (11 Elite, Harvard Apparatus, MA, USA) (Figure 1c). A flow rate of 10 µL·min^{-1} was applied unless otherwise stated, equating a flow speed of 1.984 mm·s^{-1} in the spiral channel. Before insertion into the chip device, the surface of a carbon coated screen-printed electrode (SPCE) with a working area of 11.34 mm^2 was modified with the amine-functionalized graphene oxide (NH_2-GO) by drop casting. Ten microliters of NH_2-GO suspension prepared in deionized water (1 mg·mL^{-1}) was sonicated for 30 min prior to pipetting onto the working area of the electrode, and left to air dry for ~30 min [31]. The electrodes were clamped between the PMMA plates prior to use. Electrochemical measurements via cyclic voltammetry (CV) were recorded on a PalmSens3 potentiostat

(Netherlands), interfaced with a computer using PSTrace software. The screen-printed electrode was coupled to the potentiostat via DRP-CAC (Metrohm, Runcorn, UK) cable connectors.

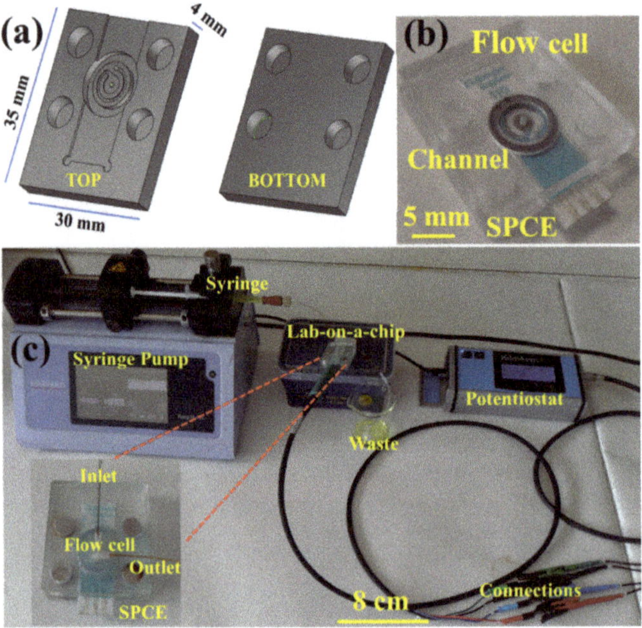

Figure 1. (**a**, **b**) Details of the microfluidic flow cell design. The top poly(methyl methacrylate) (PMMA) plate featured a groove to house the carbon-coated screen-printed electrode (SPCE), a spiral channel to move liquid over the working electrode surface with inlet an outlet holes, as well as four circular grooves to house the NdFeB magnets for clamping. The bottom plate was flat apart from the four magnet holding recesses. (**c**) Photograph of the experimental setup showing the syringe pump and collection vessel for fluid movement through the microfluidic flow cell with integrated SPCE electrode connected to a small potentiostat for recording of electrochemical measurements.

2.3. Electrochemical Measurements

Cyclic voltammetry (CV) was used for electrochemical characterization of the electrodes along with ferritin sensing. Ferritin antibodies (Fer$_{Ab}$) were immobilized on NH$_2$-GO modified SPCE for ferritin (Fer) sensing. For immobilization of the ferritin antibodies, 10 µL of antibody solution (10 µg·mL^{-1}) prepared in phosphate buffer was drop-casted on the NH$_2$-GO modified SPCE and left for air dry for about 30 min. It is expected that the amine groups present on the NH$_2$-GO will bind to the carboxyl groups present in the Fc region of the antibody, therefore creating a strong binding force for the antibody to remain on the electrode surface. For CV measurements, the phosphate buffer consisting of 1.0 mM Fe(CN)$_6^{3-}$/Fe(CN)$_6^{4-}$ as redox marker was used as an electrolyte and voltammograms were run in the potential window of −0.5 V to 0.7 V. The electrochemical characterization of the electrodes involved measuring the electrochemical response of the SPCEs after each modification; bare SPCE, NH$_2$-GO@SPCE, Fer$_{Ab}$/NH$_2$-GO@SPCE, and Fer/Fer$_{Ab}$/NH$_2$-GO@SPCE. This was performed to ensure electrochemical activity of the sequentially modified electrodes, which is a prerequisite for any electrochemical sensing. For electrochemical characterization, the effect of scan rate (0.02 V·s^{-1} to 0.10 V·s^{-1}) and flow rate (0.0 and 40 µL·min^{-1}) on the response of the developed lab-on-a-chip embedded SPCE was also studied. The stability of the NH$_2$-GO-modified electrode was verified by recording its cyclic voltammograms for 96 cycles from −0.5 V to 0.5 V at a scan rate of 0.04 V·s^{-1}.

For ferritin quantification, different concentrations of ferritin were prepared in a redox probe containing buffer. A redox probe containing buffer without any ferritin was used as a blank. Solutions were loaded into the syringe and pumped through the microfluidic device at 10 µL·min^{-1}. This flow rate had been optimized and is discussed below. In between measurements, the electrode surface was washed with phosphate buffer for 10 min. The effect of buffer pH (pH 5.5 to pH 8.5) and interfering agents (hemoglobin, BSA, and myoglobin) on the CV signal were also studied. For this, the pH of the redox probe containing buffer was adjusted with hydrochloric acid and sodium hydroxide. The interferents were added to the sample solutions at a concentration of 500 ng·mL^{-1}.

To check the reusability of the sensor, regeneration studies were performed. For this, after pumping a certain concentration of ferritin over the electrode and recording its response, a solution of 30 mM EDTA was introduced to remove the bound antigen. Then, PBS was pumped through the device for washing, followed by pumping of ferritin analyte solution. This cycle was repeated five times. The sensor performance was further validated by comparing the results with those obtained from a commercial ELISA kit. For serum studies, human serum samples were aliquoted and kept in the freezer until use and were thawed prior to use. A known concentration of ferritin was spiked into the serum to obtain samples with different ferritin concentration and the current response was recorded as before. All the experiments were performed at room temperature (~20 °C).

2.4. Computer Simulations

Computer simulations were carried out in COMSOL Multiphysics 5.2 (COMSOL AB). Mesh convergence assessments were performed on simulation geometries to determine the appropriate mesh settings in each of the simulations. COMSOL's physics for laminar flow and transport of diluted chemical species were employed for the simulations. The computed flows were calculated based on the assumption of a non-compressible fluid and with a no-slip boundary condition imposed on the wall domains. The transport of ferritin in the simulation was described by the generic diffusion equation (Equation (1)) with R as a reaction term, $\vec{u}$ as the velocity field, and c is the concentration of the species.

$$\frac{dc}{dt} = \nabla \cdot (D\nabla c) - \nabla \cdot (\vec{u}c) + R \tag{1}$$

3. Results and Discussion

3.1. Characterization of Amine-Functionalized Graphene Oxide (NH_2-GO)

The amine-functionalized graphene oxide was characterized via spectroscopy, X-ray diffraction analysis, and electron microcopy as laid out in the experimental section. The detailed results of UV/Vis spectra (Figure S1), FTIR spectra (Figure S2), Raman spectra (Figure S3), XRD pattern (Figure S4), and TEM images (Figure S5) are discussed in the Supplementary Information.

3.2. Computer Simulations

COMSOL was used to study the flow profile and changes in the concentration of the liquid inside the flow cell. For the simulations, the probe points were set at the inlet port as well as spread across the spiral channel in the flow cell (Figure 2). The simulation showed the evolution over time of the concentration at five distinct points on the electrode surface. The results demonstrate that a concentration of 50% is attained at the center surface point following approximately 17 s with 100% being achieved at 60 s. In comparison, the two peripheral surface points closest to the outlet attained 50% concentration at ~47 s, whereas the two furthest away peripheral surface points took more than 90 s.

Furthermore, it is worth noting from the modeling that the spiral channel, intended to effectively guide the flow over the counter and reference electrode, did not fully achieve this aim. This is likely to be due to the gap of ~500 µm in our design between the top and bottom plates of the magnetically clamped assembly, caused by the O-ring seal. The spiral channel is thus located above a 500 µm high circular chamber above the electrode. The design and thus guidance of the flow could be improved by the application of a thinner O-ring, a stronger clamping force via stronger magnets or a change to nuts and bolts to reduce the height of this gap. This would also reduce diffusion distances and diffusion times significantly and increase the sensitivity and speed of the sensor response.

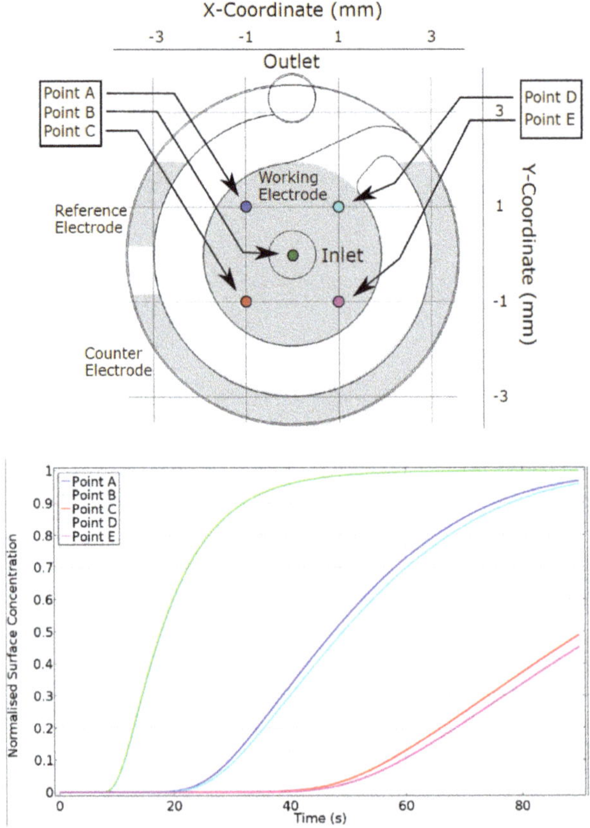

Figure 2. COMSOL simulations showing (**a**) the simulation geometry with five simulation probe points. The location of the reference, working and counter electrode at the bottom of the flow cell are shown in grey. (**b**) Evolution of concentration over time (normalized) at the five simulation points.

3.3. Electrochemical Characterization

Effect of Scan Rate

The effect of scan rate on electrode response was studied for a bare SPCE, as well as for SPCEs functionalized with NH_2-GO@SPCE and Fer_{Ab}/NH_2-GO@SPCE. The scan rate was varied from 0.02 V·s^{-1} to 0.10 V·s^{-1}. The obtained cyclic voltammograms, as well as plots of current versus scan rate and current versus square root of scan rate, are shown in Figure S6. The cyclic voltammograms showed oxidation and reduction peaks at 0.3 V·s^{-1} and −0.1 V·s^{-1}, respectively, corresponding to the conversion of Fe^{3+} to Fe^{2+} and vice versa (Figure S6a,d,g). With increased scan rate, the current increased in all the electrodes (Figure S6b,e,h), while the redox potential changed in position. The increase in current signal with increase in scan rate, can be accounted for decrease in diffusion layer at the electrode surface, as a result of which more current flows [32,33]. The R^2 value for the plots of current versus square root of scan rate was observed to be higher as compared to current versus scan rate, as expected by the Randles–Sevcik equation, which states that for reversible electrochemical reactions, the current increases linearly with the square root of the scan rate. This can be extended to the state that our system has free diffusing species and the electroactive species are not limited to the electrode surface [32,33].

Effect of Flow Rate

The effect of flow rate through the channels on the electrode signal was studied next. A phosphate buffer containing redox probe was pumped at different flow rates (0.0 and 40 μL·min^{-1}) through the device with a non-functionalized SPCE. The I_{pa}/I_{pc} ratios were calculated and the observed trend is shown in Figure S7. The ratio of I_{pa}/I_{pc} is linked to the reversibility of the reactions taking place at the electrode surface; a ratio close to 1 is expected for a highly reversible process. It is clearly evident from Figure S7 that at flow rate of 10 μL·min^{-1}, the ratio for the anodic to the cathodic current is closest to 1. Thus, this flow rate was chosen for further experimentation purposes.

Stability of NH$_2$-GO Coating

The stability of the NH$_2$-GO coating on the electrode was checked by continuously pumping the redox marker over the modified electrode and 48 scans were run. For up to 25 scans, the response was constant (Figure 3a); thereafter, a slight decrease in electrochemical response started to appear. On reaching the 48th scan, a 6.7% decline in current was observed. This suggests that the NH$_2$-GO coating on the electrode is sturdy and stable, and thus potentially suitable as a biosensor for continuous detection of an analyte over an extended period of time. The stability of the Fer$_{Ab}$/NH$_2$-GO was not evaluated. This can be attributed to fact that the amine groups present on the NH$_2$-GO were able to covalently bind to the carboxyl groups on the antibody. This ensured a strong binding force and hence the check on stability of the surface after antibody immobilization was not considered. This is in line with the work done by Gupta et al. wherein they used NH$_2$-GO as a platform for the direct immobilization of anti-*E. coli* antibodies [34].

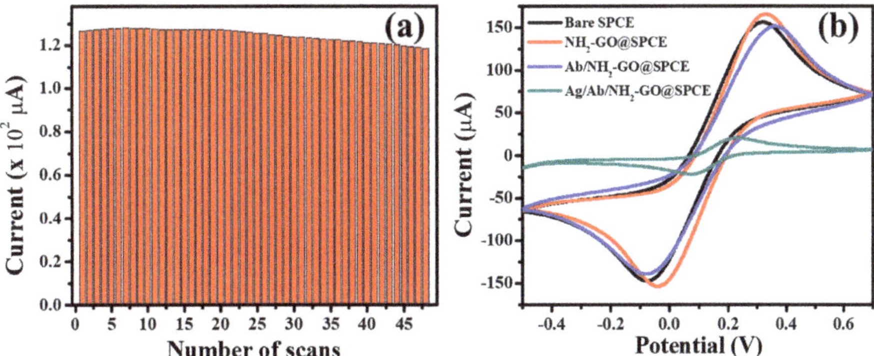

Figure 3. (a) Effect of number of scans on stability of NH$_2$-GO-modified SPCE. (b) Cyclic Voltammogram (CV) response with varying flow rates and Cyclic Voltammogram of sequentially modified electrode (Bare SPCE, NH$_2$-GO@SPCE, Fer$_{Ab}$/NH$_2$-GO@SPCE, and Fer/Fer$_{Ab}$/NH$_2$-GO@SPCE, respectively) in buffer containing redox marker at a scan rate of 0.04 V·s^{-1}.

Effect of Functionalization and Antibody Immobilization

To study the effect of functionalization and antibody immobilization, cyclic voltammograms were recorded for all the sequentially modified electrodes, i.e., bare SPCE, NH$_2$-GO@SPCE, Fer$_{Ab}$/NH$_2$-GO@SPCE and Fer/Fer$_{Ab}$/NH$_2$-GO@SPCE (Figure 3b). The electrochemical response of bare SPCE increased when modified with NH$_2$-GO, which can be explained by the role of lone pair of electrons present on the nitrogen atom. This easily available lone pair of electrons participates in the electron transfer processes taking place at the electrode surface resulting in an increase in the conduction of the system. The introduction of ferritin antibodies on the electrode however resulted in a decrease in current, which is attributed to the insulating effect caused by biological molecules [35]. Though we could not calculate the antibody density on the electrode surface, work done by Chouda et al. reports an antibody density of 4.8×10^{12} Ab/cm^2 for an immunosensor for *Staphylococcus aureus*. The authors used charge transfer resistance values before and after antibody immobilization to calculate the antibody coverage on the electrode [36]. On the addition of antigen, an immunocomplex forms between antigen and antibody, which reduces the current significantly, further hindering the flow of current [37].

Electrochemical Detection of Ferritin and Sensor Performance

The microfluidic flow cell was next tested for quantitative electrochemical analysis of ferritin. The Fer/Fer$_{Ab}$/NH$_2$-GO@SPCE was fitted into the flow cell and different concentrations of ferritin prepared in phosphate buffer consisting of 1.0 mM Fe(CN)$_6{}^{3-}$/Fe(CN)$_6{}^{4-}$ were pumped at 10 μL min^{-1} over the electrode whilst the cyclic voltammograms were recorded (Figure 4a). The current decreased as the concentration of ferritin was increased (Figure 4b) as expected from the insulating effect of the immune complex [35]. The linearity range obtained was 7.81 to 500 ng·mL^{-1}, which covers the clinically relevant range. An R^2 value of 0.996 was obtained with the limit of detection (LOD) of 0.413 ng·mL^{-1}. The LOD, a value which indicates the lowest quantity of an analyte, differentiable from the absence of that substance, was calculated using the formula: LOD = 3.3 × (Standard deviation/slope). As per the IUPAC nomenclature, the LOD is defined as the minimal detectable value as the mean blank value plus three times the standard deviation [38–41].

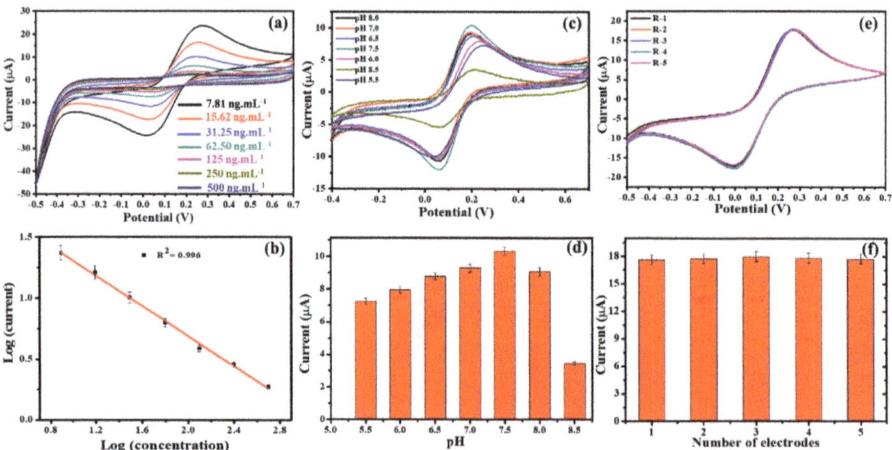

Figure 4. (a) Cyclic voltammograms from Fer/Fer$_{Ab}$/NH$_2$-GO@SPCE in presence of ferritin (7.81–500 ng·mL^{-1}). (**b**) Logarithmic plot of ferritin concentration versus obtained current. (**c**) Cyclic voltammograms and (**d**) bar chart of obtained currents when varying the pH between pH 5.5 and pH 8.5. (**e**) Cyclic voltammograms and (**f**) bar charts of currents obtained from five separately prepared electrodes to confirm reproducibility.

Further, the effect of pH on the response of the lab-on-a-chip-based immunosensor was studied (Figure 4c,d). As the pH was changed from acidic (pH 5.5) to neutral (pH 7.5), the current response increased, with the maximum response observed at pH 7.5. When moving towards the alkaline pH range, i.e., pH 8.0 and pH 8.5, the current signal decreased, which could be attributed to denaturation and instability of the proteins at non-physiological conditions [35]. Ferritin has a pI of ~5.5, thus the increasing negative charge at the higher pH values will increase electrostatic repulsion between the participating moieties and hence the current decreases [42]. The same trend with pH variation was also observed by Garg et al. [19]. The sensor reproducibility was verified by preparing five electrodes in the same way and recording their response (Figure 4e,f). It was found that all the electrodes gave a similar response with a relative standard deviation of 0.74%.

The selectivity of the immunosensor chip device towards ferritin was confirmed by interference studies with oxygen binding proteins such hemoglobin, myoglobin, and bovine serum albumin (Figure 5a,b). It was observed that these interferents had a negligible effect on the sensor readout. The reusability of the sensor was evaluated using regeneration following EDTA and PBS wash, as described in the experimental section. As can be seen from Figure 5c,d, after five cycles of regeneration, the current response dropped to ~11% as compared to the initial point. This hints at the possibility of reusing the sensor device for a number of scans before a change of electrode would be required.

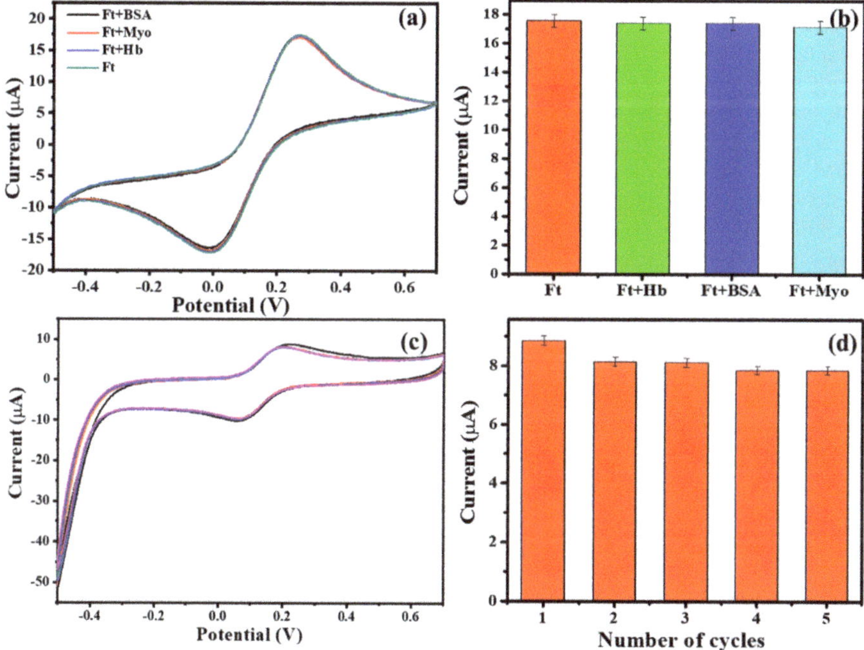

Figure 5. (a) Cyclic voltammograms and (b) bar chart of currents of obtained from ferritin in the absence and presence of interferents, i.e., hemoglobin (Hb), myoglobin (Myo), and bovine serum albumin (BSA). (c) Cyclic voltammograms and (d) bar chart of currents of obtained five cycles of ferritin analysis and regeneration.

The practical applicability of the biosensor device was investigated with ferritin spiked into human serum at different concentrations, i.e., 31.25 ng·mL^{-1}, 62.50 ng·mL^{-1}, and 125 ng·mL^{-1}. The obtained signal was converted to a quantitative readout via the calibration curve as shown in Figure 4b. The current obtained from these spiked serum samples was used to calculate the concentration of ferritin in those samples by using the linear equation obtained from the standard calibration curve. This resulted in a deviation of ~5–10% with potential for further optimization (Table 1).

Table 1. Results for ferritin spiked into human serum samples.

Concentration Added (ng·mL^{-1})	Concentration Found (ng·mL^{-1})	% Found
31.25	34.47	110.30
62.50	70.91	113.45
125	120.27	96.21

For further evaluation, sensor response was compared to that of a commercially available ELISA kit. The linear range in case of the ELISA was 0 to 1500 ng·mL^{-1} with an R^2 value of 0.966 (Figure S8). Though the linear range was higher than the present developed method, the major disadvantage for the commercially available ELISA method is the use of secondary conjugated antibodies and a chromogenic substrate to generate a readable signal. Moreover, for microplate readings used in ELISA tests, a microplate reader is required, increasing the cost of the overall method. The lab-on-chip device reported in the present manuscript on the other hand, uses only primary antibodies specific to the ferritin antigen for the bio-recognition reactions. The handheld potentiostat employed with lab-on-chip is not only portable but is also economical.

The current immunosensor performance is comparable with the previously reported microfluidic or flow-based electrochemical sensors for ferritin. Ko et al. designed and fabricated a PDMS/PMMA-laminated microfluidic device having an integrated gold-based electrode in the chip. The electrode was immobilized with ferritin antigen and two sets of antibodies, anti-ferritin IgG and anti-IgG antibody labeled with HRP, were used. The system relied on the precipitation reaction to generate the signal. The work however suffered from the drawback of non-reusability of the electrode, which hinders its usage as a point-of-care device [22]. The work by Reymond et al. on a microfluidic-based ferritin immunoassay used an eight channel electrochemical microchip modified with magnetic and non-magnetic microspheres for antibody immobilization. However, the research did not cover the required clinical range of ferritin quantification [43]. In another article by Song et al., the use of cotton thread as a microfluidic circuit was suggested, where three sets of antibodies were used in the sensor fabrication. Though the linear range achieved by this sensor is very high (5 to 5000 ng·mL^{-1}), the use of multiple antibodies and a complex system increases the overall cost of the system [12]. The most recent report for the detection of ferritin in a microfluidic fashion is the use of a paper-based electrode system. The electrode was functionalized with graphene and further modified with EDC/NHS for antibody immobilization to take place. This sensor performance was better than in our study in terms of both linear range as well as the obtained LOD value. The main disadvantages of this work are the multiple electrode preparation steps and the inability to change the electrode in the same microfluidic device [44].

Based on the literature available for the microfluidic-based electrochemical detection of ferritin, we can state that most of these were based on multiple steps for electrode functionalization and employed different chemistries for antibody immobilization. Moreover, the systems lacked easy replacement of electrodes. The immunosensor reported in the present manuscript is advantageous over other reported immunosensors for ferritin in terms of covering the clinical ferritin range, and one step process for both functionalization and immobilization. The amine functionalized graphene oxide plays a dual role of electrode surface modifier and as well as immobilization matrix. The major highlight of the current method is the continuous electrochemical detection of ferritin as compared to the static-based systems mentioned in the introduction section. The current system has the microfluidic channels manufactured into the top chip which allows easy electrode swapping, thereby giving flexibility in terms of usage. We were able to demonstrate a dynamic method for the sensing of ferritin which can be further improved upon, to develop a system for clinical settings. This device can be useful in a small clinical setting with a lightly trained skilled technician. This system with further improvements in the design (to make portable) can be used as a point-of-care technology. Moreover, the system is open ended and can also be used for sensing of other analytes by changing the bioreceptor.

4. Conclusions

An electrochemical microfluidic device for the sensitive determination of ferritin is demonstrated. The microfluidic device enabled ferritin estimation in continuous flow, which will be helpful for continuous monitoring. Furthermore, the amine functionalization of SPCE, integrated with the lab-on-chip device, provided a platform for both electrode surface modification and immuno-sensing to take place, thus simplifying the overall analysis process. The lab-on-chip with integrated NH_2-GO-functionalized SPCE is capable of sensing the ferritin in the range 7.81 to 500 ng·mL^{-1}, which covers the clinically relevant range and is found to be robust against pH changes and interferents. Validation of the device was performed by comparing its response with the results from a standard ELISA kit. The sensor response with spiked human serum samples showed promising results for consideration for use in real time applications that can be operated even by nurses in small/local healthcare facilities without requiring sophisticated machinery. This is a better alternative as compared to a central facility where samples would have to be transported. Moreover, the device is open ended in terms of applicability as simply by changing the bio-receptor, it can also be used for other biomarkers.

Supplementary Materials: The following are available online at http://www.mdpi.com/2079-6374/10/8/91/s1, Figure S1: UV/vis absorption of graphene oxide (GO), NH_2-BDC and amine functionalized graphene oxide (NH_2-GO), Figure S2: FTIR spectra of graphene oxide (GO), NH_2-BDC, and amine functionalized graphene oxide (NH_2-GO), Figure S3: Raman spectroscopy of graphene oxide (GO) and amine functionalized graphene oxide (NH_2-GO), Figure S4: XRD analysis of graphene oxide (GO), amine BDC (NH_2-BDC), and amine functionalized graphene oxide (NH_2-GO), Figure S5: TEM images of NH_2-GO, Figure S6: Effect of scan rate studies, Figure S7: Effect of flow rate, Figure S8: Calibration curve from ELISA kit

Author Contributions: M.G.: Conceptualization, Methodology, Data curation, Writing—Original draft preparation, and Funding acquisition. M.G.C.: Methodology, Data curation, and Writing–Original draft preparation. A.I.: Methodology, Data curation, and Writing—Original draft preparation. A.L.S.: Conceptualization, Supervision, and Writing—Reviewing and Editing. N.P.: Conceptualization, Supervision, and Writing—Reviewing and Editing. S.S.: Conceptualization, Supervision, and Writing—Reviewing and Editing. All authors have read and agreed to the published version of the manuscript.

Funding: This research was funded by a Newton-Bhabha Fund PhD Placement 2019 Award, grant number [428798537].

Acknowledgments: The authors acknowledge the Director, CSIR-CSIO for his constant support and encouragement. Mayank Garg would like to extend warm thanks to British Council, U.K. and Department of Biotechnology, Govt. of India for the selection in the Newton-Bhabha Ph.D. Placement Fellowship 2018–19. Mayank Garg also acknowledges the SRF-GATE fellowship from Council of Scientific and Industrial Research (CSIR-HRDG), New Delhi, India.

Conflicts of Interest: The authors declare no conflicts of interest.

Ethics: Necessary ethical clearance was taken prior performing experiments on human serum.

References

1. Kaur, K. Anaemia 'a silent killer'among women in India: Present scenario. *Eur. J. Zool. Res.* **2014**, *3*, 32–36.
2. Winzerling, J.J.; Pham, D.Q.D. 4.10-Ferritin. In *Comprehensive Molecular Insect Science*; Gilbert, L.I., Ed.; Elsevier: Amsterdam, the Netherlands, 2005; pp. 341–356.
3. Winzerling, J.J.; Pham, D.Q.D. Ferritin☆. In *Reference Module in Life Sciences*; Elsevier: Amsterdam, the Netherlands, 2017.
4. Ho, C.-H.; Yuan, C.-C.; Yeh, S.-H. Serum ferritin levels and their significance in normal full-term pregnant women. *Int. J. Gynecol. Obstet.* **1987**, *25*, 291–295. [CrossRef]
5. Kadoglou, N.P.E.; Biddulph, J.P.; Rafnsson, S.B.; Trivella, M.; Nihoyannopoulos, P.; Demakakos, P. The association of ferritin with cardiovascular and all-cause mortality in community-dwellers: The English longitudinal study of ageing. *PLoS ONE* **2017**, *12*, e0178994. [CrossRef] [PubMed]
6. Nakanishi, T.; Kuragano, T.; Nanami, M.; Otaki, Y.; Nonoguchi, H.; Hasuike, Y. Importance of Ferritin for Optimizing Anemia Therapy in Chronic Kidney Disease. *Am. J. Nephrol.* **2010**, *32*, 439–446. [CrossRef]
7. Evensen, K.J.; Swaak, T.J.G.; Nossent, J.C. Increased ferritin response in adult Still's disease: Specificity and relationship to outcome. *Scand. J. Rheumatol.* **2007**, *36*, 107–110. [CrossRef]

8. Emmenegger, U.; Frey, U.; Reimers, A.; Fux, C.; Semela, D.; Cottagnoud, P.; Spaeth, P.J.; Neftel, K. Hyperferritinemia as indicator for intravenous immunoglobulin treatment in reactive macrophage activation syndromes. *Am. J. Hematol.* **2001**, *68*, 4–10. [CrossRef]

9. Alkhateeb, A.A.; Connor, J.R. The significance of ferritin in cancer: Anti-oxidation, inflammation and tumorigenesis. *Biochim. Biophys. Acta (BBA) Rev. Cancer* **2013**, *1836*, 245–254. [CrossRef]

10. Barnett, M.D.; Gordon, Y.B.; A Amess, J.; Mollin, D.L. Measurement of ferritin in serum by radioimmunoassay. *J. Clin. Pathol.* **1978**, *31*, 742–748. [CrossRef]

11. Wang, J. CHAPTER 3—Electrochemical glucose biosensors. In *Electrochemical Sensors, Biosensors and their Biomedical Applications*; Zhang, X., Ju, H., Wang, J., Eds.; Academic Press: San Diego, CA, USA, 2008; pp. 57–69.

12. Song, T.-T.; Wang, W.; Meng, L.-L.; Liu, Y.; Jia, X.-B.; Mao, X. Electrochemical detection of human ferritin based on gold nanorod reporter probe and cotton thread immunoassay device. *Chin. Chem. Lett.* **2017**, *28*, 226–230. [CrossRef]

13. Matysiak-Brynda, E.; Wagner, B.; Bystrzejewski, M.; Grudzinski, I.P.; Nowicka, A.M. The importance of antibody orientation in the electrochemical detection of ferritin. *Biosens. Bioelectron.* **2018**, *109*, 83–89. [CrossRef]

14. Wang, X.; Tao, G.; Meng, Y. Nanogold hollow microsphere-based electrochemical immunosensor for the detection of ferritin in human serum. *Microchim. Acta* **2009**, *167*, 147. [CrossRef]

15. Zhang, X.; Wang, S.; Hu, M.; Xiao, Y. An immunosensor for ferritin based on agarose hydrogel. *Biosens. Bioelectron.* **2006**, *21*, 2180–2183. [CrossRef]

16. Yang, X.; Yuan, R.; Chai, Y.; Zhuo, Y.; Hong, C.; Liu, Z.; Su, H. Porous redox-active Cu_2O–SiO_2 nanostructured film: Preparation, characterization and application for a label-free amperometric ferritin immunosensor. *Talanta* **2009**, *78*, 596–601. [CrossRef] [PubMed]

17. Ren, J.; Tang, D.; Su, B.; Tang, J.; Chen, G. Glucose oxidase-doped magnetic silica nanostrutures as labels for localized signal amplification of electrochemical immunosensors. *Nanoscale* **2010**, *2*, 1244–1249. [CrossRef] [PubMed]

18. Yen, L.-C.; Pan, T.-M.; Lee, C.-H.; Chao, T.-S. Label-free and real-time detection of ferritin using a horn-like polycrystalline-silicon nanowire field-effect transistor biosensor. *Sens. Actuators B Chem.* **2016**, *230*, 398–404. [CrossRef]

19. Garg, M.; Chatterjee, M.; Sharma, A.L.; Singh, S. Label-free approach for electrochemical ferritin sensing using biosurfactant stabilized tungsten disulfide quantum dots. *Biosens. Bioelectron.* **2020**, *151*, 111979. [CrossRef] [PubMed]

20. Kudr, J.; Zitka, O.; Klimanek, M.; Vrba, R.; Adam, V. Microfluidic electrochemical devices for pollution analysis—A review. *Sens. Actuators B Chem.* **2017**, *246*, 578–590. [CrossRef]

21. Fragoso, A.; Latta, D.; Laboria, N.; Von Germar, F.; Hansen-Hagge, T.E.; Kemmner, W.; Gärtner, C.; Klemm, R.; Drese, K.S.; Sullivan, C.O. Integrated microfluidic platform for the electrochemical detection of breast cancer markers in patient serum samples. *Lab Chip* **2011**, *11*, 625–631. [CrossRef]

22. Soo Ko, J.; Yoon, H.C.; Yang, H.; Pyo, H.-B.; Chung, K.H.; Kim, S.J.; Kim, Y.T. A polymer-based microfluidic device for immunosensing biochips. *Lab Chip* **2003**, *3*, 106–113. [CrossRef]

23. Hayat, A.; Marty, J.L. Disposable screen printed electrochemical sensors: Tools for environmental monitoring. *Sensors* **2014**, *14*, 10432–10453. [CrossRef]

24. Yamanaka, K.; Vestergaard, M.d.C.; Tamiya, E. Printable Electrochemical Biosensors: A Focus on Screen-Printed Electrodes and Their Application. *Sensors* **2016**, *16*, 1761. [CrossRef] [PubMed]

25. Pereira, S.V.; Bertolino, F.A.; Fernández-Baldo, M.A.; Messina, G.A.; Salinas, E.; Sanz, M.I.; Raba, J. A microfluidic device based on a screen-printed carbon electrode with electrodeposited gold nanoparticles for the detection of IgG anti-Trypanosoma cruziantibodies. *Analyst* **2011**, *136*, 4745–4751. [CrossRef] [PubMed]

26. Eletxigerra, U.; Martinez-Perdiguero, J.; Merino, S. Disposable microfluidic immuno-biochip for rapid electrochemical detection of tumor necrosis factor alpha biomarker. *Sens. Actuators B Chem.* **2015**, *221*, 1406–1411. [CrossRef]

27. Vasiliadou, R.; Esfahani, M.M.N.; Brown, N.J.; Welham, K.J. A Disposable Microfluidic Device with a Screen Printed Electrode for Mimicking Phase II Metabolism. *Sensors* **2016**, *16*, 1418. [CrossRef] [PubMed]

28. Chen, S.; Wang, Z.; Cui, X.; Jiang, L.; Zhi, Y.; Ding, X.; Nie, Z.; Zhou, P.; Cui, D. Microfluidic Device Directly Fabricated on Screen-Printed Electrodes for Ultrasensitive Electrochemical Sensing of PSA. *Nanoscale Res. Lett.* **2019**, *14*, 71. [CrossRef] [PubMed]

29. Damiati, S.; Peacock, M.; Leonhardt, S.; Damiati, L.; Baghdadi, M.A.; Becker, H.; Kodzius, R.; Schuster, B. Embedded Disposable Functionalized Electrochemical Biosensor with a 3D-Printed Flow Cell for Detection of Hepatic Oval Cells (HOCs). *Genes* **2018**, *9*, 89. [CrossRef]

30. Patinglag, L.; Esfahani, M.M.N.; Ragunathan, K.; He, P.; Brown, N.J.; Archibald, S.J.; Pamme, N.; Tarn, M.D. On-chip electrochemical detection of glucose towards the miniaturised quality control of carbohydrate-based radiotracers. *Analyst* **2020**. [CrossRef]

31. Singh, S.; Kumar, N.; Kumar, M.; Jyoti; Agarwal, A.; Mizaikoff, B. Electrochemical sensing and remediation of 4-nitrophenol using bio-synthesized copper oxide nanoparticles. *Chem. Eng. J.* **2017**, *313*, 283–292. [CrossRef]

32. Bard, A.J.; Faulkner, L.R. *Electrochemical Methods: Fundamentals and Applications*; Wiley: New York, NY, USA, 1980; Volume 2.

33. Elgrishi, N.; Rountree, K.J.; McCarthy, B.D.; Rountree, E.; Eisenhart, T.T.; Dempsey, J.L. A Practical Beginner's Guide to Cyclic Voltammetry. *J. Chem. Educ.* **2018**, *95*, 197–206. [CrossRef]

34. Gupta, A.; Bhardwaj, S.K.; Sharma, A.L.; Deep, A. A graphene electrode functionalized with aminoterephthalic acid for impedimetric immunosensing of Escherichia coli. *Mikrochim. Acta* **2019**, *186*, 800. [CrossRef]

35. Kukkar, M.; Singh, S.; Kumar, N.; Tuteja, S.K.; Kim, K.-H.; Deep, A. Molybdenum disulfide quantum dot based highly sensitive impedimetric immunoassay for prostate specific antigen. *Microchim. Acta* **2017**, *184*, 4647–4654. [CrossRef]

36. Chrouda, A.; Braiek, M.; Rokbani, K.B.; Bakhrouf, A.; Maaref, A.; Jaffrezic-Renault, N. An Immunosensor for Pathogenic *Staphylococcus aureus* Based on Antibody Modified Aminophenyl-Au Electrode. *ISRN Electrochem.* **2013**, *2013*, 367872. [CrossRef]

37. Kukkar, M.; Tuteja, S.K.; Kumar, P.; Kim, K.-H.; Bhadwal, A.S.; Deep, A. A novel approach for amine derivatization of MoS2 nanosheets and their application toward label-free immunosensor. *Anal. Biochem.* **2018**, *555*, 1–8. [CrossRef] [PubMed]

38. Gauglitz, G. Analytical evaluation of sensor measurements. *Anal. Bioanal. Chem.* **2018**, *410*, 5–13. [CrossRef]

39. IUPAC. Compendium of Analytical Nomenclature (Orange Book). 2002. Available online: http://media. iupac.org/publications/analytical_compendium/ (accessed on 27 July 2020).

40. Oleneva, E.; Khaydukova, M.; Ashina, J.; Yaroshenko, I.; Jahatspanian, I.; Legin, A.; Kirsanov, D. A Simple Procedure to Assess Limit of Detection for Multisensor Systems. *Sensors* **2019**, *19*, 1359. [CrossRef]

41. de Vicente, J.; Lavín, Á.; Holgado, M.; Laguna, M.F.; Casquel, R.; Santamaría, B.; Quintero, S.; Hernandez, A.L.; Ramírez, Y.; Bolanos, M.H. The uncertainty and limit of detection in biosensors from immunoassays. *Meas. Sci. Technol.* **2020**, *31*, 044004. [CrossRef]

42. He, D.; Marles-Wright, J. Ferritin family proteins and their use in bionanotechnology. *New Biotechnol.* **2015**, *32*, 651–657. [CrossRef]

43. Reymond, F.; Vollet, C.; Jendelova, P.; Horák, D. Fabrication and characterization of tosyl-activated magnetic and nonmagnetic monodisperse microspheres for use in microfluic-based ferritin immunoassay. *Biotechnol. Prog.* **2013**, *29*, 532–542. [CrossRef]

44. Boonkaew, S.; Teengam, P.; Jampasa, S.; Rengpipat, S.; Siangproh, W.; Chailapakul, O. Cost-effective paper-based electrochemical immunosensor using a label-free assay for sensitive detection of ferritin. *Analyst* **2020**, *145*, 5019–5026. [CrossRef]

Review

A Review of Capillary Pressure Control Valves in Microfluidics

Shaoxi Wang [1,†], Xiafeng Zhang [1,†], Cong Ma [2,3], Sheng Yan [4], David Inglis [5] and Shilun Feng [2,5,6,*]

[1] School of Microelectronics, Northwestern Polytechnical University, Xi'an 710072, China; shxwang@nwpu.edu.cn (S.W.); zhangxiafeng@mail.nwpu.edu.cn (X.Z.)

[2] State Key Laboratory of Transducer Technology, Shanghai Institute of Microsystem and Information Technology, Chinese Academy of Sciences, Shanghai 200050, China; macong1@mail.sim.ac.cn

[3] School of Information Science and Technology, ShanghaiTech University, Shanghai 201210, China

[4] Institute for Advanced Study, Shenzhen University, Shenzhen 518060, China; shengyan@szu.edu.cn

[5] School of Engineering, Faculty of Science and Engineering, Macquarie University, Sydney, NSW 2109, Australia; david.inglis@mq.edu.au

[6] School of Electrical and Electronic Engineering, Nanyang Technological University, Singapore 639798, Singapore

* Correspondence: shilun.feng@mail.sim.ac.cn

† These two authors contribute equally.

Abstract: Microfluidics offer microenvironments for reagent delivery, handling, mixing, reaction, and detection, but often demand the affiliated equipment for liquid control for these functions. As a helpful tool, the capillary pressure control valve (CPCV) has become popular to avoid using affiliated equipment. Liquid can be handled in a controlled manner by using the bubble pressure effects. In this paper, we analyze and categorize the CPCVs via three determining parameters: surface tension, contact angle, and microchannel shape. Finally, a few application scenarios and impacts of CPCV are listed, which includes how CPVC simplify automation of microfluidic networks, work with other driving modes; make extensive use of microfluidics by open channel, and sampling and delivery with controlled manners. The authors hope this review will help the development and use of the CPCV in microfluidic fields in both research and industry.

Keywords: capillary pressure control valve (CPCV); microfluidics; passive valve

Citation: Wang, S.; Zhang, X.; Ma, C.; Yan, S.; Inglis, D.; Feng, S. A Review of Capillary Pressure Control Valves in Microfluidics. *Biosensors* **2021**, *11*, 405. https://doi.org/10.3390/bios11100405

Received: 8 September 2021

Accepted: 14 October 2021

Published: 19 October 2021

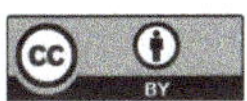

1. Introduction

Microfluidic technology has made great progress in the past two decades, and is a significant feature in a wide range of scientific and industrial work [1,2]. Microfluidic processes have strong potential in biomedical applications due to their small volume of samples and reagents, high throughput, and potential for automation. After decades of development, microfluidic devices have moved partly from the laboratory to practical applications, such as cell separation [3,4], analytical reactions and detections [5,6], immunoassays [7–9], as well as polymerase chain reaction (PCR) [10–12].

As a part of controlling and regulating liquid flow in microfluidics, micro-valves are essential. They can also be reliable and inexpensive [13]. With the increasing complexity and scale of microfluidic systems, research into microvalve designs has grown and various types have been demonstrated [14]. The function of these valves can be divided into stop valves [15], check valves [16,17], delay valves [18], retention valves [14], trigger valves [19], and siphon valves [20]. In addition to the purpose of the valve, we can categorize valves by the mechanism of actuation, which can be either active or passive. Active valves require an external energy source such as electrostatic [21–23], electromagnetic [24,25], pneumatic [26], hydraulic or photothermal [27–29]. These energy sources can control fluid flow through the deformation of a boundary, as in electromechanical [30] and pneumatic valves [31]. The energy may also change the state of a boundary, for example, by melting ice [10], wax [32,33], or a hydrogel [34]. These valves are best suited for repeatedly administering the liquid, controlling the pressure of the liquid, or pumping the liquid, but their operation requires peripheral actuators that limit their use. In contrast, the passive valve does not require additional driving equipment, which is more conducive to the integration and miniaturization of equipment. Existing passive valves include capillary pressure control valves [35], capillary burst valves [36], siphon valves [20], retention valves [14], flap valves [37], and check valves [38–40].

In an active valve, flow stops until the barrier is removed by some external force. In a passive valve the forward flow is stopped by a change in Laplace pressure. The capillary pressure control valve (CPCV) is a kind of passive valve that relies on the Laplace pressure generated by the change of the liquid front meniscus concavity. It may also be referred to as the Laplace pressure control valve or capillary valve. CPCVs do not require external equipment and rely solely on Laplace pressure at the liquid interfaces, which depends on the surface tension, contact angle and channel shape. CPCVs have been widely used as flow control valves in various microfluidic systems, which further enhances the performance of micro devices and expands the functions available in an integrated microsystem [14,41].

Surface tension changes little for a given combination of fluid and gas at a constant temperature, so the CPCV can be divided based on their actuating mechanism: either contact angle changes or geometric changes. In the first group, the Laplace pressure is different in two adjacent locations because of hydrophilic and hydrophobic patterning [42–45]. In the second group, the Laplace pressure is different in two adjacent locations because of a change in geometry, typically an expansion, which changes the radius of curvature [15,46,47].

The stopping principle of these valves is that the liquid will be blocked due to a Laplace pressure change caused by a sudden change in the liquid front meniscus. To open it and restore flow, the existing meniscus needs to be rebuilt to remove additional pressure [36]. There are essentially two ways to do this. The first is to increase the fluid pressure beyond the bubble pressure, usually by increasing the pumping pressure. In a Lab on Disk system the pressure is increased by increasing the spin speed [48]. Through theoretical calculation and experimental analysis, a large number of articles have reported the relationship between break pressure and valve channel characteristics, including valve channel geometry shape and material characteristics [36,46,49–51]. The second way to open the valve is to bring fluid in from the stopped side of the valve. The merged meniscus can then pass the stop valve. This is known as a trigger valve [19,52].

A. Olanrewaju et al. [14] have recently reviewed capillary microfluidics and described the capillary networks. Among them, part of the contents are the collections of the CPCV examples including stop valve, capillary trigger valve, capillary soft valve, capillary retention valve, retention burst valve, and delay valve. Moreover, the physics theories governing capillary flow are described, and the concepts of capillary circuits are also introduced. In this review, we focus on CPCV descriptions through the whole article. We discussed the specific fundamental physics theories related to the CPCV in detail; categorized them based on the theory elements with new applications of each kind of valve; discussed the potential of real applications in industrial and academic fields. We analyzed the CPCV examples from three fundamental elements: surface tension, contact angle and different microchannel shape. We hope it can give a direction describing the specific phenomena theoretically and easily for other academic and industrial control valves research, which can have more useful applications.

There are four parts in this paper. Part 2 is for the categories of CPCV and different three determining parameters of Laplace pressure, including surface tension, contact angle as well as microchannel shape. Part 3 provides a review for different developments and applications of CPCV. Part 4 is for the conclusions and outlook.

2. Categories of CPCV

With the expansion of microfluidic technology, CPCV is more attractive as a kind of passive valve, the threshold pressure of the valve is affected only by the geometry and liquid properties of the device. As no moving parts are involved, these valves are easier to manufacture and less likely to clog than moving valves.

The bubble pressure (or Laplace pressure) is the difference in pressure between the inside and the outside of a curved surface, such as bubbles or droplets. The relationship between Laplace pressure (P) and surface tension is described by Young–Laplace equation as follows:

$$\Delta P \equiv P_{inside} - P_{outside} = \gamma(\frac{1}{R_1} + \frac{1}{R_2}) \tag{1}$$

The pressure is larger on the concave side of the meniscus (gas–liquid interface) than on the convex side. Where γ is the surface tension of the liquid. R_1 and R_2 are the principal radii of curvature of the meniscus related to the contact angle and microchannel shape, where $R = \frac{\gamma}{cos\theta_c}$. The CPCV is categorized from the following aspects: surface tension γ, contact angle θ_c, and its channel shape.

2.1. Surface Tension (γ)

Surface tension is defined in a very pragmatic way: the tension between any two adjacent parts of a liquid surface that interact perpendicular to their unit length boundary is called surface tension (γ). Using an arbitrary line to divide the liquid level into two parts, surface tension can be understood as the pull of a molecule on one side on the other side per unit length. At each point on both sides of the line, there is a surface tension perpendicular to the line and tangent to the surface.

The surface tension of water/mineral oil can be calculated according to:

$$\gamma[\text{mN/m}] = 51.83 - 0.103T \tag{2}$$

where T is temperature in Celsius. If the room temperature T is 22 °C, surface tension is around 49.56 mN/m.

The surface tension can be caused by either the metallic bonds or the hydrogen bonds, while the former one has a larger effect. Surface tension may be altered locally through changing surface energy, such as addition of surfactants [53] or by applied sunlight or even lasers [54] or magnetic fields [55].

2.2. Contact Angle (θ)

In 1805, Thomas Young defined the contact angle at the solid–liquid–gas three-phase boundary. As shown in Figure 1A, the contact angle θ_c is the included angle from the solid–liquid interface through the liquid interior to the gas–liquid interface. The contact angle represents a method of showing liquid–solid adhesion.

The relation between interfacial tension and contact angle is given by Young's equation:

$$\gamma_{SG} = \gamma_{SL} + \gamma_{LG}cos\theta_c \tag{3}$$

where γ_{SG}, γ_{SL} and γ_{LG} are the interfacial tensions between solid and gas, solid and liquid, and liquid and gas, respectively.

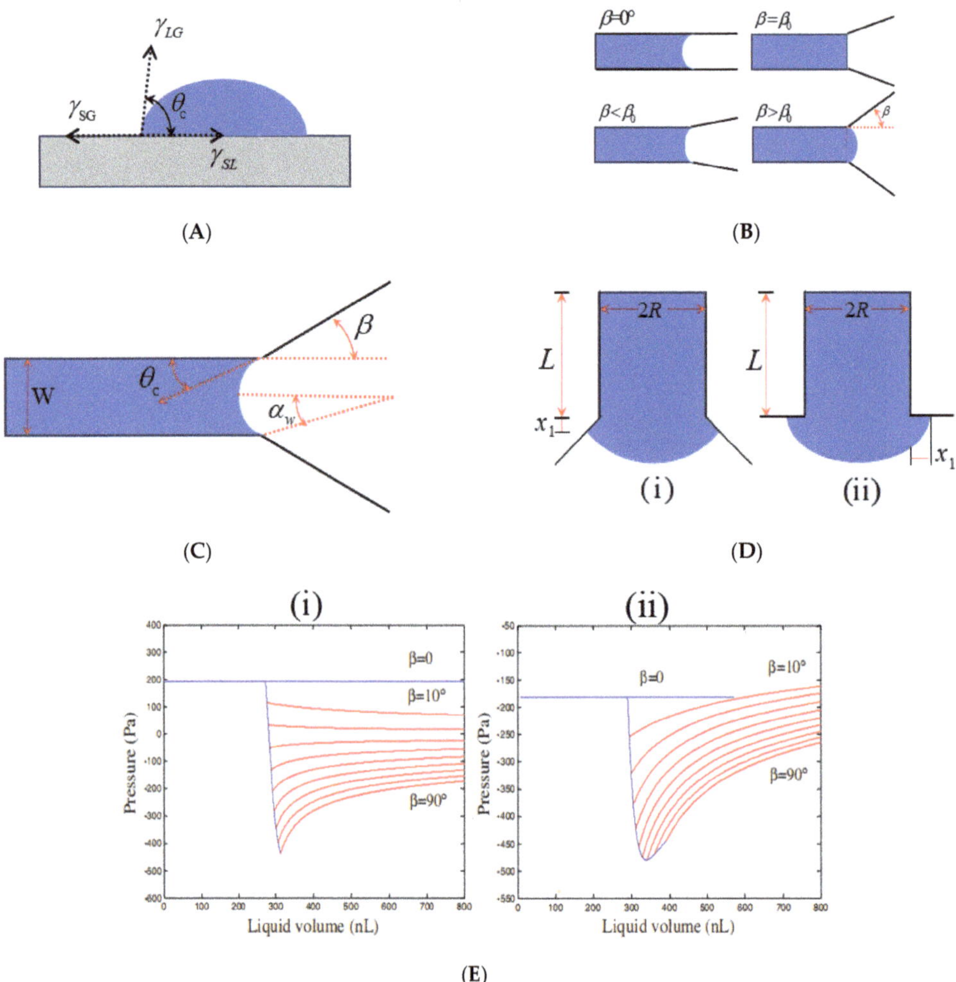

Figure 1. CPCV parameters and their relationships. (**A**) Definition of contact angle. (**B**) The relationship between the angle of triangle expansion β and the liquid level meniscus. (**C**) The common geometrical parameters in CPCV boundary pressure model are given. (**D**) Meniscus of axisymmetric channel in expansion regime bursting into divergent section, where advancement of liquid is described mainly by wetted length x_1 for (i) $\beta < 90°$ and (ii) $\beta = 90°$. (**E**) Variation in liquid pressure with liquid volume for hydrophilic and hydrophobic channels of R = 0.3 mm with expansion angles ranging from $\beta = 0$ to $90°$, (i) hydrophilic channels interfacial properties $\gamma = 0.072$ N/m and $\theta_c = 66°$; (ii) hydrophobic channels interfacial properties $\gamma = 0.072$ N/m and $\theta_c = 112°$. Reprinted with permission from [56]. Copyright (2008) The Japan Society of Applied Physics.

2.2.1. Hydrophobic/Hydrophilic

Since γ_{SG} and γ_{LG} are constant, it is known from the Young's equation that the contact angle θ_c decreases accordingly when the surface tension between solid and liquid γ_{SL} decreases, leading to the increase of adhesion energy standing for hydrophilic characteristics. Additionally, vice versa, the contact angle increases standing for the hydrophobic characteristics. A surface with droplet water contact angle greater than 90° is hydrophobic, but less than 90° is deemed hydrophilic. The affinity between a material and water is described by the term hydrophilic/hydrophobic. Materials with polar groups usually have great affinity for water and can attract water molecules to be easily wetted. On the contrary, hydrophobic materials tend to be non-polar with no affinity for water and are not easily wetted. For hydrophobic materials, when water comes into contact with the surface of the material, the contact angle is generally greater than $90°$. Droplets that are large enough to experience gravitational forces that exceed the capillary forces are likely to break up into small droplets or beads. The opposite is true for hydrophilic materials, where small droplets will aggregate into a film. As the water molecule is polar, materials with polarizable surface groups tend to be hydrophilic, having contact angles that are less than $90°$.

Common hydrophilic substances are aluminum, zinc, and other metals and their oxides, glass and mica, quartz, talc, calcite, quartz, and many other minerals. In addition, the single and associating -OH polar groups on the surface of the material, which can form hydrogen bonds with water molecules, are hydrophilic. The hydrophobic group is mainly represented by $-NO_2$, Si-H, and Si-CHx groups, and the Si-F group also exists in a small amount [57], such as paraffin, Teflon (PTFE), polyamide (PA), PC (polycarbonate), PAN (polyacrylonitrile), fluorinated polyethylene, fluorocarbon wax, polyolefin, polyester, fluoro-free acrylates, and so on. Modifying the surface of the material to chang the hydrophilic and hydrophobic nature can change the contact angle, which is an important method to construct the surface tension valve.

Due to the capillarity property generated by surface tension, water is pulled into a microchannel with a hydrophilic surface inside, but it meets the stop barrier at the hydrophobic surfaces. Taking advantage of hydrophobic–hydrophilic interface (hydrophobic—more hydrophobic; hydrophilic—more hydrophilic) effect can be used to manage the flow of liquid in the microchannel [58].

2.2.2. Material Properties

Material selection is the first step in the fabrication of microfluidic chips, which directly affects the function of chips and determines the processing and production methods of subsequent chips. Chips made of different materials have different production costs, processing difficulty and specific processing methods. The selection of chip materials is also related to the observation and detection of subsequent experiments. Different chip materials have a direct impact on the difficulty of optical detection in subsequent experiments due to their different light transmittance. CPCV is an important part of a chip, and different properties of different materials also directly affect the design and production of CPCV.

The materials for making microfluidic chips generally include silicon, glass and a variety of polymer materials, for example, polymethylmethacrylate (PMMA), polydimethylsiloxane (PDMS), polystyrene (PS), and Polycarbonate (PC), etc. Table 1 shows the advantages and weakness of common materials used to fabricate microfluidic chips, where the contact angles with water were summarized.

It can be seen from Table 1 that, compared with silicon and glass, low cost and easy processing are advantages of polymer materials, so polymer materials are mostly used in the production of microfluidic chips. PDMS is a widely used polymer material for microfluidic chip processing. PDMS is chemically inert and non-toxic; it has excellent light permeability and is convenient for optical detection; with good plasticity, it is easy to process; it is convenient for surface modification and bonding; it can be repeated many times for mass production. In addition, most of the polymer materials are naturally hydrophobic, such as PDMS (contact angle: $113.5 \pm 2°$ [59]) and PMMA (contact angle: $97°$ [60]), and the surface performance is not stable. After hydrophilic treatment, there is the phenomenon of hydrophobic recovery, which will gradually lose hydrophilicity and return to hydrophobic state. This is very unfavorable for CPCV constructed by changing surface hydrophilicity and hydrophobicity, and seriously affects the application expansion of CPCV.

Table 1. Properties of different materials for preparing microfluidic chips.

Materials	Advantages	Weaknesses	Contact Angles with Water (°)
Silicon	Chemical inertness Smooth surface Mature technology Easy to mass production	Fragile High cost Opaque Complex surface chemistry	Silica (super-hydrophobic): 161 [61] Silica (conventional): 24.5 [61] Silicon (FDTS): 113.7 ± 3.1 [62] Silicon (DDMS):105 [63]
Glass	Good electroosmotic Good optical properties Easy for surface treatment	Fragile, high cost Bonding difficult Difficulty in a large aspect ratio	Glass (uncoated): 68.5 [64] Glass (2% APTES): 40 [65] Glass (MSNPs-CVD): 175 ± 2 [66] Glass (MSNPs-Sol): 158 ± 2 [66]
Polymer	Variety and low cost Transmission of light Easy to process and form Cheap mass production	Low heat-resistant Low thermal conductivity	PDMS: 113.5 ± 2 [59] PMMA: 97 [60] PDMS (APTES+MA): 60 [67]

2.2.3. Surface Treatment Methods

The most commonly used materials for microfluidic devices, and therefore for surface tension valves, are glass and PDMS. The chemical stability of PDMS is very good, so its surface modification is quite difficult. In addition, the surface properties of PDMS are unstable. Even if the surface of PDMS becomes hydrophilic through modification, the surface will gradually lose hydrophilicity and return to the hydrophobic state, a phenomenon known as hydrophobic recovery [68,69]. The mechanism of PDMS surface instability is not yet clear. Most surface modification methods have not solved the problem of PDMS hydrophobic recovery. Common surface treatment methods include plasma treatment [70,71], covalent surfactants treatment [72], and Sol–gel coating [73,74].

Plasma treatment: this surface treatment method is currently the most commonly used method for surface modification of PDMS [69]. Plasma is an ionized gaseous substance composed of atoms deprived of some electrons and positive and negative ions generated after the ionization of atomic groups. It is often regarded as the fourth state of matter in addition to solid, liquid, and gas. In plasma treatment, oxygen plasma is used to react with the surface of the material to expose its hydrophilic chemical functional groups. When a surface is treated with hydrophilic plasma, it increases the surface energy of the object. When the surface energy of an object is high, its adhesion is high. However, plasma surface modification of PDMS can restore its hydrophobic properties within a few minutes, as exposed hydrophilic groups recombine with uncured hydrophobic polymer chains, causing the surface to lose hydrophilic properties. Maintaining the surface of the material in water immediately after plasma treatment or using solvent extraction to remove the uncured polymer can effectively slow down or prevent this recovery [70,71].

Surfactants treatment: surfactant molecules have charged "heads" and hydrophobic "tails" that can easily adsorb onto the hydrophobic surface and change its surface properties. Amphiphilic surfactant molecules enter the microchannel by running the buffer, the hydrophobic tail is physically adsorbed on the PDMS surface, and the hydrophilic surfactant head is extended into the buffer, making the PDMS surface hydrophilic in situ, to achieve the purpose of changing the surface properties of PDMS. This method not only reduces the cost of surface modification, but also makes it more rapid and simple. As the surfactant does not form a strong covalent bond with primary PDMS only by weak hydrogen bond binding, desorption can occur, and excessive surfactants are needed to dynamically supplement the desorption substances in the process, so as not to have a negative impact on the surface performance.

Nowadays, a large number of surfactants are applied to PDMS surface modification. For hydrophilic treatment, there is poly ethylene glycol methacrylate (PEGMA), 2-hydroxy ethyl methacrylate (HEMA), O2 plasma, tween-20 [75], 3-aminopropyl triethoxysilane (APTES) [67,76], polyvinyl alcohol (PVA) [72], polyvinylpyrrolidone (PVP) [77],

Pluronic [78], etc. For hydrophobic treatment, there is octadecyltrichlorosilane (OTS) [79], Sigmacoat [80], Fluorosilane coupling agent, Saline, etc.

Sol–gel coating: in this method, the "solution" and "sol" of compounds containing high chemical active components are cured into a "gel" state through a series of treatments. Sol–gel technology can rapidly and repeatedly construct a large number of ordered hydrophilic and hydrophobic surface structures on nanoscale surface, which is a promising surface modification method at present [81–85]. The Sol–gel coating can be produced by electrophoretic deposition, impregnation, and sputtering. At present, the deposition coating mainly uses silica sol, which is composed of alkoxy compound and its composite material with metal salt solution. Hydrophobic coatings can be formed directly by introducing a hydrophobic agent (hexamethyl-disilazane, trimethylchlorosilane) into the aerosol and curing it on the material surface. In recent years, the preparation of hydrophobic and super-hydrophobic surfaces has received extensive attention, among which the use of organosilicon fluoro compounds, especially containing hydrolytic groups, is the key development direction [86–88]. Fluorinated compounds can be used not only as modification additives in the preparation of Sol–gel composites used for coating formation, but also as the main components of hydrophobic coatings [89].

According to different materials and application scenarios, a large number of surface modification methods have been proposed. Surface modification is a popular research field, especially super-hydrophobic surfaces (which provide a wet angle of more than 150 degrees), including Nano-surface [86,87,90], which has been a new research hotspot in the past two decades [74,89,91]. The surface modification method of PDMS mentioned above is also applicable to common microfluidic materials such as plexiglass. For example, hydrophobic glass can be obtained by preparing the surface of glass by Sol–gel method [92]. There are also other methods, such as ultraviolet (UV) treatment [93,94], chemical vapor deposition (CVD) [95,96], self-assembled monolayers (SAMs) coatings, etc.

By changing the contact angle of the material through any of these methods, the Laplace pressure of the liquid interface is changed. This shift in Laplace pressure is essential for the manufacture and design of the CPCV.

2.2.4. Partial Hydrophilic/Hydrophobic Treatment

By coating the parylene layer and etching it with a designed pattern, the chip surface can be hydrophobic. While the bottom of the microcavity is still hydrophilic, the parylene layer also can be peeled off to restore surface hydrophilicity (Figure 2A). In another article, silica pillars were coated with Cr and Teflon on top (red). During the lift off process, only the top of silica pillars recovered hydrophilicity because Cr was only coated on the top of pillars (Figure 2B). Local hydrophilic/hydrophobic property also can be obtained by transferring the coating from plane to rugged structure (Figure 2C). To extract aqueous droplets from oil, some local hydrophilic/hydrophobic structures were built. In one work, Sigmacote (Sigma Aldrich, Burlington, MA, USA) was used to render the channels hydrophobic upon further baking at 120 °C for 1 h. After that, the capillaries were brought back to a hydrophilic state by stripping the silane layer. This is done by flowing 2-propanol at both the inlet (10 kPa) and the outlet (−100 kPa) through the 50 μm channels while drawing 0.1 M potassium hydroxide through the capillaries. In a similar work, a piece of membrane was bonded with a droplets generation chip made of cyclic olefin copolymer (COC) to extract aqueous droplets from oil (Figure 2D).

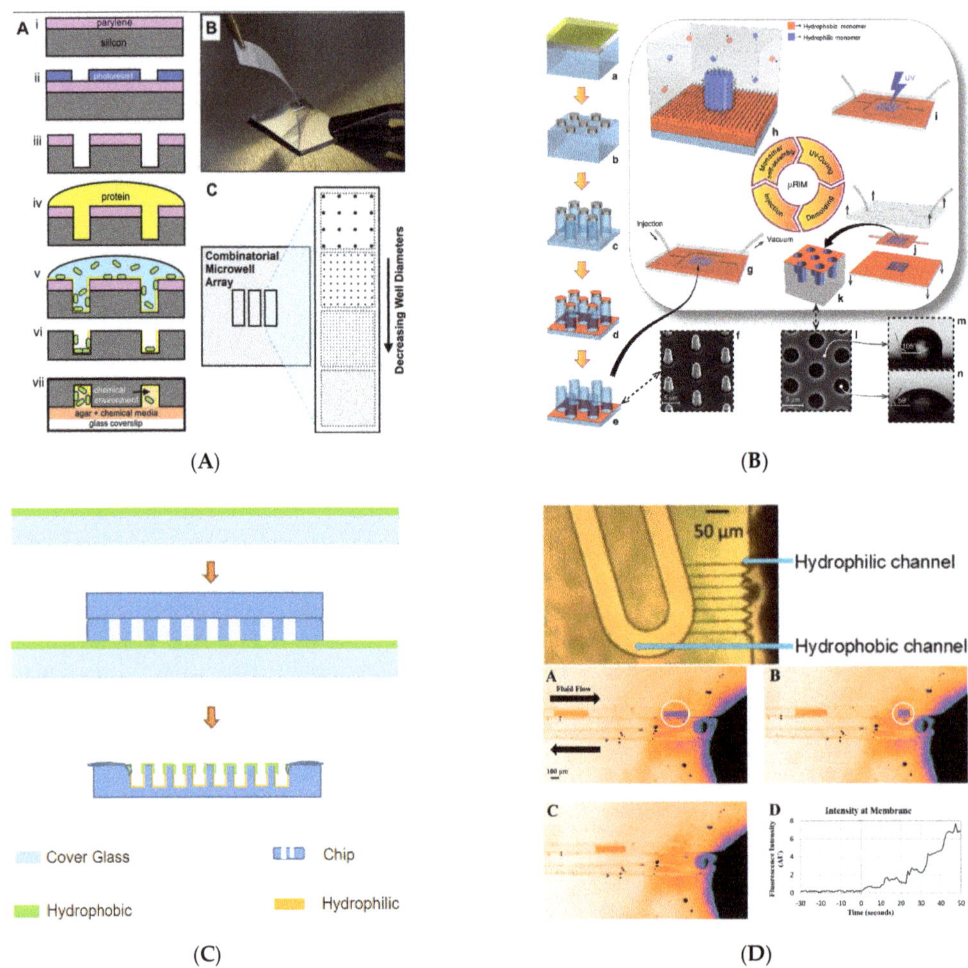

Figure 2. Some examples of surface treatment. (**A**) Microwell arrays for bacteria stochastic assembly. Reprinted with permission from [97]. (**B**) Hydrophilic in-hydrophobic femtolitre-well arrays. Blue color indicates hydrophilic surfaces; red color hydrophobic surfaces. Reprinted with permission from [98]. (**C**) Spin coating liquid hydrophobic liquid to make surface and part of chip wall hydrophobic. (**D**) Local hydrophilic/hydrophobic treatment silicon chip used in sampling (the top half part) and hydrophilic membrane used in microfluidic droplet extraction (the lower half part). Reprinted with permission from [80,99].

2.3. Channel Shape

2.3.1. Straight Microchannel

The principal radii of curvature R from Equation (1) may be different depending on the shape of the microchannel. If the microchannel is a closed cylinder with radius r, the fluid boundary meniscus in the channel is relatively static, and the two principal radii of curvature of the meniscus are equal, substituting $R = r/cos\theta_c$ into Equation (1), the Laplace pressure of the cylindrical channel can be obtained as follows:

$$\Delta P = \frac{2\gamma}{r}cos\theta_c \tag{4}$$

where θ_c is the contact angle for the fluids at the solid boundary. Due to the extensive use of simple molding technology, a large number of microfluidic devices are rectangular channels. The equation can be modified to a 3D model as [50]:

$$\Delta P = 2\gamma(\frac{1}{w} + \frac{1}{h})cos\theta_c \tag{5}$$

where, the curvature radius of the height h direction in the microchannel is $R_h = h/(2cos\theta_c)$, and the width w direction is $R_w = w/(2cos\theta_c)$.

2.3.2. Shape Change Microchannel

In a closed hydrophilic microchannel, by gradually increasing the cross section of the channel, the meniscus in front of the liquid will be eliminated and the additional Laplace pressure generated by the contact angle can be offset. The details of several gradually widening walls are summarized below.

In the rectangular channel, the gradual expansion of the channel can be carried out along the two directions of width and height. In microchannel, the triangular expansion angle β_0 without attachment pressure should meet the following requirements:

$$\Delta P = \frac{2\gamma}{h}(\frac{cos\theta_c - \frac{\alpha}{sin\alpha}sin\beta}{cos\beta + \frac{sin\beta}{sin\alpha}[\frac{\alpha}{sin\alpha} - cos\alpha]})\Delta P = 2\gamma(\frac{1}{w} + \frac{1}{h})cos\theta_c \tag{6}$$

The angle of triangular expansion is β. When $\beta < \beta_0$, the liquid level meniscus is concave, and liquid advances, wetting the surface. When $\beta > \beta_0$, the liquid level meniscus protrudes, and Laplace pressure creates a pressure barrier, stopping the flow of liquid (Figure 1B).

If the microchannel height h is much bigger than the width w, then the meniscus can be regarded as one-dimensional. The meniscus contour is considered as a section of sphere with an arc angle of 2α. Man et al. give the pressure barrier P caused by 2D geometric change in the case of channel mutation [47]:

$$\Delta P = \frac{2\gamma}{h}(\frac{cos\theta_c - \frac{\alpha}{sin\alpha}sin\beta}{cos\beta + \frac{sin\beta}{sin\alpha}[\frac{\alpha}{sin\alpha} - cos\alpha]}) \tag{7}$$

The expression derived by Chen et al. gives the pressure model of the geometric expansion of the micro-channel in the 3D model [100]:

$$\Delta P = \frac{2\gamma}{w}[(-\frac{w}{h})cos\theta_c + (\frac{cos\theta_c - \frac{\alpha_w sin\beta}{sin\alpha_w}}{-cos\beta + \frac{sin\beta}{sin\alpha_w}(\frac{\alpha_w}{sin\alpha_w} - cos\alpha_w)})] \tag{8}$$

Figure 1C is a top view of the model. β is the angle of expansion in the width w direction. The depth h remains constant in the whole channel. The α_w is circular arcs angle in width directions.

Jerry M et al. undertook an extensive analysis of the axisymmetric channel pressure barrier with sudden expansion (by suddenly increasing the cross section of the channel), along the channel direction and gave the expression [56]:

$$\Delta P = \frac{-2\gamma}{R + x_1 tan\beta}[\frac{cos\theta_c - \frac{2sin\beta}{1+cos\alpha}}{cos\beta - \frac{sin\beta}{sin^3\alpha}(2 - 3cos\alpha + cos^3\alpha)}] \tag{9}$$

As shown in Figure 1D, the axially symmetric channel radius of the column is R, and the wetting length is x_1. The meniscus shape, defined by $\alpha = min(\pi/2 - \theta_c - \beta, -\pi)$, is held invariant throughout this regime. When $\beta = 90°$, the meniscus shape is described by $\alpha = -\theta_c$. The formula is as follows:

$$\Delta P = \frac{2\gamma sin\alpha}{R + x_1} \tag{10}$$

Figure 1E shows the liquid pressure in hydrophilic (Figure 1E(i)) and hydrophobic (Figure 1E(ii)) microchannels with a radius of 300 μm as a function of liquid volume for expansion angles in the range $\beta = 0–90°$. For a straight channel ($\beta = 0$), the pressure in hydrophilic microchannels is maintained at a positive constant of 195 Pa driving the liquid downstream without being blocked, the pressure in hydrophobic microchannels is negative (-195 Pa) indicating that an external force is required to propel the liquid forward. As the meniscus arrives at the edge where sudden expansion in the cross section begins, the pressure drops rapidly to reach a minimum, followed by a steep increase. The liquid is blocked by the pressure barrier formed at the sudden opening and an increase in liquid volume needs an extra force. In other words, a sudden expansion of the microchannel can act as a valve to control the flow of liquid in the channel. In hydrophilic pipe, if β is small ($\beta < \beta_0$), the falling pressure of the liquid is still positive, and the Laplace pressure continues to push the liquid forward, it is a valveless state. Note that except for the channels with large expansion angles ($\beta + \theta_c > \pi$), threshold pressure increases with an increase in expansion angle.

2.4. Examples of the CPCV

In microfluidics, a series of valves with unique structure and function are manufactured using the principle above. There are many examples to show the control valve applications [101–111]. According to the different functions of valves, CPCV includes stop valve, trigger valve, and delay valve [112].

2.4.1. Stop Valves

The purpose of the stop valve is to temporarily block the flow of liquid in the capillary. A stop valve can be formed by reasonably changing the structure of the meniscus in the microchannel to generate the additional pressure we need and block the liquid flow. There are two main ways: the first kind of valve depends on changes in the hydrophilic and hydrophobic properties of the microchannel surface, such as the hydrophilic microchannel surface which is coated with hydrophobic materials [45,113,114]. The second type of capillary valve is fairly simple to manufacture and does not require additional surface treatment to alter the microchannel hydrophilicity. The liquid is simply stopped when the microchannel cross section changes abruptly (hydrophilic microchannels expand suddenly or hydrophobic microchannels shrink suddenly) [46]. These stop valves can stop the flow of liquid in microchannels without external forces and have been successfully used in the partially wetted regime.

Surface treatment type stop valve. Go Takei et al. designed and manufactured a four-step wettability passive surface tension stop valve with a pressure barrier of 6.8–12.5 kPa based on the characteristics of titanium dioxide nanoparticles, and proposed a method for microfluidics to control the surface wettability of microchannels [113]. The super-hydrophobic surface can be obtained by hydrophobic treatment on the rough surface of titanium dioxide. The surface wettability changes from super-hydrophobic to super-hydrophilic by photocatalytic decomposition of hydrophobic molecules under controlled light irradiation. The wettability pattern of the channel surface can be flexibly drawn, exposure can be controlled by graphics, and wettability can be adjusted over a wide range. Yingxue Zhang et al. used a surface tension stop valve that can be used in collecting skin sweat. The valve is positioned at the interface between the single-opening liquid collection chamber and the microfluidic channel to ensure the collection of sweat and reduce the evaporation of sweat [45]. As shown in Figure 3A, when sweat enters the hydrophilic microchannel, the stop valve forms a pressure barrier to prevent sweat from entering the back channel and forcing sweat into the collecting chamber. When the sweat fills the chamber completely along the wall, the stop valve will not work and the liquid passes over the stop valve.

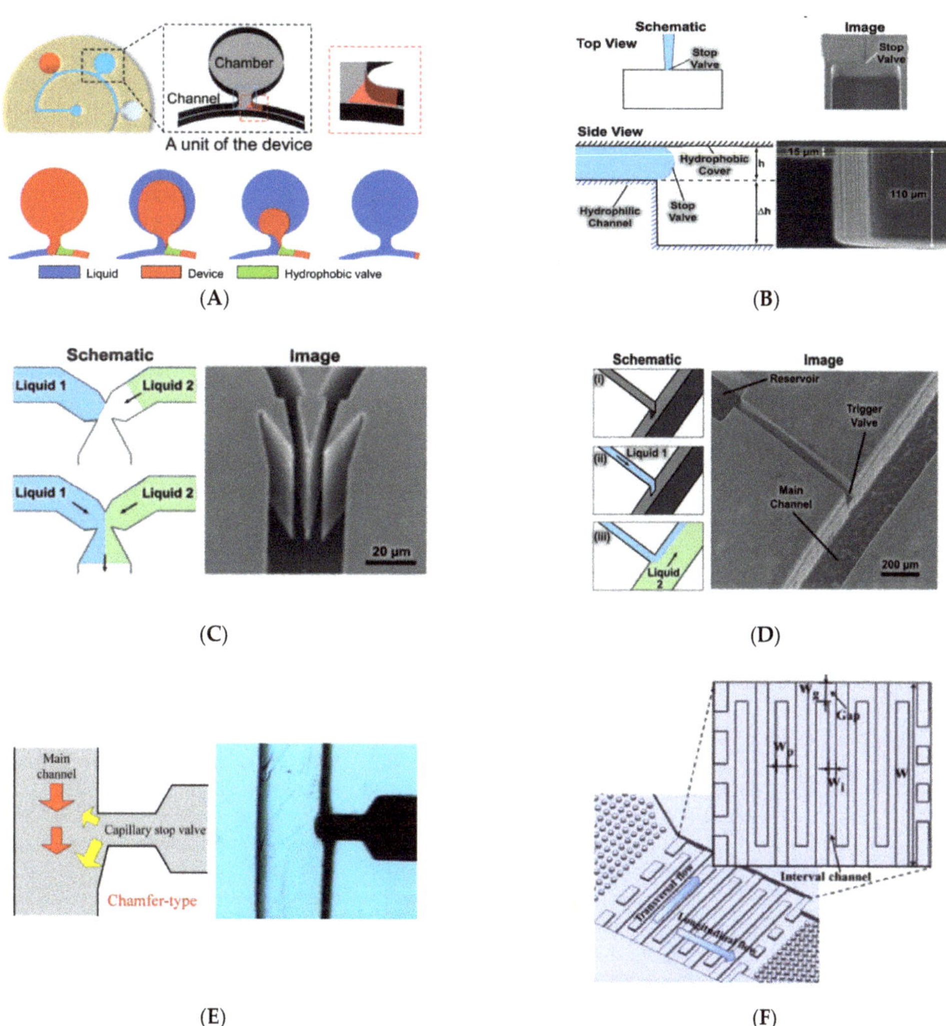

Figure 3. Structures of CPCV. (**A**) Surface treatment type stop valve. Reprinted with permission from [45]. Copyright 2020 Royal Society of Chemistry. (**B**) Cross section expanding type stop valve. (**C**) Single layer trigger valve. (**D**) Stair-step trigger valve. ((**B–D**) reprinted with permission from [14]. Copyright 2018 Royal Society of Chemistry.) (**E**) Chamfer-type trigger valve. Reprinted with permission from [15]. (**F**) Comb-like delay valve. Reprinted with permission from [18].

Cross section expanding type stop valve. The liquid flow in the microchannel is stopped by a sudden expansion in the channel geometry without external intervention. The normal stop valve stops the liquid only by expanding the width of the microchannel. This single-layer method is very simple and reliable in active hydrophobic systems. A two-level stop valve was designed and manufactured by Gliere and Delattre using hydrophilic microchannel and hydrophobic PDMS covers, and the threshold pressure and microchannel size were theoretically analyzed and experimentally verified (Figure 3B). A two-level stop valve with channel cross section extending along the width and depth was prepared by reactive ion etching technique. Tests on the 15 μm × 15 μm valve confirmed that the valve can block the buffer normally and it opens at a burst pressure within the range of 1–10 kPa [46].

2.4.2. Trigger Valves

Stop valves can be easily converted into trigger valves by cross-setting two microchannels. The flow of the first microchannel stops flowing at the intersection of channels, and the stopped liquid flow can be restored when the second liquid reaches the intersection.

The trigger valve can be assembled by stop valves to allow the stopped liquid to resume flow due to the subsequent trigger liquid [115,116]. For microfluidic applications requiring accurate control of reaction time and liquid flow, ensuring that multiple reagents reach the reaction chamber simultaneously is the core of the problem. If the reagents that should have reached the cross junction of the channel do not arrive at the same time, bubbles are often generated between multiple reagents because the inlet and outlet channels are directly connected to the reaction chamber. These bubbles may change the planned flow path of the reagent or directly truncate the channel [52]. Considering the principle of the stop valve at the junction of multiple channels, the trigger valve is designed to ensure that a single fluid stops at the valve, and the flow will be resumed only after the arrival of the second liquid, which can effectively avoid the bubble problem existing in the microchannel.

The geometry of the trigger valve determines its function. The basic structure consists of two inlets and one outlet in a shape similar to the Y-junction (Figure 3C). When the liquid from the two inlet reaches the connection point alone, the Laplace pressure of the liquid increases due to the abrupt change of channel cross section, and the Y-junction acts as a geometric stop valve. However, when more than one inlet liquid is present at the junction, the valve allows the liquid to pass through the exit, because liquid contact rebuilds the original liquid meniscus, and the Laplace pressure of the newly formed meniscus is not enough to stop the liquid. These trigger valves can hold liquid in an inlet for an average of 15 min, but require a very high aspect ratio to successfully stop the liquid [14].

Yaw-Jen Chang et al. introduced the chamfered side structure into the trigger valve [15], as shown in Figure 3E. As a flow control device, there will be a countercurrent phenomenon in the commonly used T-type trigger valve, and the use of chamfer can alleviate this phenomenon. The chamfered side of the valve not only holds the reagent in place, but also increases the contact area between the reagents. Due to chamfering, the stop reagent meniscus is more likely to be triggered by reconstruction, and the reagent is also more likely to be pulled into the main channel.

The stair-step triggered valve with lower aspect ratios designed and discussed by Zhang Lei et al. is shown in Figure 3D. Instead of two microchannels narrowing into a point, the stair-step trigger valve microchannels meet at an orthogonal intersection. The stair-trigger valve is sufficient to stop the liquid for more than 30 min and is more reliable than the single level trigger valve. However, a single level trigger valve can stop either liquid until both are present while the stair-trigger valve can only stop one liquid, which seems more useful. The valve channel, with sudden expansion along the width and depth of the channel, is fabricated on silicon using a two-step etching process. The two entrances of the stair-step trigger valve are non-equivalent. The function of the stair-step trigger valve is to stop the flow of the liquid from the cut-off passage and allow the liquid from the triggered passage to pass through the stepped structure normally [19]. Chen, Xi et al. used polyethylene glycol (PEG) and other materials to modify the surface to form the stair-step liquid trigger valve, and further evaluated and discussed the reliability of the valve [117].

2.4.3. Delay Valves

For some microfluidic systems, it is necessary to accurately control the arrival time of reagents. This can be accomplished with delay valves.

The delay valve controls the liquid flow time by increasing the flow resistance which decreases the liquid velocity or increasing the channel length to increase the wetting length [18,42]. In order to reduce the velocity of the liquid, it is possible to add hydrophobic marks in the channel to increase the resistance.

The moving speed of the filling liquid front in the wide microchannel is lower than that in the narrow microchannel. M. Zimmermann et al. achieved two parallel flow paths by simply combining the smaller channel with the larger one. At the same time, they used the guide structure to prefabricate the sequence of liquid entering each area, so as to achieve the purpose of delay [112].

Ji Won Suk et al. proposed a simple method for making the delay valve using an array of hydrophobic patterns to control the liquid velocity to reach the purpose of delay. The whole microchannel is composed of a hydrophilic floor and a hydrophobic roof. In order to reduce the liquid flow rate, hydrophobic patches are created on the floor. By adjusting the location, number, and spacing of hydrophobic patches, the flow rate of liquid throughout the channel network can be customized [42].

The comb-like delay valve proposed by Jingmin Li et al. (Figure 3F) provides a delay range from tens of seconds to several minutes, providing accurate time control for the reaction between samples and reagents. The delay time of the delay valve can be changed by adjusting the number and layout of the comb protrusions. As the simple structure of the valve and the minimum line width of the comb protrusions is over 100 μm, it can be easily manufactured in stainless steel molds and used in mass production [18]. The capillary stop valve, capillary retention valve, also belong to this part.

3. Applications and Impacts of CPCV

With the diversification of microfluidic chip application scenarios, the function of microfluidic chip is more and more powerful, and the role of valves is more and more important. As one of the most unique valves to control the flow of microfluidic equipment, the CPCV is valued for its simple design and customizable nature. CPCV can be easily integrated into different applications by designing microchannel structures or modifying hydrophilicity and will shine with the development of microfluidic technology. A brief overview of valve applications in different scenarios is listed as below, which includes but is not limited to making easy automation of microfluidic networks; making broader use of microfluidics by open channel; sampling and delivery in a controlled manner.

3.1. Simplifying Automation of Microfluidic Networks

The use of CPCV greatly expands the application of microfluidic networks and increases the automation of microfluidic chips [14]. It has accelerated the automation of microfluidic chips combined with the point of care testing (POCT) instrument for the automated detection of disease, which is easily broadly applied and commercialized.

3.1.1. Microfluidic Networks

A simple microfluidic network was constructed by Olanrewaju, A. O et al., in order to achieve rapid and simple bacterial detection. The trigger valve is used to isolate the preloaded sample and the two test reagents, and the buffer solution is added to trigger the automatic analysis of the bacterial test without human intervention [118].

With the expansion of microfluidic networks, bubble-free filling is a key prerequisite. By using CPCV to optimize the entire microfluidics network and pre-determine the flow of liquid through the device, bubbles within the microfluidic structure can be effectively avoided. As shown in the Figure 4A, CPCV can be set at each node of the microfluidic network, such as the merging or splitting of channels and the inlet and outlet of the assembly chamber. By setting valves at the nodes, the flow of the nodes in the network can be controlled and the reagents filling order for the entire network can be customized. With the cascade of valves, the driving pressure of the whole network increases continuously, and it needs to be established through several stages to realize the bubble-free filling of the hydrophobic microfluidic device [119].

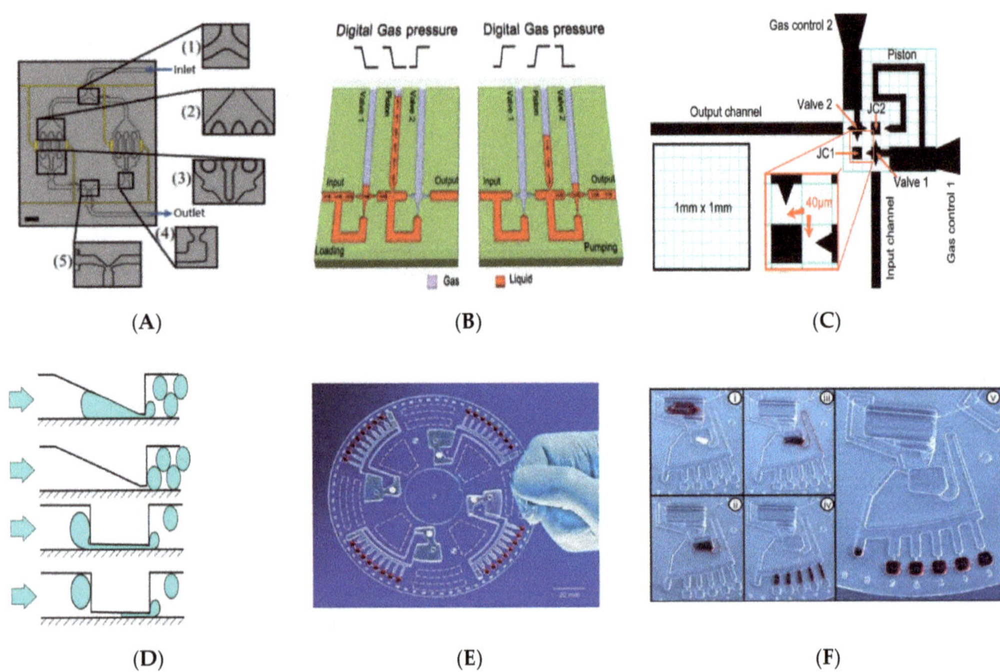

Figure 4. Application of CPCV in different microfluidic scenarios. (**A**) CPCV is used to realize the bubble-free microfluidic network for filling. Reprinted with permission from [119]. Copyright 2014 American Institute of Physics. (**B**) Micropump using CPCV components. Reprinted with permission from [120]. Copyright 2011 Royal Society of Chemistry. (**C**) The optimized micro pump with an area of one square millimeter. Reprinted with permission from [121]. (**D**) Structure of droplet generator using CPCV principle. Reprinted with permission from [122]. Copyright 2017 Royal Society of Chemistry. (**E**) A panoramic view of the lab on disk chip. Reprinted with permission from [123]. Copyright 2010 Royal Society of Chemistry. (**F**) The process of working in a lab on a disk chip. Reprinted with permission from [124]. Copyright 2010 Royal Society of Chemistry.

3.1.2. Micropump

Professor Peng applied the principle of CPCV to assemble a micro surface tension pump [120], the pump can pump liquid at a speed of 10 nL·s^{-1} into a 300 μm wide microchannel by pneumatic control. As shown in Figure 4B, during the pump loading process, valve 1 (inlet valve) is opened at low pressure and valve 2 (outlet valve) is closed at high pressure, as Laplace pressure fluid passes through valve 1 into the intermediate piston section. During pumping, valve 1 is closed and valve 2 is opened, using high pressure to pump the liquid from the piston through valve 2. In 2016, Professor Peng optimized the above pump (Figure 4C). By combining the pneumatic control of the input valve and the piston into one channel and simplifying the channel, the original three-way control was reduced to a two-way control and the compressed pump area was reduced to 1 square millimeter, which greatly enhanced the fault-tolerance and recoverability of the pump [121].

3.1.3. Droplet Generation

Accurate generation of micro droplets is a hot topic for researchers in the microfluidic laboratory. Zhu, Pingan and Wang, Liqiu reviewed in detail the various active and passive methods now used for droplet generation [125]. The T-shaped structure is the most common droplet generation method. When mutually insoluble fluids meet at the intersection of vertical T-shaped channel, under the action of pressure and shear force, the continuous phase truncates the dispersed phase, which can simply disperse the dispersed phase droplets in the continuous oil phase. In addition, CPCV-like structures are widely used to generate micro droplets [122,126–128]. For example, the structure (Figure 4D) used by Postek, W. et al., through a gradually narrowed microchannel or slit, can further disperse larger droplets to produce nanoscale droplets whose size is largely determined by the geometry of the device and which do not require additional external phase flows [122].

3.1.4. One Example: Lab on a Disk

In the last 10 years, the field of centrifugal microfluidics has experienced tremendous development, and its powerful and high-performance liquid handling capabilities have led to the widespread use of modular, multipurpose lab-on-a-disc platforms in the life sciences [41,48,129,130]. CPCV is often used to separate the measuring chamber from the reaction chamber in centrifugal microfluidics [123,131], such as the work by Lutz, Sascha et al. [124]. As shown in the Figure 4E,F, after adding the specimen and sealing the inlet hole, the disc will rotate at a frequency of 6.6 Hz, the sample will be calibrated in five adjacent metering chambers, and the excess liquid will be collected in an adjacent waste reservoir located behind the metering chambers, during which the CPCV prevents the sample from entering the reaction well. When unidirectional and alternating rotation frequencies between 6.6 and 27 Hz are applied, the CPCV is opened, and the sample enters the reaction well until it is filled. Then, the system enters a stable equilibrium state, and the rotation frequency will not affect the position of the equable sample. CPCV with similar structure is widely used in clinical chemistry, immunodiagnostics and protein analysis, cell processing [129], medical diagnosis [132], and other fields.

3.2. Various Valve Actuation Mode

As a kind of passive valve, CPCV can stop the liquid in the valve without external force. That is, the valve is in the closed state. However, opening the valve to allow the fluid to pass normally requires additional force to break the meniscus in order to completely wet the channel. This additional force can be introduced in a number of ways, the most commonly mentioned being the capillary burst valve [36]. There are many microfluidic devices that use external pressure directly to open the valve [114]. This approach is simple and straightforward but requires additional equipment such as injection pumps and does not reflect the driving advantages of the CPCV. The following are brief introductions of some other driving common modes of CPCV.

3.2.1. External Forces

Using Laplace pressure to directly drive liquid (such as water in the hydrophilic channel, oily liquid in the hydrophobic channel) to rebuild and reconstruct the meniscus to open the valve is the most ideal method to open CPCV. This completely passive way does not require external forces or any external equipment, and only depends on the channel structure and the liquid properties, which is easy to implement and extremely conducive to the miniaturization of equipment. Due to the deterministic structure and passive triggering, these trigger valves lack flexibility in opening. When the valve is fully open, it needs to drain all the fluid in the channel and then close the valve to restore it to its original state, which lacks reusability. However, its disposable features, due to its low cost, are more promising in real-time diagnosis, biological analysis, and other aspects.

Centrifugal force can be used not only to open the CPCV, but also to control the flow direction of the centrifugal microfluid after opening the valve. For symmetrical structures, as shown in Figure 5A, the liquid is driven into the selected chamber by changing the direction of rotation [133]. Amin Kazemzadeh et al. designed an asymmetric structure (Figure 5B), which can control the flow direction of centrifugal fluid without changing the rotation direction and without using external sources or surface treatment. The device is related to the speed or frequency of rotation. At low speeds, because of the asymmetric structure of the valve, the liquid is directed in one direction along the microchannel. At high speeds, the liquid is directed in the opposite direction by Coriolis forces [134].

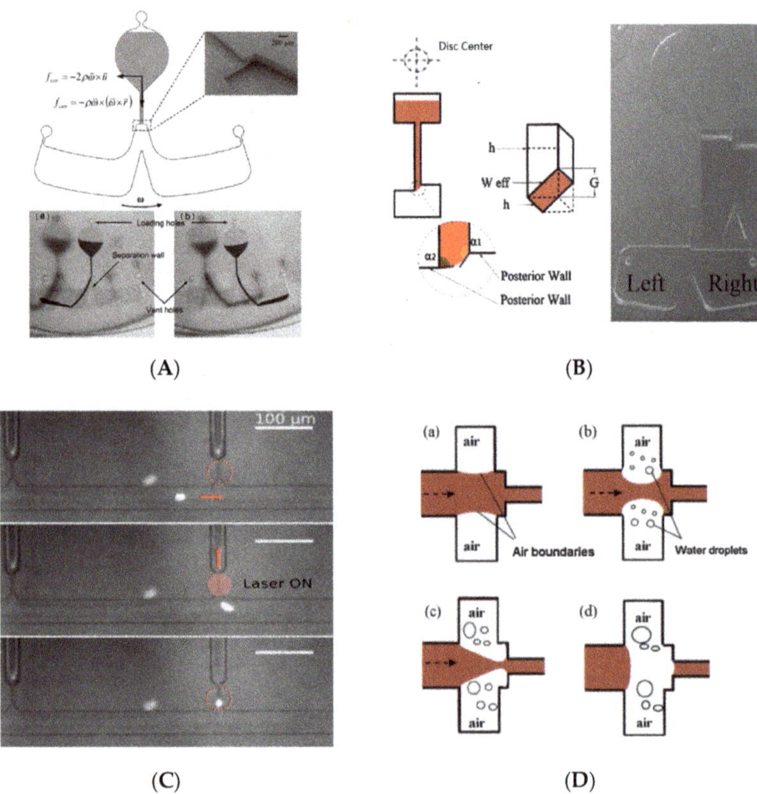

(A)

(B)

(C)

(D)

Figure 5. Different modes of valve actuation. (**A**) Centrifugal force drives CPCV by changing the direction of rotation to allow the liquid to enter different chambers. Reprinted with permission from [133]. (**B**) Asymmetric structure, centrifugal force driven, changing the speed of liquid into different chambers. Reprinted with permission from [134]. (**C**) The laser drive is used to capture particles. Reprinted with permission from [135]. Copyright 2017 American Institute of Physics. (**D**) An example using thermally driven CPCV. Reprinted with permission from [136].

We know that the surface tension forming the meniscus creates a pressure barrier that can stop the capillary flow. As a valve, it is a good idea to use the centrifugal force generated by spinning to break through the pressure barrier. For example, Lab-on-CD devices make the chip rotate through a simple platform. They use the Laplace barrier generated by surface tension to prevent the liquid flow in the microchannel, and use the centrifugal force generated by rotation to drive the liquid to break through the pressure barrier and control the liquid flow [137,138]. CPCV will generate different threshold pressure barriers with different channel sizes and shapes. When CPCV structure is fixed, the threshold speed of the Lab-on-CD device can be calculated by CPCV position, reagent density, and reagent column length in the microchannel. When the rotating speed of the platform exceeds the threshold value, the centrifugal force generated is greater than the barrier pressure, the meniscus burst liquid enters the expansion space, and the valve is opened. According to the change of interface energy, Jerry M et al. give a simple formula to calculate the critical rotational speed and burst pressure to overcome the CPCV threshold pressure at the sudden expansion of the rectangular channel section [50]. By setting the CPCV of different thresholds and accurately controlling the rotation speed, it is possible to integrate a variety of functions into Lab-on-CD devices, such as flow sequencing, cascade micro-mixing, capillary metering, and so on [139,140]. For example, the sample separator designed by Leu, T. S. et al. using density gradient centrifugation to segment the samples liquid in the microstructure. The different density segments required by the sample can be simply separated by rotating at different speeds [141].

3.2.2. By Changing Surface Tension

Londe, G et al. made thermos-sensitive CPCV using a switchable thermos-sensitive polymer [142]. At room temperature, the surface of thermos-sensitive polymer is hydrophilic and allows water to flow. When the temperature exceeds 65 degrees Celsius, the surface becomes hydrophobic, thus inhibiting water flow. Recently, L. Li et al. grafted thermo-responsive polymer Poly(Nisopropylacrylamide) (PNIPAm) onto PDMS. When the channel temperature increased from 20 to 37 °C, the channel surface changes from hydrophilic to hydrophobic, forming a CPCV [143]. Thus, the temperature control valve can regulate sample flow more flexibly. Due to the interrelationship between fluid surface tension and temperature, increasing the temperature can make the CPCV meniscus rupture. Johan Eriksen et al. embedded the near-infrared absorption dye film into a sealed microfluidic device and used laser local heating to open a single valve [135], shown in Figure 5C.

Wei Xu et al. proposed a basic valve design using trapped air to control flow. The small concave structure was made in the micro channel at the same position as the CPCV section expansion, as shown in Figure 5D. When the liquid filled the micro channel, due to the hydrophobicity of the manufacturing material PDMS, the surface tension limited the fluid into the concave chamber, and the hydrophobic chamber intercepted the air to form air bubbles. By increasing the chip temperature, the bubbles gradually expanded and moved into the main channel, limiting the flow of fluid in the channel. Finally, the bubble completely truncated the main channel and blocked the flow of liquid [136].

3.3. Make Broader Use of Microfluidics by Open Channel

Open channel microfluidics devices usually lack at least one limiting liquid physical sidewall or cap. By applying the hydrophilic and hydrophobic properties of the channel and the surface tension of the liquid, the liquid is fixed to flow along a specific path [144,145]. The open channel microfluidic device is a special application of CPCV using the theory from Section 2.4.1. Compared with traditional closed microfluidic structure, the open channel microfluidic structure has obvious advantages and disadvantages. As the top is open or semi-open, it facilitates the entry of sample reagents and eliminates the air-bubble problem in traditional closed wall microsystems. At the same time, since there is no cover at the top, it is easier to manufacture. However, the top opening also greatly increases sample evaporation and the risk of reagent contamination. In addition, the open structure greatly restricts the use of the drive system [146,147].

V.A. Papadimitriou et al. used CPCV to realize the fast and convenient immobilization of antibodies after chip bonding. This method can automatically and conveniently shape the liquid in a closed chip. The basic structure is shown in Figure 6A, a deep channel is placed across another shallow channel, the deep channel is filled with reagents containing antibodies, and the shallow channel perfusion is used to detect the samples. The reagent fills the deep channel rapidly due to capillary force, and when it reaches the intersection of the deep and shallow channels, the upper part of the reagent in the deep channel stops flowing because of the sudden opening of the channel, which is bound by Laplace pressure. The lower part of the reagent is free to flow through the trench below the shallow channel. When the reagent passes through the trench through the shallow channel, the capillary force pulls the reagent to fill the whole deep channel. Throughout the process, the reagents are fixed to the bottom and vertical sides of the intersection of the deep and shallow channels, which functions like a globe valve [148].

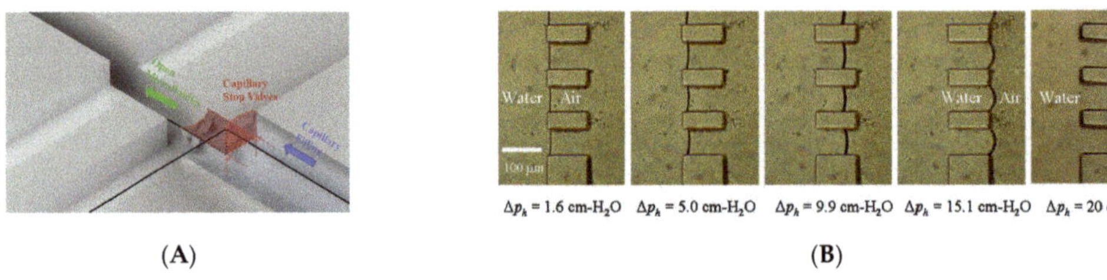

(A)   (B)

Figure 6. Application of CPCV in different microfluidic scenarios. (**A**) Use of CPCV to quick rivet antibody with open channel. Reprinted with permission from [148]. (**B**) Use of CPCV to build virtual channels. Reprinted with permission from [149]. Copyright 2011 American Institute of Physics.

Hsuan-hong Lai et al. used the characteristics of CPCV to design the virtual walls with a range of feature sizes [149]. As shown in Figure 6B, the virtual wall is composed of a row of rectangular PDMS micro-columns, and the two rows of parallel virtual walls constitute the channel, and then the device is divided into an intermediate liquid channel and an air chamber on both sides. When the liquid is filled, because of the Laplace pressure of the liquid, the liquid is confined within the middle channel, and the air is squeezed into chambers on both sides of the channel. When the hydraulic pressure is enhanced, the liquid between the micro-columns will advance towards the air chamber on both sides; when the pressure decreases, the liquid will contract towards the middle channel; when the pressure exceeds the threshold, the liquid passes through the virtual wall composed of PDMS micro-columns and completely fills the air chamber. To restore the virtual wall, the device needs to be cleared of all fluids and started at low pressure.

3.4. Sampling and Delivery with Controlled Manner

Sample collection and delivery are two indispensable functions of microfluidic chips. The chip contact tip must first enter into the sample and then sample or deliver and transport along a given route in a controlled manner, which is the basic guarantee for the implementation of a microfluidic probe. CPCV can help to achieve these functions based on the theory from theory 3.1.1. and sometimes with theory 3.1.3. It can transport and stop the flow as well as protect the sampling or delivery microenvironment at the chip contact tip by forming the safe barrier of hydrophilic features where only water phase can go through rather than oil/gas phases in a controlled manner.

Choi, Jungil et al. designed a skin microfluidics device with accurate sampling capability, as shown in Figure 7A, using a set of carefully designed CPCVs to guide sweat to fill the micro reservoir in a sequential flow [150]. The initial liquid arriving at CPCV1 and CPCV2 meets them in a closed state. When the breakthrough pressure of CPCV1 is reached or exceeded, CPCV1 opens, allowing fluid to enter the chamber. After this chamber is filled completely, CPCV2 is broken down under sufficient pressure. The breakthrough pressure of CPCV2 is lower than that of CPCV3. Through this process, all chambers are sequentially filled with fluid at pressures greater than the CPCV2 breakthrough pressure. Then, the chip is removed from the skin and placed into the centrifuge. CPCV3 is turned on by centrifugal force and the sweat is moved to the reaction chamber behind the micro reservoir for experimental analysis. This design of CPCV satisfies that the pressure generated by sweat glands exceeds the breakthrough pressure of CPCV1 and CPCV2 that ensures sweat can enter and fully fill the micro reservoir. As the pressure generated by centrifugation exceeds the breakthrough pressure of CPCV3, sweat can enter the reaction chamber.

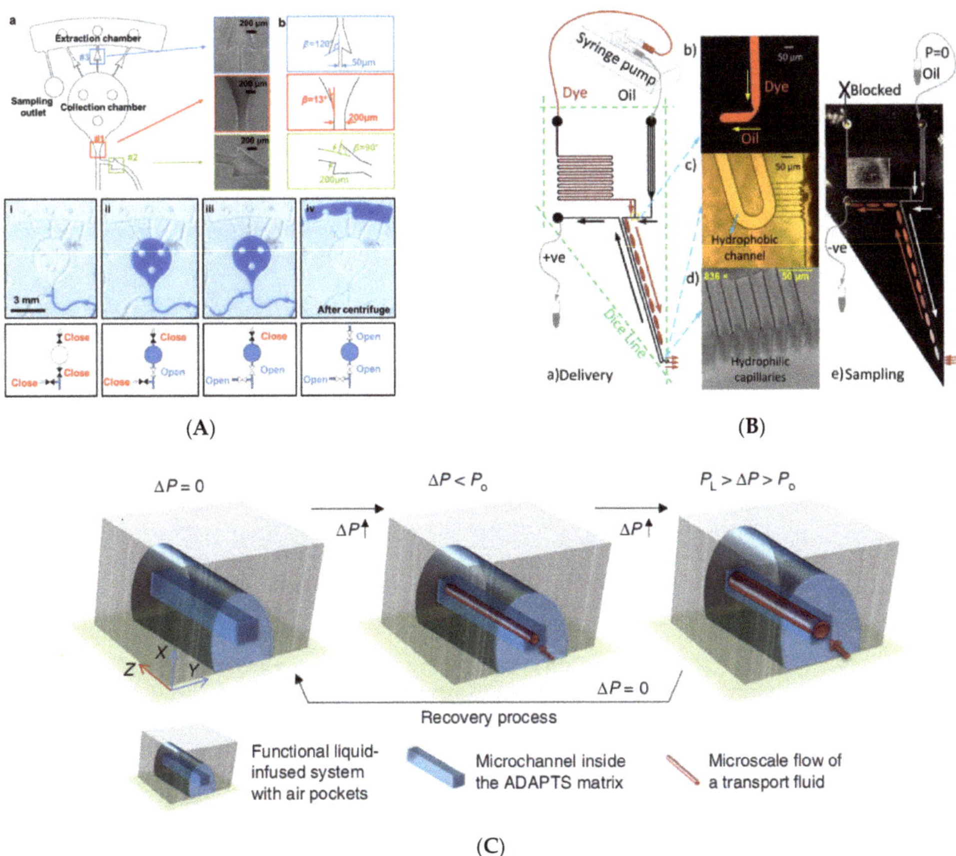

Figure 7. Use of CPCV in sampling and delivery. (**A**) A sweat sampling microfluidic device that adheres to the skin. Reprinted with permission from [150]. Copyright 2017 Wiley-VCH. (**B**) A needle-shaped liquid sampling and delivery device. Reprinted with permission from [80]. Copyright 2017 American Institute of Physics. (**C**) Surface tension applied in an adaptive air/liquid pocket transport system (ADAPTS) where the microscale flow (shown in red) is in the square microchannel (shown in dark blue). Reprinted with permission from [151].

Based on the principle of CPCV, a needle-shaped microfluidic device is proposed, as shown in Figure 7B, which can provide the sampling and delivery of nanoliter solution in droplets [80], where CPCV serves as the safe barrier for a safe sampling and delivery microenvironment. The device consists of a T-junction forming droplets. The needle was formed by a parallel hydrophobic channel, and the sampling tip was formed by seven hydrophilic microchannels (similar to CPCV). During the delivery process, droplets dispersed in the oil phase produced by the T-junction move along the channel under pressure at the needle tip, where the tip (seven capillaries) is hydrophilic and the droplets can pass through with a pressure gradient close to zero but the mineral oil requires 19 kPa pressure. During the sampling process, a negative pressure is added to the device to allow the liquid to enter the device from the tip, create droplets in the oil phase and move with the pressure gradient. This needle was applied for H_2O_2 detection [152]. The similar principles have been widely applied by Feng on the membrane on the plastic chip [99]. It has been further reviewed by this type of principle used in the sampling probe [153].

Xu, Hou et al. presented an adaptive air/liquid pocket transport system (ADAPTS) [151] with solid porous structures and liquid/gas interfaces, which are suitable for low-pressure applications. Initially, functional liquid is filled in the porous structures. When the transport fluid enters, it displaces the functional liquid to open a flow path in the microchannel (Figure 7C). By then, the applied pressure (ΔP) is bigger than the threshold pressure (Po). The functional liquid can refill the microchannel after removal of the pressure and recover to the original state immediately.

4. Conclusions and Outlook

In this paper, the principle, composition, and application of CPCV were briefly reviewed. In microfluidics, CPCV is used as flow controller and forms important parts of microfluidic chips. CPCVs are generated by additional pressure due to meniscus variations in microchannels. The characteristics produced by it reflect its advantages and disadvantages. The CPCV does not need additional equipment; the switching pressure is convenient, controllable, and only related to its structure; it is simple to manufacture, convenient to integrate and put on scale production; it has a good prospects in all branches of micro-flow control (biomedical testing, wearable devices, etc.). The disadvantages are also relatively obvious, the valve is more passive and there is a lack of flexibility. If you want to close the valve after opening the valve, you need to drain all the liquid in the channel and restore it to dry, which lacks reusability. In Section 3, Xu Hou has put effort into this, while it can only be used for low pressure applications, but it is still an improvement. The shortcoming is also the direction of further research in the future, which needs to be jointly promoted by the industry.

We hope the CPCV can make automation easy and merge into the point of care instrument, which can help with the commercialization of microfluidics chips, precise medicine and POCT detections, especially in this COVID-19 virus outbreak period. CPCV can facilitate broader use of open channels in microfluidics by eliminating the air-bubble problem in traditional closed wall microsystems. CPCV can be widely used in the bionic microfluidics; CPCV can be widely used in the sampling probes to avoid the gas/oil pollution to the sampling/delivery microenvironment.

In this review, we focused on fundamental physics theories in detail, categorized them based on the theory elements with new applications of each kind, and examined the potential of real applications in industrial and academic fields. We analyzed and categorized the examples from three fundamental elements: surface tension, contact angle, and different microchannel shape. We hope this review can give a direction for using the CACV in various academic and industrial research, which can provide broader applications.

Author Contributions: Conceptualization, S.F. and D.I.; writing—original draft preparation, S.W. and X.Z.; writing—review and editing, C.M., S.Y. and D.I.; supervision, S.F. All authors have read and agreed to the published version of the manuscript.

Funding: This work was financially supported by Technology Development Program of Taicang (No.TC2019DYDS07) and Technology Development Program of Xi'an (No.201805042YD 20CG26).

Institutional Review Board Statement: Not applicable.

Informed Consent Statement: Not applicable.

Conflicts of Interest: The authors declare no competing interest.

References

1. Thorsen, T.; Maerkl, S.J.; Quake, S.R. Microfluidic large-scale integration. *Science* **2002**, *298*, 580–584. [CrossRef] [PubMed]
2. Whitesides, G.M. The origins and the future of microfluidics. *Nature* **2006**, *442*, 368–373. [CrossRef] [PubMed]
3. Feng, S.; Skelley, A.M.; Anwer, A.G.; Liu, G.; Inglis, D.W. Maximizing particle concentration in deterministic lateral displacement arrays. *Biomicrofluidics* **2017**, *11*, 024121. [CrossRef] [PubMed]
4. Inglis, D.; Vernekar, R.; Krüger, T.; Feng, S. The fluidic resistance of an array of obstacles and a method for improving boundaries in deterministic lateral displacement arrays. *Microfluid. Nanofluid.* **2020**, *24*, 18. [CrossRef]
5. Liu, G.; Cao, C.; Ni, S.; Feng, S.; Wei, H. On-chip structure-switching aptamer-modified magnetic nanobeads for the continuous monitoring of interferon-gamma ex vivo. *Microsyst. Nanoeng.* **2019**, *5*, 35. [CrossRef] [PubMed]

6. Kanitthamniyom, P.; Zhou, A.; Feng, S.; Liu, A.; Vasoo, S.; Zhang, Y. A 3D-printed modular magnetic digital microfluidic architecture for on-demand bioanalysis. *Microsyst. Nanoeng.* **2020**, *6*, 48. [CrossRef] [PubMed]
7. Ghodbane, M.; Stucky, E.C.; Maguire, T.J.; Schloss, R.S.; Shreiber, D.I.; Zahn, J.D.; Yarmush, M.L. Development and validation of a microfluidic immunoassay capable of multiplexing parallel samples in microliter volumes. *Lab Chip* **2015**, *15*, 3211–3221. [CrossRef]
8. Han, S.W.; Jang, E.; Koh, W.-G. Microfluidic-based multiplex immunoassay system integrated with an array of QD-encoded microbeads. *Sens. Actuators B-Chem.* **2015**, *209*, 242–251. [CrossRef]
9. Soares, R.R.G.; Ramadas, D.; Chu, V.; Aires-Barros, M.R.; Conde, J.P.; Viana, A.S.; Cascalheira, A.C. An ultrarapid and regenerable microfluidic immunoassay coupled with integrated photosensors for point-of-use detection of ochratoxin A. *Sens. Actuators B-Chem.* **2016**, *235*, 554–562. [CrossRef]
10. Amasia, M.; Cozzens, M.; Madou, M.J. Centrifugal microfluidic platform for rapid PCR amplification using integrated thermo-electric heating and ice-valving. *Sens. Actuators B-Chem.* **2012**, *161*, 1191–1197. [CrossRef]
11. Kieu The Loan, T.; Wu, W.; Lee, N.Y. Planar poly(dimethylsiloxane) (PDMS)-glass hybrid microdevice for a flow-through polymerase chain reaction (PCR) employing a single heater assisted by an intermediate metal alloy layer for temperature gradient formation. *Sens. Actuators B-Chem.* **2014**, *190*, 177–184. [CrossRef]
12. Tachibana, H.; Saito, M.; Tsuji, K.; Yamanaka, K.; Le Quynh, H.; Tamiya, E. Self-propelled continuous-flow PCR in capillary-driven microfluidic device: Microfluidic behavior and DNA amplification. *Sens. Actuators B-Chem.* **2015**, *206*, 303–310. [CrossRef]
13. Idota, N.; Kikuchi, A.; Kobayashi, J.; Sakai, K.; Okano, T. Microfluidic valves comprising nanolayered thermoresponsive polymer-grafted capillaries. *Adv. Mater.* **2005**, *17*, 2723–2727. [CrossRef]
14. Olanrewaju, A.; Beaugrand, M.; Yafia, M.; Juncker, D. Capillary microfluidics in microchannels: From microfluidic networks to capillaric circuits. *Lab Chip* **2018**, *18*, 2323–2347. [CrossRef] [PubMed]
15. Chang, Y.-J.; Lin, Y.-T.; Liao, C.-C. Chamfer-Type Capillary Stop Valve and Its Microfluidic Application to Blood Typing Tests. *SLAS Technol.* **2019**, *24*, 188–195. [CrossRef] [PubMed]
16. Al-Faqheri, W.; Ibrahim, F.; Thio, T.H.G.; Aeinehvand, M.M.; Arof, H.; Madou, M. Development of novel passive check valves for the microfluidic CD platform. *Sens. Actuators A-Phys.* **2015**, *222*, 245–254. [CrossRef]
17. Kim, D.; Beebe, D.J. A bi-polymer micro one-way valve. *Sens. Actuators A-Phys.* **2007**, *136*, 426–433. [CrossRef]
18. Li, J.; Liang, C.; Zhang, B.; Liu, C. A comblike time-valve used in capillary-driven microfluidic devices. *Microelectron. Eng.* **2017**, *173*, 48–53. [CrossRef]
19. Zhang, L.; Jones, B.; Majeed, B.; Nishiyama, Y.; Okumura, Y.; Stakenborg, T. Study on stair-step liquid triggered capillary valve for microfluidic systems. *J. Micromech. Microeng.* **2018**, *28*, 065005. [CrossRef]
20. Siegrist, J.; Gorkin, R.; Clime, L.; Roy, E.; Peytavi, R.; Kido, H.; Bergeron, M.; Veres, T.; Madou, M. Serial siphon valving for centrifugal microfluidic platforms. *Microfluid. Nanofluid.* **2010**, *9*, 55–63. [CrossRef]
21. Desai, A.V.; Tice, J.D.; Apblett, C.A.; Kenis, P.J.A. Design considerations for electrostatic microvalves with applications in poly(dimethylsiloxane)-based microfluidics. *Lab Chip* **2012**, *12*, 1078–1088. [CrossRef]
22. Li, H.Q.; Roberts, D.C.; Steyn, J.L.; Turner, K.T.; Yaglioglu, O.; Hagood, N.W.; Spearing, S.M.; Schmidt, M.A. Fabrication of a high frequency piezoelectric microvalve. *Sens. Actuators A-Phys.* **2004**, *111*, 51–56. [CrossRef]
23. Tice, J.D.; Desai, A.V.; Bassett, T.A.; Apblett, C.A.; Kenis, P.J.A. Control of pressure-driven components in integrated microfluidic devices using an on-chip electrostatic microvalve. *RSC Adv.* **2014**, *4*, 51593–51602. [CrossRef]
24. Hartshorne, H.; Backhouse, C.J.; Lee, W.E. Ferrofluid-based microchip pump and valve. *Sens. Actuators B-Chem.* **2004**, *99*, 592–600. [CrossRef]
25. Luharuka, R.; LeBlanc, S.; Bintoro, J.S.; Berthelot, Y.H.; Hesketh, P.J. Simulated and experimental dynamic response characterization of an electromagnetic microvalve. *Sens. Actuators A-Phys.* **2008**, *143*, 399–408. [CrossRef]
26. Rich, C.A.; Wise, K.D. A high-flow thermopneumatic microvalve with improved efficiency and integrated state sensing. *J. Microelectromechan. Syst.* **2003**, *12*, 201–208. [CrossRef]
27. Huang, C.; Tsou, C. The implementation of a thermal bubble actuated microfluidic chip with microvalve, micropump and micromixer. *Sens. Actuators A-Phys.* **2014**, *210*, 147–156. [CrossRef]
28. Liu, R.H.; Bonanno, J.; Yang, J.N.; Lenigk, R.; Grodzinski, P. Single-use, thermally actuated paraffin valves for microfluidic applications. *Sens. Actuators B-Chem.* **2004**, *98*, 328–336. [CrossRef]
29. Zahra, A.; Scipinotti, R.; Caputo, D.; Nascetti, A.; de Cesare, G. Design and fabrication of microfluidics system integrated with temperature actuated microvalve. *Sens. Actuators A-Phys.* **2015**, *236*, 206–213. [CrossRef]
30. Casals-Terre, J.; Duch, M.; Plaza, J.A.; Esteve, J.; Perez-Castillejos, R.; Valles, E.; Gomez, E. Design, fabrication and characterization of an externally actuated ON/OFF microvalve. *Sens. Actuators A-Phys.* **2008**, *147*, 600–606. [CrossRef]
31. Kong, M.C.R.; Salin, E.D. Pneumatic Flow Switching on Centrifugal Microfluidic Platforms in Motion. *Anal. Chem.* **2011**, *83*, 1148–1151. [CrossRef]
32. Al-Faqheri, W.; Ibrahim, F.; Thio, T.H.G.; Moebius, J.; Joseph, K.; Arof, H.; Madou, M. Vacuum/Compression Valving (VCV) Using Parrafin-Wax on a Centrifugal Microfluidic CD Platform. *PLoS ONE* **2013**, *8*, e58523. [CrossRef]
33. Abi-Samra, K.; Hanson, R.; Madou, M.; Gorkin, R.A., III. Infrared controlled waxes for liquid handling and storage on a CD-microfluidic platform. *Lab Chip* **2011**, *11*, 723–726. [CrossRef] [PubMed]

34. Sugiura, S.; Szilagyi, A.; Sumaru, K.; Hattori, K.; Takagi, T.; Filipcsei, G.; Zrinyi, M.; Kanamori, T. On-demand microfluidic control by micropatterned light irradiation of a photoresponsive hydrogel sheet. *Lab Chip* **2009**, *9*, 196–198. [CrossRef] [PubMed]

35. Ducree, J.; Haeberle, S.; Lutz, S.; Pausch, S.; von Stetten, F.; Zengerle, R. The centrifugal microfluidic bio-disk platform. *J. Micromech. Microeng.* **2007**, *17*, S103–S115. [CrossRef]

36. Cho, H.; Kim, H.-Y.; Kang, J.Y.; Kim, T.S. How the capillary burst microvalve works. *J. Colloid Interface Sci.* **2007**, *306*, 379–385. [CrossRef] [PubMed]

37. Thio, T.; Nozari, A.A.; Soin, N.; Kahar, M.K.B.A.; Dawal, S.Z.M.; Samra, K.A.; Madou, M.; Ibrahim, F. Hybrid Capillary-Flap Valve for Vapor Control in Point-of-Care Microfluidic CD. In Proceedings of the 5th Kuala Lumpur International Conference on Biomedical Engineering 2011, Kuala Lumpur, Malaysia, 20–23 June 2011; Volume 35, pp. 578–581.

38. Brask, A.; Snakenborg, D.; Kutter, J.P.; Bruus, H. AC electroosmotic pump with bubble-free palladium electrodes and rectifying polymer membrane valves. *Lab Chip* **2006**, *6*, 280–288. [CrossRef] [PubMed]

39. Nguyen, N.T.; Truong, T.Q.; Wong, K.K.; Ho, S.S.; Low, C.L.N. Micro check valves for integration into polymeric microfluidic devices. *J. Micromech. Microeng.* **2004**, *14*, 69–75. [CrossRef]

40. Ni, J.; Huang, F.; Wang, B.; Li, B.; Lin, Q. A planar PDMS micropump using in-contact minimized-leakage check valves. *J. Micromech. Microeng.* **2010**, *20*, 095033. [CrossRef]

41. Strohmeier, O.; Keller, M.; Schwemmer, F.; Zehnle, S.; Mark, D.; von Stetten, F.; Zengerle, R.; Paust, N. Centrifugal microfluidic platforms: Advanced unit operations and applications. *Chem. Soc. Rev.* **2015**, *44*, 6187–6229. [CrossRef]

42. Suk, J.W.; Cho, J.-H. Capillary flow control using hydrophobic patterns. *J. Micromech. Microeng.* **2007**, *17*, N11–N15. [CrossRef]

43. Andersson, H.; van der Wijngaart, W.; Griss, P.; Niklaus, F.; Stemme, G. Hydrophobic valves of plasma deposited octafluorocyclobutane in DRIE channels. *Sens. Actuators B-Chem.* **2001**, *75*, 136–141. [CrossRef]

44. Lu, C.; Xie, Y.; Yang, Y.; Cheng, M.M.C.; Koh, C.-G.; Bai, Y.; Lee, L.J. New valve and bonding designs for microfluidic biochips containing proteins. *Anal. Chem.* **2007**, *79*, 994–1001. [CrossRef]

45. Zhang, Y.; Chen, Y.; Huang, J.; Liu, Y.; Peng, J.; Chen, S.; Song, K.; Ouyang, X.; Cheng, H.; Wang, X. Skin-interfaced microfluidic devices with one-opening chambers and hydrophobic valves for sweat collection and analysis. *Lab Chip* **2020**, *20*, 2635–2645. [CrossRef]

46. Gliere, A.; Delattre, C. Modeling and fabrication of capillary stop valves for planar microfluidic systems. *Sens. Actuators A-Phys.* **2006**, *130*, 601–608. [CrossRef]

47. Man, P.F.; Mastrangelo, C.H.; Burns, M.A.; Burke, D.T. Microfabricated capillarity-driven stop valve and sample injector. In Proceedings of the Proceedings MEMS 98. IEEE. Eleventh Annual International Workshop on Micro Electro Mechanical Systems, Heideberg, Germany, 25–29 January 1998; pp. 45–50. [CrossRef]

48. Kong, L.X.; Perebikovsky, A.; Moebius, J.; Kulinsky, L.; Madou, M. Lab-on-a-CD: A Fully Integrated Molecular Diagnostic System. *J. Lab. Autom.* **2016**, *21*, 323–355. [CrossRef] [PubMed]

49. Liu, M.; Zhang, J.; Liu, Y.; Lau, W.M.; Yang, J. Modeling of flow burst, flow timing in Lab-on-a-CD systems and its application in digital chemical analysis. *Chem. Eng. Technol.* **2008**, *31*, 1328–1335. [CrossRef]

50. Chen, J.M.; Huang, P.-C.; Lin, M.-G. Analysis and experiment of capillary valves for microfluidics on a rotating disk. *Microfluid. Nanofluid.* **2008**, *4*, 427–437. [CrossRef]

51. Zhang, H.; Hong Hanh, T.; Chung, B.H.; Lee, N.Y. Solid-phase based on-chip DNA purification through a valve-free stepwise injection of multiple reagents employing centrifugal force combined with a hydrophobic capillary barrier pressure. *Analyst* **2013**, *138*, 1750–1757. [CrossRef]

52. Melin, J.; Roxhed, N.; Gimenez, G.; Griss, P.; van der Wijngaart, W.; Stemme, G. A liquid-triggered liquid microvalve for on-chip flow control. *Sens. Actuators B-Chem.* **2004**, *100*, 463–468. [CrossRef]

53. Rosen, M.J.; Kunjappu, J.T. *Surfactants and Interfacial Phenomena*; John Wiley & Sons: Hoboken, NJ, USA, 2012; ISBN 9780470541944. [CrossRef]

54. Kumar, K.; Knie, C.; Bléger, D.; Peletier, M.A.; Friedrich, H.; Hecht, S.; Broer, D.J.; Debije, M.G.; Schenning, A.P.H.J. A chaotic self-oscillating sunlight-driven polymer actuator. *Nat. Commun.* **2016**, *7*, 11975. [CrossRef]

55. Amiri, M.; Dadkhah, A.A. On reduction in the surface tension of water due to magnetic treatment. *Colloids Surf. A Physicochem. Eng. Asp.* **2006**, *278*, 252–255. [CrossRef]

56. Chen, J.M.; Chen, C.-Y.; Liu, C.-H. Pressure barrier in an axisymmetric capillary microchannel with sudden expansion. *Jpn. J. Appl. Phys.* **2008**, *47*, 1683–1689. [CrossRef]

57. Grundner, M.; Jacob, H. Investigations on hydrophilic and hydrophobic silicon (100) wafer surfaces by X-ray photoelectron and high-resolution electron-energy loss-spectroscopy. *Appl. Phys. A-Mater.* **1986**, *39*, 73–82. [CrossRef]

58. Zhao, B.; Moore, J.S.; Beebe, D.J. Surface-directed liquid flow inside microchannels. *Science* **2001**, *291*, 1023–1026. [CrossRef]

59. Mata, A.; Fleischman, A.J.; Roy, S. Characterization of polydimethylsiloxane (PDMS) properties for biomedical micro/nanosystems. *Biomed. Microdevices* **2005**, *7*, 281–293. [CrossRef]

60. Tihan, T.G.; Ionita, M.D.; Popescu, R.G.; Iordachescu, D. Effect of hydrophilic-hydrophobic balance on biocompatibility of poly(methyl methacrylate) (PMMA)-hydroxyapatite (HA) composites. *Mater. Chem. Phys.* **2009**, *118*, 265–269. [CrossRef]

61. Chi, F.; Liu, D.; Wu, H.; Lei, J. Mechanically robust and self-cleaning antireflection coatings from nanoscale binding of hydrophobic silica nanoparticles. *Sol. Energy Mater. Sol. Cells* **2019**, *200*, 109939. [CrossRef]

62. Larsen, S.T.; Andersen, N.K.; Sogaard, E.; Taboryski, R. Structure Irregularity Impedes Drop Roll-Off at Superhydrophobic Surfaces. *Langmuir* **2014**, *30*, 5041–5045. [CrossRef]
63. Zhong, J.; Chinn, J.; Roberts, C.B.; Ashurst, W.R. Vapor-Phase Deposited Chlorosilane-Based Self-Assembled Monolayers on Various Substrates for Thermal Stability Analysis. *Ind. Eng. Chem. Res.* **2017**, *56*, 5239–5252. [CrossRef]
64. Li, D.W.; Wang, H.Y.; Liu, Y.; Wei, D.S.; Zhao, Z.X. Large-scale fabrication of durable and robust super-hydrophobic spray coatings with excellent repairable and anti-corrosion performance. *Chem. Eng. J.* **2019**, *367*, 169–179. [CrossRef]
65. Chaudhary, S.; Kamra, T.; Uddin, K.M.A.; Snezhkova, O.; Jayawardena, H.S.N.; Yan, M.; Montelius, L.; Schnadt, J.; Ye, L. Controlled short-linkage assembly of functional nano-objects. *Appl. Surf. Sci.* **2014**, *300*, 22–28. [CrossRef]
66. Rezaei, S.; Manoucheri, I.; Moradian, R.; Pourabbas, B. One-step chemical vapor deposition and modification of silica nanoparticles at the lowest possible temperature and superhydrophobic surface fabrication. *Chem. Eng. J.* **2014**, *252*, 11–16. [CrossRef]
67. Cordeiro, A.L.; Zschoche, S.; Janke, A.; Nitschke, M.; Werner, C. Functionalization of Poly(dimethylsiloxane) Surfaces with Maleic Anhydride Copolymer Films. *Langmuir* **2009**, *25*, 1509–1517. [CrossRef] [PubMed]
68. Wong, I.; Ho, C.-M. Surface molecular property modifications for poly(dimethylsiloxane) (PDMS) based microfluidic devices. *Microfluid. Nanofluid.* **2009**, *7*, 291–306. [CrossRef] [PubMed]
69. Zhou, J.; Ellis, A.V.; Voelcker, N.H. Recent developments in PDMS surface modification for microfluidic devices. *Electrophoresis* **2010**, *31*, 2–16. [CrossRef]
70. Vickers, J.A.; Caulum, M.M.; Henry, C.S. Generation of hydrophilic poly(dimethylsiloxane) for high-performance microchip electrophoresis. *Anal. Chem.* **2006**, *78*, 7446–7452. [CrossRef] [PubMed]
71. Peterson, S.L.; McDonald, A.; Gourley, P.L.; Sasaki, D.Y. Poly(dimethylsiloxane) thin films as biocompatible coatings for microfluidic devices: Cell culture and flow studies with glial cells. *J. Biomed. Mater. Res. Part A* **2005**, *72A*, 10–18. [CrossRef]
72. Trantidou, T.; Elani, Y.; Parsons, E.; Ces, O. Hydrophilic surface modification of PDMS for droplet microfluidics using a simple, quick, and robust method via PVA deposition. *Microsyst. Nanoeng.* **2017**, *3*, 16091. [CrossRef]
73. Abate, A.R.; Lee, D.; Do, T.; Holtze, C.; Weitz, D.A. Glass coating for PDMS microfluidic channels by sol–gel methods. *Lab Chip* **2008**, *8*, 516–518. [CrossRef]
74. Khamova, T.V.; Shilova, O.A.; Krasil'nikova, L.N.; Ladilina, E.Y.; Lyubova, T.S.; Baten'kin, M.A.; Kruchinina, I.Y. Sol–gel synthesis and study of the hydrophobicity of coatings prepared using modified aerosils. *Glass Phys. Chem.* **2016**, *42*, 194–201. [CrossRef]
75. Guo, W.; Hansson, J.; van der Wijngaart, W. Synthetic Paper Separates Plasma from Whole Blood with Low Protein Loss. *Anal. Chem.* **2020**, *92*, 6194–6199. [CrossRef] [PubMed]
76. Lee, M.; Lopez-Martinez, M.J.; Baraket, A.; Zine, N.; Esteve, J.; Plaza, J.A.; Jaffrezic-Renault, N.; Errachid, A. Polymer micromixers bonded to thermoplastic films combining soft-lithography with plasma and aptes treatment processes. *J. Polym. Sci. Part A-Polym. Chem.* **2013**, *51*, 59–70. [CrossRef]
77. Kim, J.A.; Lee, J.Y.; Seong, S.; Cha, S.H.; Lee, S.H.; Kim, J.J.; Park, T.H. Fabrication and characterization of a PDMS-glass hybrid continuous-flow PCR chip. *Biochem. Eng. J.* **2006**, *29*, 91–97. [CrossRef]
78. Hitzbleck, M.; Delamarche, E. Advanced Capillary Soft Valves for Flow Control in Self-Driven Microfluidics. *Micromachines* **2013**, *4*, 1–8. [CrossRef]
79. Zhou, S.; Li, M.; Tang, Q.; Song, Z.; Tong, Y.; Liu, Y. Deposition of Pentacene Thin Film on Polydimethylsiloxane Elastic Dielectric Layer for Flexible Thin-Film Transistors. *IEEE Electron Device Lett.* **2017**, *38*, 1031–1034. [CrossRef]
80. Feng, S.; Liu, G.; Jiang, L.; Zhu, Y.; Goldys, E.; Inglis, D. A Microfluidic Needle For Sampling And Delivery Of Chemical Signals By Segmented Flows. *Appl. Phys. Lett.* **2017**, *111*, 183702. [CrossRef]
81. Shirtcliffe, N.; McHale, G.; Newton, M.; Perry, C. Intrinsically superhydrophobic organosilica sol–gel foams. *Langmuir* **2003**, *19*, 5626–5631. [CrossRef]
82. Rao, A.V.; Latthe, S.S.; Nadargi, D.Y.; Hirashima, H.; Ganesan, V. Preparation of MTMS based transparent superhydrophobic silica films by sol–gel method. *J. Colloid Interface Sci.* **2009**, *332*, 484–490.
83. Basu, B.J.; Hariprakash, V.; Aruna, S.; Lakshmi, R.; Manasa, J.; Shruthi, B. Effect of microstructure and surface roughness on the wettability of superhydrophobic sol–gel nanocomposite coatings. *J. Sol-Gel Sci. Technol.* **2010**, *56*, 278–286. [CrossRef]
84. Ogihara, H.; Katayama, T.; Saji, T. One-step electrophoretic deposition for the preparation of superhydrophobic silica particle/trimethylsiloxysilicate composite coatings. *J. Colloid Interface Sci.* **2011**, *362*, 560–566. [CrossRef]
85. Mahadik, S.A.; Mahadik, D.; Kavale, M.; Parale, V.; Wagh, P.; Barshilia, H.C.; Gupta, S.C.; Hegde, N.; Rao, A.V. Thermally stable and transparent superhydrophobic sol–gel coatings by spray method. *J. Sol-Gel Sci. Technol.* **2012**, *63*, 580–586. [CrossRef]
86. Graham, P.; Stone, M.; Thorpe, A.; Nevell, T.G.; Tsibouklis, J. Fluoropolymers with very low surface energy characteristics. *J. Fluor. Chem.* **2000**, *104*, 29–36. [CrossRef]
87. Thorpe, A.A.; Peters, V.; Smith, J.R.; Nevell, T.G.; Tsibouklis, J. Poly(methylpropenoxyfluoroalkylsiloxane)s: A class of fluoropolymers capable of inhibiting bacterial adhesion onto surfaces. *J. Fluor. Chem.* **2000**, *104*, 37–45. [CrossRef]
88. Yarosh, A.A.; Krukovsky, S.P.; Pryakhina, T.A.; Kotov, V.M.; Zavin, B.G.; Sakharov, A.M. Synthesis of water- and oil-repellent organofluorosilicon compounds. *Mendeleev Commun.* **2006**, *16*, 190–192. [CrossRef]
89. Li, X.-M.; Reinhoudt, D.; Crego-Calama, M. What do we need for a superhydrophobic surface? A review on the recent progress in the preparation of superhydrophobic surfaces. *Chem. Soc. Rev.* **2007**, *36*, 1350–1368. [CrossRef]
90. Zhang, Q.; He, M.; Chen, J.; Wang, J.; Song, Y.; Jiang, L. Anti-icing surfaces based on enhanced self-propelled jumping of condensed water microdroplets. *Chem. Commun.* **2013**, *49*, 4516–4518. [CrossRef]

91. Cho, S.J.; An, T.; Kim, J.Y.; Sung, J.; Lim, G. Superhydrophobic nanostructured silicon surfaces with controllable broadband reflectance. *Chem. Commun.* **2011**, *47*, 6108–6110. [CrossRef]

92. Ali, U.; Abd Karim, K.J.B.; Buang, N.A. A Review of the Properties and Applications of Poly (Methyl Methacrylate) (PMMA). *Polym. Rev.* **2015**, *55*, 678–705. [CrossRef]

93. Efimenko, K.; Wallace, W.E.; Genzer, J. Surface modification of Sylgard-184 poly(dimethyl siloxane) networks by ultraviolet and ultraviolet/ozone treatment. *J. Colloid Interface Sci.* **2002**, *254*, 306–315. [CrossRef]

94. Berdichevsky, Y.; Khandurina, J.; Guttman, A.; Lo, Y.H. UV/ozone modification of poly(dimethylsiloxane) microfluidic channels. *Sens. Actuators B-Chem.* **2004**, *97*, 402–408. [CrossRef]

95. Chen, H.-Y.; McClelland, A.A.; Chen, Z.; Lahann, J. Solventless adhesive bonding using reactive polymer coatings. *Anal. Chem.* **2008**, *80*, 4119–4124. [CrossRef] [PubMed]

96. Chen, H.Y.; Lahann, J. Fabrication of discontinuous surface patterns within microfluidic channels using photodefinable vapor-based polymer coatings. *Anal. Chem.* **2005**, *77*, 6909–6914. [CrossRef] [PubMed]

97. Hansen, R.H.; Timm, A.C.; Timm, C.M.; Bible, A.N.; Morrell-Falvey, J.L.; Pelletier, D.A.; Simpson, M.L.; Doktycz, M.J.; Retterer, S.T. Stochastic Assembly of Bacteria in Microwell Arrays Reveals the Importance of Confinement in Community Development. *PLoS ONE* **2016**, *11*, e0155080. [CrossRef]

98. Zandi Shafagh, R.; Decrop, D.; Ven, K.; Vanderbeke, A.; Hanusa, R.; Breukers, J.; Pardon, G.; Haraldsson, T.; Lammertyn, J.; van der Wijngaart, W. Reaction injection molding of hydrophilic-in-hydrophobic femtolitre-well arrays. *Microsyst. Nanoeng.* **2019**, *5*, 25. [CrossRef]

99. Feng, S.; Nguyen, M.N.; Inglis, D.W. Microfluidic Droplet Extraction by Hydrophilic Membrane. *Micromachines* **2017**, *8*, 331. [CrossRef]

100. Mohammed, M.I.; Abraham, E.; Desmulliez, M.P.Y. Rapid laser prototyping of valves for microfluidic autonomous systems. *J. Micromech. Microeng.* **2013**, *23*, 035034. [CrossRef]

101. Weigl, B.; Klein, G. Surface Tension Valves for Microfluidic Applications. International Patent WO0190614, 13 June 2002.

102. Wang, M.Y.; Wan, Z.; Surangalikar, H.; Wu, G.; Ata, E. Micromachined Electrowetting Microfluidic Valve. U.S. Patent US20080257438, 23 October 2008.

103. Field, L.A.; Schiaffino, S.; Barth, P.W.; Hoen, S.T.; Kawamura, N.A.; Donald, D.K.; Robertson, C.R.; Servaites, J.D. Bubble Valve and Bubble Valve-Based Pressure Regulator. U.S. Patent US6062681, 16 May 2000.

104. Krulevitch, P.A.; Benett, W.J.; Rose, K.A.; Hamilton, J.; Maghribi, M. Low Power Integrated Pumping and Valving Arrays for Microfluidic Systems. International Patent WO03027508, 3 April 2003.

105. Larsson, O.; Tiensuu, A.-L. Hydrophobic Barriers. U.S. Patent US2007059216, 15 March 2007.

106. Lee, L.; Lu, C.; Juang, Y.-J. Valve for Microfluidic Chips. U.S. Patent US2007113908, 24 May 2007.

107. Kozicki, M.N. Programmable Surface Control Devices and Method of Making Same. European Patent EP1440485, 21 June 2006.

108. Shartle, R.; Besemer, D.; Gorin, M. Capillary Stop-Flow Junction Having Improved Stability against Accidental Fluid Flow. U.S. Patent US5230866, 27 July 1993.

109. Banerjee, D.; Faulstich, K.; Lau, A.; Ulmanella, U.; Xie, J. Device Including a Dissolvable Structure for Flow Control. U.S. Patent US2006093528, 4 May 2006.

110. Chung, K.H.; Ko, J.S.; Yoon, H.C.; Yang, H.S.; Pyo, H.B.; Kim, S.J.; Kim, Y.T. Device for Controlling Fluid Using Surface Tension. U.S. Patent US7445754, 4 November 2008.

111. Gerhardt, G.C.; Bouvier, E.S.; Dourdeville, T. Fluid Flow Control Freeze/Thaw Valve for Narrow Bore Capillaries or Microfluidic Devices. European Patent EP1446601, 26 March 2008.

112. Zimmermann, M.; Hunziker, P.; Delamarche, E. Valves for autonomous capillary systems. *Microfluid. Nanofluid.* **2008**, *5*, 395–402. [CrossRef]

113. Takei, G.; Nonogi, M.; Hibara, A.; Kitamori, T.; Kim, H.-B. Tuning microchannel wettability and fabrication of multiple-step Laplace valves. *Lab Chip* **2007**, *7*, 596–602. [CrossRef]

114. Ellinas, K.; Tserepi, A.; Gogolides, E. Superhydrophobic, passive microvalves with controllable opening threshold: Exploiting plasma nanotextured microfluidics for a programmable flow switchboard. *Microfluid. Nanofluid.* **2014**, *17*, 489–498. [CrossRef]

115. Olanrewaju, A.O.; Robillard, A.; Dagher, M.; Juncker, D. Autonomous microfluidic capillaric circuits replicated from 3D-printed molds. *Lab Chip* **2016**, *16*, 3804–3814. [CrossRef] [PubMed]

116. Safavieh, R.; Juncker, D. Capillarics: Pre-programmed, self-powered microfluidic circuits built from capillary elements. *Lab Chip* **2013**, *13*, 4180–4189. [CrossRef] [PubMed]

117. Chen, X.; Chen, S.; Zhang, Y.; Yang, H. Study on Functionality and Surface Modification of a Stair-Step Liquid-Triggered Valve for On-Chip Flow Control. *Micromachines* **2020**, *11*, 690. [CrossRef] [PubMed]

118. Olanrewaju, A.O.; Ng, A.; DeCorwin-Martin, P.; Robillard, A.; Juncker, D. Microfluidic Capillaric Circuit for Rapid and Facile Bacteria Detection. *Anal. Chem.* **2017**, *89*, 6846–6853. [CrossRef]

119. Hagmeyer, B.; Zechnall, F.; Stelzle, M. Towards plug and play filling of microfluidic devices by utilizing networks of capillary stop valves. *Biomicrofluidics* **2014**, *8*, 056501. [CrossRef]

120. Peng, X.Y. A micro surface tension pump (MISPU) in a glass microchip. *Lab Chip* **2011**, *11*, 132–138. [CrossRef]

121. Peng, X.Y. A One-Square-Millimeter Compact Hollow Structure for Microfluidic Pumping on an All-Glass Chip. *Micromachines* **2016**, *7*, 63. [CrossRef]

122. Postek, W.; Kaminski, T.S.; Garstecki, P. A passive microfluidic system based on step emulsification allows the generation of libraries of nanoliter-sized droplets from microliter droplets of varying and known concentrations of a sample. *Lab Chip* **2017**, *17*, 1323–1331. [CrossRef]

123. Focke, M.; Stumpf, F.; Faltin, B.; Reith, P.; Bamarni, D.; Wadle, S.; Mueller, C.; Reinecke, H.; Schrenzel, J.; Francois, P.; et al. Microstructuring of polymer films for sensitive genotyping by real-time PCR on a centrifugal microfluidic platform. *Lab Chip* **2010**, *10*, 2519–2526. [CrossRef]

124. Lutz, S.; Weber, P.; Focke, M.; Faltin, B.; Hoffmann, J.; Mueller, C.; Mark, D.; Roth, G.; Munday, P.; Armes, N.; et al. Microfluidic lab-on-a-foil for nucleic acid analysis based on isothermal recombinase polymerase amplification (RPA). *Lab Chip* **2010**, *10*, 887–893. [CrossRef]

125. Zhu, P.; Wang, L. Passive and active droplet generation with microfluidics: A review. *Lab Chip* **2017**, *17*, 34–75. [CrossRef] [PubMed]

126. Mittal, N.; Cohen, C.; Bibette, J.; Bremond, N. Dynamics of step-emulsification: From a single to a collection of emulsion droplet generators. *Phys. Fluids* **2014**, *26*, 082109. [CrossRef]

127. Amstad, E.; Chemama, M.; Eggersdorfer, M.; Arriaga, L.; Brenner, M.; Weitz, D. Robust scalable high throughput production of monodisperse drops. *Lab Chip* **2016**, *16*, 4163–4172. [CrossRef] [PubMed]

128. Dangla, R.; Kayi, S.C.; Baroud, C.N. Droplet microfluidics driven by gradients of confinement. *Proc. Natl. Acad. Sci. USA* **2013**, *110*, 853–858. [CrossRef] [PubMed]

129. Burger, R.; Kirby, D.; Glynn, M.; Nwankire, C.; O'Sullivan, M.; Siegrist, J.; Kinahan, D.; Aguirre, G.; Kijanka, G.; Gorkin, R.A.; et al. Centrifugal microfluidics for cell analysis. *Curr. Opin. Chem. Biol.* **2012**, *16*, 409–414. [CrossRef]

130. Mark, D.; Haeberle, S.; Roth, G.; von Stetten, F.; Zengerle, R. Microfluidic lab-on-a-chip platforms: Requirements, characteristics and applications. *Chem. Soc. Rev.* **2010**, *39*, 1153–1182. [CrossRef]

131. Mark, D.; Weber, P.; Lutz, S.; Focke, M.; Zengerle, R.; von Stetten, F. Aliquoting on the centrifugal microfluidic platform based on centrifugo-pneumatic valves. *Microfluid. Nanofluid.* **2011**, *10*, 1279–1288. [CrossRef]

132. Arshavsky-Graham, S.; Segal, E. Lab-on-a-Chip Devices for Point-of-Care Medical Diagnostics. *Adv. Biochem. Eng./Biotechnol.* **2020**, 1–19. [CrossRef]

133. Kim, J.; Kido, H.; Rangel, R.H.; Madou, M.J. Passive flow switching valves on a centrifugal microfluidic platform. *Sens. Actuators B-Chem.* **2008**, *128*, 613–621. [CrossRef]

134. Kazemzadeh, A.; Ganesan, P.; Ibrahim, F.; Aeinehvand, M.M.; Kulinsky, L.; Madou, M.J. Gating valve on spinning microfluidic platforms: A flow switch/control concept. *Sens. Actuators B-Chem.* **2014**, *204*, 149–158. [CrossRef]

135. Eriksen, J.; Bilenberg, B.; Kristensen, A.; Marie, R. Optothermally actuated capillary burst valve. *Rev. Sci. Instrum.* **2017**, *88*, 045101. [CrossRef] [PubMed]

136. Xu, W.; Wu, L.L.; Zhang, Y.; Xue, H.; Li, G.-P.; Bachman, M. A vapor based microfluidic flow regulator. *Sens. Actuators B-Chem.* **2009**, *142*, 355–361. [CrossRef] [PubMed]

137. Madou, M.; Zoval, J.; Jia, G.; Kido, H.; Kim, J.; Kim, N. Lab on a CD. *Annu. Rev. Biomed. Eng.* **2006**, *8*, 601–628. [CrossRef]

138. Wu, H.-C.; Shih, C.-H.; Chen, W.-H.; Lin, M.-H. Design and Analysis of a Robust Sequential Flow Control Using Burst Valves on the Centrifugal Platform. *J. Nanosci. Nanotechnol.* **2016**, *16*, 12602–12608. [CrossRef]

139. Madou, M.J.; Lee, L.J.; Daunert, S.; Lai, S.; Shih, C.-H. Design and Fabrication of CD-like Microfluidic Platforms for Diagnostics: Microfluidic Functions. *Biomed. Microdevices* **2001**, *3*, 245–254. [CrossRef]

140. Zoval, J.V.; Madou, M.J. Centrifuge-based fluidic platforms. *Proc. IEEE* **2004**, *92*, 140–153. [CrossRef]

141. Leu, T.S.; Chang, P.Y. Pressure barrier of capillary stop valves in micro sample separators. *Sens. Actuators A-Phys.* **2004**, *115*, 508–515. [CrossRef]

142. Londe, G.; Chunder, A.; Wesser, A.; Zhai, L.; Cho, H.J. Microfluidic valves based on superhydrophobic nanostructures and switchable thermosensitive surface for lab-on-a-chip (LOC) systems. *Sens. Actuators B-Chem.* **2008**, *132*, 431–438. [CrossRef]

143. Li, L.; Westerbeek, E.Y.; Vollenbroek, J.C.; de Beer, S.; Shui, L.; Odijk, M.; Eijkel, J.C.T. Autonomous capillary microfluidic devices with constant flow rate and temperature-controlled valving. *Soft Matter.* **2021**. [CrossRef]

144. Bouaidat, S.; Hansen, O.; Bruus, H.; Berendsen, C.; Bau-Madsen, N.K.; Thomsen, P.; Wolff, A.; Jonsmann, J. Surface-directed capillary system; theory, experiments and applications. *Lab Chip* **2005**, *5*, 827–836. [CrossRef]

145. Lam, P.; Wynne, K.J.; Wnek, G.E. Surface-tension-confined microfluidics. *Langmuir* **2002**, *18*, 948–951. [CrossRef]

146. You, I.; Yun, N.; Lee, H. Surface-Tension-Confined Microfluidics and Their Applications. *ChemPhysChem* **2013**, *14*, 471–481. [CrossRef] [PubMed]

147. Wang, L.; Niu, C.; Zhang, C.; Wang, Z. Steady flow of pressure-driven water-in-oil droplets in closed-open-closed microchannels. *AIP Adv.* **2019**, *9*, 125040. [CrossRef]

148. Papadimitriou, V.A.; Segerink, L.I.; van den Berg, A.; Eijkel, J.C.T. 3D capillary stop valves for versatile patterning inside microfluidic chips. *Anal. Chim. Acta* **2018**, *1000*, 232–238. [CrossRef] [PubMed]

149. Lai, H.-H.; Xu, W.; Allbritton, N.L. Use of a virtual wall valve in polydimethylsiloxane microfluidic devices for bioanalytical applications. *Biomicrofluidics* **2011**, *5*, 024105. [CrossRef]

150. Choi, J.; Kang, D.; Han, S.; Kim, S.B.; Rogers, J.A. Thin, Soft, Skin-Mounted Microfluidic Networks with Capillary Bursting Valves for Chrono-Sampling of Sweat. *Adv. Healthc. Mater.* **2017**, *6*, 1601355. [CrossRef] [PubMed]

151. Hou, X.; Li, J.; Tesler, A.B.; Yao, Y.; Wang, M.; Min, L.; Sheng, Z.; Aizenberg, J. Dynamic air/liquid pockets for guiding microscale flow. *Nat. Commun.* **2018**, *9*, 733. [CrossRef]
152. Feng, S.; Clement, S.; Zhu, Y.; Goldys, E.M.; Inglis, D.W. Microfabricated needle for hydrogen peroxide detection. *RSC Adv.* **2019**, *9*, 18176–18181. [CrossRef]
153. Feng, S.; Shirani, E.; Inglis, D.W. Droplets for Sampling and Transport of Chemical Signals in Biosensing: A Review. *Biosensors* **2019**, *9*, 80. [CrossRef]

Review

Recent Progress in Wearable Biosensors:
From Healthcare Monitoring to Sports Analytics

Shun Ye [1,2,3,†]**, Shilun Feng** [4,5,†]**, Liang Huang** [6] **and Shengtai Bian** [1,*]

1   Microfluidics Research & Innovation Laboratory, School of Sport Science, Beijing Sport University, Beijing 100084, China; sjy5271@psu.edu
2   Biomedical Engineering Department, College of Engineering, Pennsylvania State University, University Park, PA 16802, USA
3   State Key Laboratory of Microbial Resources, Institute of Microbiology, Chinese Academy of Sciences, Beijing 100101, China
4   State Key Laboratory of Transducer Technology, Shanghai Institute of Microsystem and Information Technology, Chinese Academy of Sciences, Shanghai 200050, China; shilun.feng@ntu.edu.sg
5   School of Electrical and Electronic Engineering, Nanyang Technological University, Singapore 639798, Singapore
6   School of Instrument Science and Opto–Electronics Engineering, Hefei University of Technology, Hefei 230009, China; lianghuang@hfut.edu.cn
*   Correspondence: stbian@bsu.edu.cn; Tel.: +86-188-1085-9809
†   These authors contributed equally to this work.

Received: 9 November 2020; Accepted: 13 December 2020; Published: 15 December 2020

Abstract: Recent advances in lab-on-a-chip technology establish solid foundations for wearable biosensors. These newly emerging wearable biosensors are capable of non-invasive, continuous monitoring by miniaturization of electronics and integration with microfluidics. The advent of flexible electronics, biochemical sensors, soft microfluidics, and pain-free microneedles have created new generations of wearable biosensors that explore brand-new avenues to interface with the human epidermis for monitoring physiological status. However, these devices are relatively underexplored for sports monitoring and analytics, which may be largely facilitated by the recent emergence of wearable biosensors characterized by real-time, non-invasive, and non-irritating sensing capacities. Here, we present a systematic review of wearable biosensing technologies with a focus on materials and fabrication strategies, sampling modalities, sensing modalities, as well as key analytes and wearable biosensing platforms for healthcare and sports monitoring with an emphasis on sweat and interstitial fluid biosensing. This review concludes with a summary of unresolved challenges and opportunities for future researchers interested in these technologies. With an in-depth understanding of the state-of-the-art wearable biosensing technologies, wearable biosensors for sports analytics would have a significant impact on the rapidly growing field—microfluidics for biosensing.

Keywords: wearable biosensors; biomedical microfluidics; healthcare monitoring; sports analytics

---

1. Introduction

Miniaturization of laboratory apparatus into microscale devices is a promising technology called lab-on-a-chip (LOC) [1]. About 30 years ago the concept of micro total analysis systems (μTAS) emerged from the field of semiconductor fabrication and was enhanced by microelectromechanical systems (MEMS) technologies [2–4]. The μTAS concept is to shrink an entire analytical procedure, such as cell sorting, single-cell capture, captured-cell transport, cell lysis, and intracellular analysis, into a miniaturized multifunctional chip [5–7], and nowadays its well-known synonym is called lab-on-a-chip (LOC) [3,4]. This growing field has garnered considerable attention since scaled-down biochemical analysis has several key advantages over both conventional and current laboratory benchtop methods [3,5]. These advantages are consistently demonstrated in clinical medicine, engineering, biology, and life science, etc., for example, to expedite the experimental process by embracing automation and parallelization [1,8,9]; to lower the cost by reducing the volume of expensive reagents [1,5,10,11]; to yield better interpretation of experimental results by gleaning vital information at cellular even molecular levels [12–14].

Interest in device miniaturization [15–17], combined with advances in bio-microfabrication and enabling materials [18], is motivating various microfluidic methods in which microchips can be mass-manufactured at extremely low cost via polymers (e.g., polydimethylsiloxane, PDMS) and soft lithography for microfabrication [5,19]. Microfluidics is the science of microscale devices that process and manipulate extremely low (10^{-9} to 10^{-18} L) amounts of fluids in microchannels with dimensions of tens of micrometers [10]. Conventional macroscale experimental technologies meet difficulties to deal with such low amounts of fluids, impeding their development in various fields. Conversely, microfluidic technologies begin to address numerous tough challenges, because fluid phenomena at the microscale are dramatically different from those at the macroscale [3]. For instance, capillary forces and surface tension are more dominant than gravitational forces [3], allowing for passively pumping fluids in opposition to gravity [20]. Flows at the microscale are laminar instead of turbulent, resulting in more predictable liquid handling and diffusion kinetics [5]. Based on the different phenomena behaving at the microscale, microfluidic technologies offer a sensitive, predictable, and controllable avenue for bioanalysis [21].

Despite all the attractive capacities of LOC/microfluidics devices that have enabled the widespread implementation of microchip-based systems in biology and life science [3–5,22], microfluidic technologies often only improve the performance of existing macroscale assays or provide equivalent alternatives [13]. Conversely, they have not reached their full potential due to the lack of essentially new capacities [3]. In recent years, however, LOC/microfluidics technologies begin to address some problems that have not yet been solved by current laboratory benchtop methods. An excellent example can be found in wearable/ambulatory healthcare monitoring and sports analytics harnessing skin-interfaced wearable biosensors [15,23]. Although this field is still in its infancy, the fundamentals of it are exceptionally strong: in the past decade, the wearable LOC devices gradually integrated with well-established techniques, including biocompatible materials [24,25], flexible electronics [26–30], optical/electrochemical sensors [14,26,31,32], microfluidics [21,33–35], near-field communications (NFC) [36], pain-free microneedles [37–40], as well as big data and cloud computing [14,41,42].

These above-mentioned enabling techniques establish the foundations for a new generation of wearable biosensors that directly interfaced with the human epidermis instead of rigid packages embedded in wrist straps or bands [23,43–45]. The distinguishing characteristics of the emerging wearable biosensors, lightweight, flexibility, and portability [31,36,46], have made them especially suitable for point-of-care testing (POCT). Therefore, brand-new wearable biosensors capable of real-time physiological monitoring quickly emerge, as shown in Figure 1. However, these wearable biosensors are mainly designed for health monitoring [15,34,41,45,47,48], especially, some of them are only developed to measure the physical strain/stress bending change [25,49,50]. Although many wearable devices have been deployed in sports, they are used to monitoring biophysical markers [23], such as movement [51] and cardiovascular information (e.g., blood oxygenation) [26,52,53].

Conversely, wearable biosensors are relatively underutilized resources to gain biochemical insights about athletes' health status. The POCT microfluidic devices have been used to analyze biofluids, such as urine [31,32,54], tears [46,55,56], saliva [57–60], and sweat [27,43,61,62]. However, for the applicability of wearable biosensing, not all biofluids show on-field potential.

Biofluids for physiological/pathological information acquisition are listed as follows. Their suitability for wearable biosensing is also discussed.

1. Sweat: Sweat is a biofluid produced and excreted by the eccrine glands within the epidermis, containing numerous important biomarkers that could be detected and

assessed through the non-invasive collection and biochemical analysis [27,43,45,61,62]. For example, wearable microfluidic/electrochemical sweat biosensors are able to detect various electrolytes, such as sodium ions (Na^+) [14,61,63–67], potassium ions (K^+) [14,63], calcium ions (Ca^{2+}) [32], and ammonium (NH_4^+) [68], multiple metabolites, such as glucose [14,37,69–72], lactate [14,21,63,73–75], several heavy metal species, such as zinc, iron, copper, and magnesium [61,76], and drug contents, such as Levodopa [77] (more sweat analytes can be found in Section 5.1 and Table 1). As such, sweat contains a wealth of physiologically relevant information [41,43,45] that can reflect hydration [21,27,33,74], electrolyte balance [63], exercise intensity [78], renal function [34,79], etc. For diagnostic use, wearable sweat biosensors are used to diagnose cystic fibrosis (CF), liver diseases, kidney disorders, as well as to monitor stress levels by measuring the cortisol concentrations in sweat [34,43,80]. Nonetheless, wearable biosensors capable of real-time biochemical analysis are still at an early stage of development. For instance, challenges still exist in extracting and calibrating the concentration of biomarkers in sweat due to regional variations and individual hydration status. Besides, other daunting challenges will be discussed in Section 6. In this paper, an in-depth overview of wearable sweat biosensing will be presented since sweat is the most well-studied analyte source among all six typical biofluids in the wearable biosensing field.

2. Interstitial fluid (ISF): ISF is an attractive biofluid presented in the human dermis, and has rich analytes from blood, especially through capillaries. Due to the ease of fluid exchange, lots of analytes have near levels of concentration between ISF and blood [81]. The microneedle patches have many diagnostic applications via ISF manipulation [82]. The MNs are the miniaturized replica of hypodermic needles aimed at minimally-invasive transdermal ISF biosensing [37,83] without blood sampling, which will be briefly reviewed in Section 5.3.1.

3. Saliva: Saliva has been recognized as an alternative to blood, too [57–60]. Non-invasive analysis of fluoride (F^-), Na^+, pH, and uric acid has been demonstrated [84]. The mouthguard platforms combined with electrochemical biosensors for monitoring were developed for wearable salivary monitoring of metabolites [85]. However, salivary monitoring may be affected by huge amounts of microbes (e.g., bacteria) in the oral cavity that could cause specimen contamination [86].

4. Tears: Recently, the contact lens has attracted considerable interest as a platform for in situ biosensing of tear fluid [46,56]. Research shows that glucose, Na^+, and K^+ can be found in tears and tear fluid is less complicated than blood due to the presence of the blood-tear barrier [55]. Unfortunately, tear fluid has a relatively low potential for wearable biosensing because of the low diversity of analytes and a strong reliance on proximal wireless power delivery.

5. Urine: Urinalysis has been widely used as a means to monitor the overall health status and screen various diseases due to ease of non-invasive collection and relatively large amounts [31,32,54,85]. Despite these advantages, urinalysis still has difficulties performing on-field wearable biosensing due to the difficulty of calibration of concentration levels of analytes levels, because they are strongly related to individuals' hydration levels.

6. Blood: It is known that blood analysis is usually not suitable for wearable sensing due to its invasive nature, skin-piercing for blood sampling. Only several recent studies demonstrate its utility in wearable biosensing. These studies harnessed fingernail-mounting optoelectronic biosensors to in situ continuous monitoring of vital signs [26,52,53]. A representative example [26] will be discussed in Section 4.2.4.

Overall, although the underlying values of wearable biosensing still need to be proven and demonstrated, sweat and ISF biosensing show promise that may not be recognized by researchers. Skin offers a non-invasive (sweat) and minimally-invasive (ISF) diagnostic interface rich with biological insights from our inner body [21,47]. Thus, to demonstrate the potential value of wearable biosensors for health-related and sports analytics by reviewing recent progress is required. Besides, due to the space limitations, the wearable biosensors for interrogating ISF [37,83], saliva [58], and blood [26] are discussed in a limited manner. Wearable biosensing modalities for tracking non-sweat biofluids (tears and urine) are not included, they could be found in recent reviews [15,44,81].

In this review, the latest developments in wearable biosensors will be reviewed in detail, from materials and fabrication strategies, sampling modalities, sensing modalities to key analytes and wearable biosensing platforms for health and sports monitoring with an emphasis on sweat and ISF biosensing. Afterwards, the unsolved challenges for future researchers specializing in diverse fields (e.g., biomedical/electronic engineering, biochemistry, healthcare, pharmaceutics, sports science, etc.) will also be briefly discussed. A concluding section summarizes the trend towards wearable biosensing for sports analytics and forecasts promising breakthroughs in the rapidly growing field—microfluidics for biosensing.

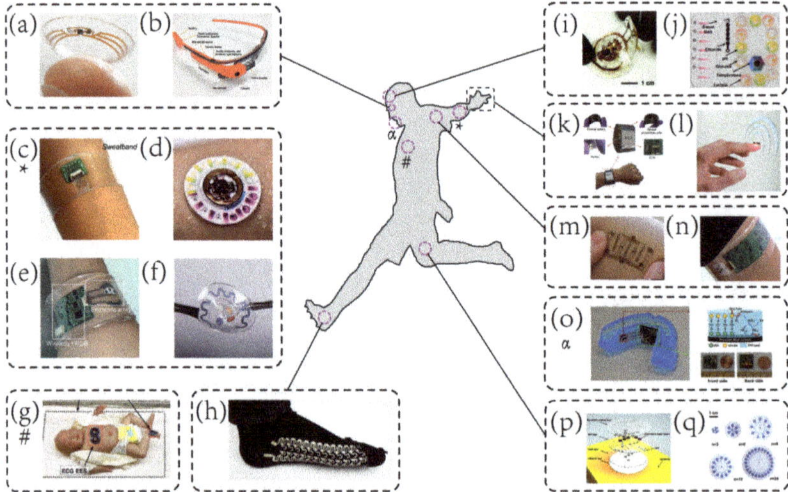

Figure 1. Representative examples of wearable biosensors for both healthcare and sports monitoring. (**a**) Contact lens sensors in ocular diagnostics [46]. Copyright 2015, Wiley. (**b**) Google glass for immunochromatographic diagnostic test analysis [87]. Copyright 2014, American Chemical Society. (**c**) A wearable microsensor array for multiplexed heavy metal monitoring [31]. Copyright 2016, American Chemical Society. (**d**) A hybrid sensor for simultaneous electrochemical, colorimetric, and volumetric analysis of sweat [73]. Copyright 2019, American Association for Advancement of Science. (**e**) A wearable sensor for autonomous sweat extraction and analysis [69]. Copyright 2017, National Academy of Sciences of United States of America (NAS). (**f**) A microfluidic device for colorimetric sensing of sweat [43]. Copyright 2018, American Association for Advancement of Science. (**g**) Binodal, wireless epidermal electronic systems with in-sensor analytics for neonatal intensive care [88]. Copyright 2019, American Association for Advancement of Science. (**h**) Wearable textile-based self-powered sensors [89]. Copyright 2016, Royal Society of Chemistry. (**i**) A microfluidic system for real-time tracking of sweat loss and electrolyte composition [27]. Copyright 2018, Wiley. (**j**) A microfluidic system for colorimetric analysis of sweat biomarkers and temperature [74]. Copyright 2019, American Chemical Society. (**k**) A smartwatch for continuous sweat glucose monitoring [72]. Copyright 2019, American Chemical Society. (**l**) A miniaturized battery-free wireless sensor for wearable pulse oximetry [26]. Copyright 2017, Wiley. (**m**) An epidermal stimulation and sensing platform for sensorimotor prosthetic control, management of lower back exertion, and electrical muscle activation [90]. Copyright 2016, Wiley. (**n**) A wearable electrochemical sensor for noninvasive simultaneous monitoring of Ca^{2+} and pH [32]. Copyright 2016, American Chemical Society. (**o**) A wearable salivary uric acid mouthguard sensor [58]. Copyright 2015, Elsevier. (**p**) A microfluidic system for time-sequenced discrete sampling and chloride analysis [91]. Copyright 2018, Wiley. (**q**) Skin-mounted microfluidic networks for chrono-sampling of sweat [63]. Copyright 2017, Wiley.

2. Materials and Fabrication Strategies

2.1. Materials

Materials for Substrate: the softness and attachability of the substrate are the main concerns for the stretchability of wearable biosensors, and directly determine the comfort of the biosensor and the reliability of long-term monitoring. Common base materials are organic materials such as polymers, silicones, and rubbers. For example, polydimethylsiloxane (PDMS) has inherent high stretchability, non-toxic, hydrophobic, and good workability. It has been used in microfluidics, prostheses, and wearable biosensors [92]. Polyethylene terephthalate (PET) is characterized by lightweight and can serve as a flexible substrate for ink-jet printed electronics [93]. Polyimide (PI) film is another commonly used substrate. It can maintain flexibility, creep resistance, and tensile strength under high temperature (up to 360 °C) and acid/alkali conditions [94]. Cellulose paper is a flexible, porous, inexpensive, recyclable, biodegradable, and biocompatible substrate material, and it has been widely used in test strips for medical diagnosis [95,96].

For attachability, substrates for skin-mounted biosensors need to meet strict requirements, including the wearable power source, prolonged wear time, and facile and painless removal after use without leaving residues. Meanwhile, they should be safe and minimize skin irritation and damage, such as epidermal stripping, blisters, skin tears, irritant contact dermatitis, allergic dermatitis, maceration, and folliculitis [97,98]. Polymeric materials, including acrylics, hydrocolloids, hydrogels, polyurethanes, rubbers, and silicones, have been used as the adhesive for skin-contact applications. Adhesives can be categorized into structural adhesives and pressure-sensitive adhesives (PSA). Materials for structural adhesives include cyanoacrylates, urethanes, epoxies, and acrylics, which lend themselves to structural applications. PSAs should deform easily under small pressures and can adhere to wet surfaces [99]. Aggressive adhesives based on acrylics, hydrocolloids, hydrogels, natural/synthetic rubbers, and polyurethanes can cause skin trauma (such as skin stripping, maceration, and allergic reactions), as well as leave residue on the skin [100]. Besides, soft silicone-based adhesives have good biocompatibility, temperature stability, chemical inertness, and environmental stability. These features make silicone-based adhesives suitable for many medical applications, including tapes, wound dressings for wearable devices, and transdermal drug delivery applications [101,102].

Materials for active element: (a) Carbon materials have been widely used in wearable sensors, and mainly include graphite, carbon nanotube (CNT), and graphene. Graphite is often used to form pencil—paper electrode, drawing patterns on the substrate by pencil painting [103]. CNT is a carbon material with extraordinary conductivity and mechanical strength [104,105]. Although a single CNT has a high sensitivity to strain, it is constructed and used on a larger scale. Therefore, CNT powders are often mixed into polymer substrates to realize wearable sensing. Compared with graphite and CNT, graphene has better electrical conductivity and mechanical properties, and has been widely used in flexible/stretchable sensors [106]. (b) Metal materials with excellent electrical conductivity have been widely used in wearable biosensors. (1) nanowires (NWs) or nanoparticles (NPs); NWs and NPs are often used as fillers to prepare piezoresistive composites and conductive ink [107,108]. For instance, AgNWs can be embedded into PDMS to build resistive-type sensors, and conductive inks with metal NPs can be cast and annealed on the substrate surface to form the electrodes for capacitive sensors. (2) Flexible or stretchable configurations with stretch structures; the flexible or stretchable configuration utilizes coiled buckled, serpentine and woven structures to make metal flexible and stretchable [109,110]. (3) Liquid metal; liquid metal utilizes its liquid state at room temperature and combines with microfluidic technology to achieve electrode fabrication [111,112]. (c) Polymers have been widely explored in sensing elements for their thermal stability, high transparency, and tunable conductivity, such as poly 3,4-ethylene dioxythiophene (PEDOT) [113] and polyvinylidene difluoride (PVDF) polymers [114].

2.2. Fabrication Strategies

According to the materials that can be used for flexible and stretchable sensors, the fabrication methods are categorized into compositing materials and pattern transferring. (a) Mixing different materials (e.g., carbon black NP, AgNW) into the composite is the simplest manufacturing method [115,116]. The active material is blended into the polymer by stirring, and then can be made into a dry elastic composite material in bulk or film form. Mixed composites have complex electromechanical features, relying on fillers and polymer substrates and doping concentration and distribution state. (b) Pattern transferring is the most common manufacturing method to produce the desired pattern of wearable biosensor geometries. It mainly includes microscale modeling, lithography, printing, and handwriting. Microscale modeling is used to prepare the substrates, electrodes, and sensing microstructures in composite materials. Lithography is a pattern transfer method that can realize various novel geometric shapes in flexible electronic products. Lithography technology specializes in manufacturing complex stretchable systems with high-precision dimensions, exquisite structure, and rich functions [117,118]. Printing can deposit and pattern many materials on various substrates at the same time without the need for complicated equipment and clean rooms. Screen printing is a typical printing technique that required masks [119,120]. This technology has been widely adopted to fabricate working electrodes in electrochemical sensors and the sensing elements in electromechanical sensors. The handwriting method is a technique in which various instruments are used to directly draw electrodes on the substrates, which has become an alternative for manufacturing low-cost, DIY sensors [96,121]. This technology gives end-users the ability to design and implement sensors based on "on-site, real-time" requirements. With a more complex structure, rollerball pens and fountain pens can be written with a variety of inks (including metallic inks, liquid metals, and organic mixtures) to generate controllable geometric shapes on many substrates. In addition to screen printing technology, inkjet printing technology is also an important technology that has been widely used in sensor fabrication [122,123]. Inkjet printing is a precise and fast film preparation technology, which sprays functional ink droplets onto paper or other substrates through nozzles. The materials used include carbon nanotube, graphene, conductive metal solution, polymer, and liquid metal, and other functional inks [124,125].

3. Sampling Modalities

To lay foundations for various biosensing modalities, including optical and electrochemical sensing, sampling modalities will be introduced. The majority of sweat analytical methodologies mainly adopt three sampling methods, direct contact, chamber collection, and multistep sampling. For example, paper-based methods, such as disposable test strips, have the devices directly contact the human epidermis to sample biofluids in just a few seconds. As shown in Figure 2a, Oncescu et al. [126] developed a smartphone-based biosensing system, which consisted of a disposable test strip for sweat/saliva sampling, a back compartment for eliminating the interference from ambient lighting conditions to improve the accuracy of colorimetric sensing, and a smartphone application for measuring biomarkers by digital image analysis and displaying the results to end-users.

Another sampling method concerns chamber collection. The microfluidic systems using this sampling method have microchannels for guiding the sweat to fill on-chip microchambers for biochemical/enzymatic reactions to yield measurable color/potential changes [34,61,74,127]. The chamber collection method for sweat sampling and subsequent in situ biomarker analysis has been widely adopted, thus examples and corresponding technical details will be thoroughly reviewed in the following sections (Sections 4.2.1 and 4.2.3).

The last sampling method, multistep sampling, also has its applications in microfluidic biosensing. In the first step, various microchambers are also exploited for in situ sweat collection. However, unlike most microfluidic systems conducting in situ analysis, these microchambers are specially designed for the temporary storage of sweat but not for on-chip analysis. Microfluidic valves, such as capillary bursting valves (CBVs) (Figure 2b) [63] and hydrophobic valves (HVs) (Figure 2c) [128], are exploited to carry out sweat collection and storage in a time-sequential manner by microfluidic networks with interconnected sets of microchambers. The CBVs and HVs open only until the microchambers set before them have filled with sweat, enabling passively guiding sweat through the microfluidic network sequentially. After sweat sample acquisition, there is another step to transport the sweat for ex-situ laboratory analysis. For instance, Choi et al. [63] developed a skin-interfaced microfluidic chip with CBVs for chrono-sampling and ex-situ biomarker analysis of sweat (Figure 2b). CBVs were exploited to collect sweat in a time-sequential manner. Afterwards, centrifugation and retrieval of collected sweat from chambers allowed for chemical analysis of sweat biomarkers (lactate, Na^+, and K^+) by mass spectrometry.

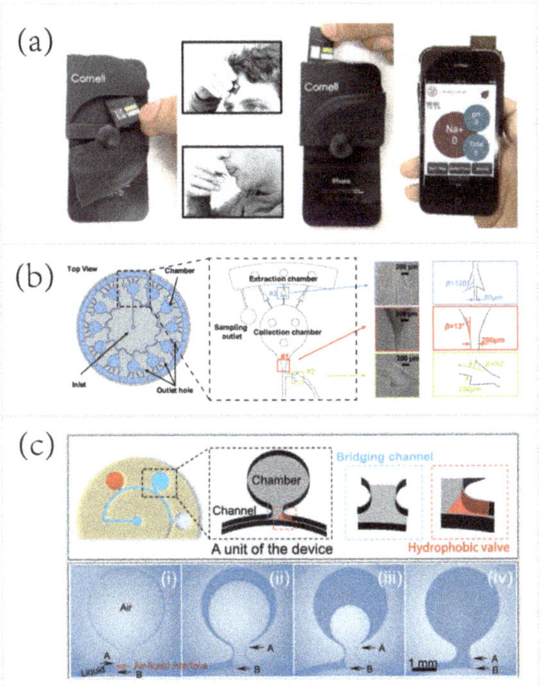

Figure 2. Representative examples of sweat sampling modalities. (**a**) A smartphone-based sweat biosensor, consisting of a disposable test strip for sweat sampling via direct contact [126]. Copyright 2013, Royal Society of Chemistry. (**b**) A microfluidic biosensor with capillary bursting valves (CBVs) for sampling sweat in a sequential manner and with an interconnected set of chambers for the temporary storage of sweat samples for subsequent ex-situ analysis of sweat biomarkers [63]. Copyright 2017, Wiley. (**c**) A skin-interfaced microfluidic biosensor with sweat collection chambers and hydrophobic valves (HVs) for chrono-sampling of sweat; (**i**)—(**iv**) A series of images during the fluid sampling for demonstrating the working principle of HVs [128]. Copyright 2020, Royal Society of Chemistry.

4. Sensing Modalities

Real-time tracking of the physiological status of patients or athletes by wearable biosensors is essential to provide insightful information about their health conditions, sports performance, cardiovascular capacities, and so on. As described previously, the progress in enabling materials and fabrication strategies further improves both mechanical features (flexible, stretchable, non-irritating) and performance (selectivity, sensitivity, durability, reproducibility, accuracy) of wearable sensors. Several microfluidic, electrochemical, and optical biosensors demonstrate greater functionality. The main two sensing modalities, including electrochemical and optical sensing, which make optimum use of these systems for wearable in situ analysis of biofluids (e.g., sweat, saliva, and blood), will be systematically reviewed in this section. Other sensing modalities will also be briefly discussed.

4.1. Electrochemical Sensing

Early developed wearable biosensors mainly focused on detecting a single parameter, such as sodium concentrations [64]. Recently, multianalyte tracking changes in biofluid chemistry becomes a widely available biosensing technique since the upsurge of the electrochemical sensing methods, including potentiometry, conductometry, and also voltammetry/amperometry [45].

4.1.1. Potentiometry

Potentiometry is a highly sensitive, selective, reproducible electrochemical sensing method that has been exploited for wearable, non-invasive monitoring of biofluids (e.g., sweat, urine, and tears). The potentiometric biosensor usually contains several ion-selective electrodes (ISEs) and a reference electrode (RE) [14]. The potential of the RE is independent of the analyte concentrations, and the potential of ISE, conversely, depends on them. Eventually, the differences between ISEs and RE are used to determine the concentrations of targeted analytes. The research group at the University of California (Berkley) developed a wearable potentiometric biosensor for in situ monitoring of Ca^{2+} and pH [32]. As shown in Figure 3a, a PET-based flexible biosensor array of ionized calcium and pH was interfaced with a flexible printed circuit board (FPCB). Measurements of the concentrations of Ca^{2+} and pH were based on ISEs, coupled with an Ag/AgCl RE. An interfacing signal conditioning circuitry was used to measure the electrical potential differences between the ISEs and a RE, that were proportional to the logarithmic concentration of respective target ions. As mentioned previously, sweat contains a wealth of electrolytes, such as Na^+, K^+, and NH_4^+. Thus, the selectivity of potentiometric biosensors is of utmost importance. To demonstrate the high selectivity of this platform, H^+, NH_4^+, Mg^{2+}, K^+, and Na^+ were added to the Ca^{2+} solution sequentially. However, as shown in Figure 3b, the potential changes due to the addition of non-targeted ions were significantly smaller than due to the Ca^{2+}. The performance test results (Figure 3b, right) also showed its high reproducibility (n = 6, RSD = 1.5%; RSD, relative standard deviation). Moreover, electrochemical sensing modalities also allow for seamless system integration, the FPCB of this biosensor integrated signal transduction, conditioning, processing, and wireless data transmission chipsets, eliminating delayed sample processing compared to traditional benchtop methods.

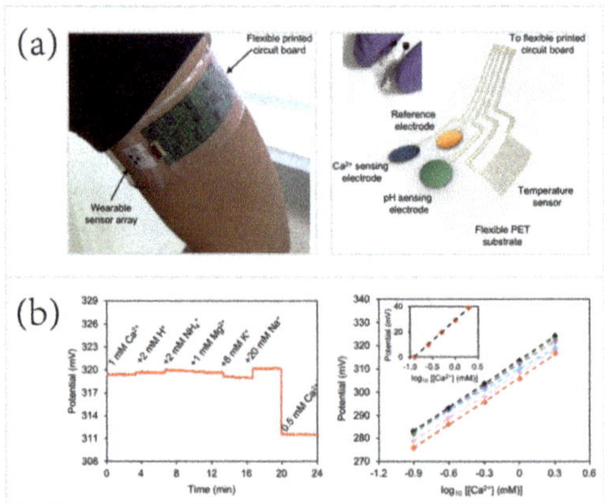

Figure 3. A fully integrated electrochemical biosensor for non-invasive in situ analysis of biofluids. (**a**) An image and a schematic illustration of a fully integrated potentiometric biosensor, consisting of Ca^{2+}, pH, and temperature sensors patterned on PET substrate (**b**) Performance test results of Ca^{2+} sensors: ion-selectivity and reproducibility (n = 6) [32]. Copyright 2016, American Chemical Society.

4.1.2. Conductometry

Conductometry is another electrochemical sensing modality, but it has not yet been applied to wearable biosensing due to the reliance on high-capacity power sources and bulky electronics [45]. Few conductometric sensors were used to analyze biofluids in the literature. Ogasawara et al. [129] reported a flexible conductimetric sensor for ocular disease diagnosis by measuring the electrical conductivity of tears. There is a significant difference ($p < 0.01$) between electrolyte concentrations of healthy individuals and keratoconjunctivitis sicca (KCS) patients, which can be used to diagnose KCS patients without ocular damage by conductometric sensing.

4.1.3. Voltammetry/Amperometry

Voltammetry and amperometry are two rapid, real-time methods for detecting a milieu of analytes, such as lactate and heavy metals [31,76,127]. Voltammetric/amperometric biosensors have at least one three-electrode configuration, which includes a working electrode (WE), an RE, and a counter electrode (CE) [45]. Especially for voltammetric sensing, a time-dependent varying voltage with respect to the RE is applied, and the response between the WE and the RE is scanned within a range to oxidize/reduce the analytes. The peak heights or amplitudes of local maxima in the measured current are exploited to determine the concentrations of the analytes. For example, Zhang et al. [127] reported a wearable electrochemical biosensor based on molecularly imprinted Ag nanowires (AgNWs) for in situ monitoring sweat lactate (Figure 4a). The majority of sweat lactate electrochemical biosensors employ enzymatic reactions for gleaning detectable electrochemical signals. However, some environmental factors (e.g., temperature) can affect the activity of enzymes.

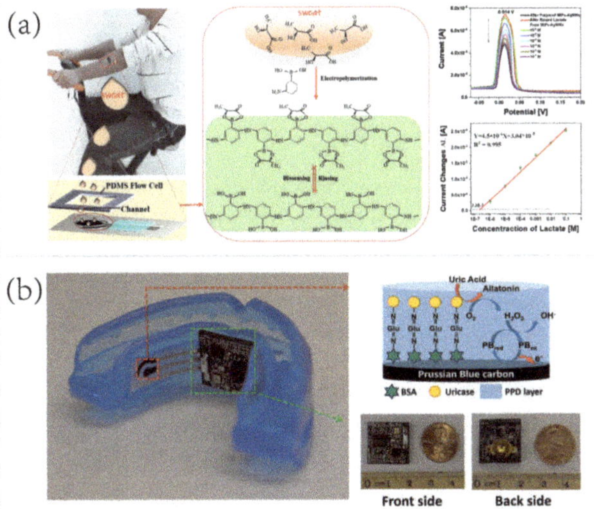

Figure 4. Wearable voltammetric/amperometric biosensors for sweat/saliva analysis. (**a**) A wearable voltammetric biosensor based on Ag nanowires (AgNWs) and molecularly imprinted polymers (MIPs) for in situ monitoring of lactate in the human sweat. A three-electrode MIPs-AgNWs biosensor mounting on the volunteer's arm; the principle of MIPs' formation and biosensing on the carbon working electrode; differential pulse voltammetry (DPV) responses and the calibration curve of MIPs-AgNWs biosensor for detection of lactate [127]. Copyright 2020, Elsevier. (**b**) A wearable amperometric mouthguard biosensor for real-time continuous salivary uric acid detection, screen printed uricase carbon working electrode, and its fabricated printed circuit board assembly (PCBA) [58]. Copyright 2015, Elsevier.

Molecularly imprinted polymers (MIPs) were introduced for in situ detection of lactate in sweat, eliminating the need for proper storage of enzymes. This novel skin-mounted biosensor consisted of a three-electrode configuration: a MIPs-AgNWs coated carbon WE for providing specific binding sites to sweat lactate molecules, an Ag/AgCl RE, and a carbon CE. As shown in Figure 4, differential pulse voltammetry (DPV) responses, especially the peak current changes, were recorded for sensing of sweat lactate at concentrations from 10^{-6} M to 0.1 M. Considerable interest in sweat lactate stems from its positive correlation to the exercise intensity [78], which can be used to evaluate the athletic performance. Thus, this AgNWs-MIPs biosensor is highly applicable for sports analytics, especially during endurance, high intensity, or high strength activities.

Similar to voltammetry for wearable sweat biosensing, amperometry has been used for electrochemical biosensing of saliva [58,84]. Mouthguard electrochemical sensors have been established for wearable biosensing of saliva analytes, such as glucose [59], lactate [84], uric acid [58], and bacteria [57]. For example, in 2015, Kim et al. [58] presented a wearable mouthguard biosensor for real-time monitoring saliva biomarker, uric acid, in a highly sensitive, selective manner (Figure 4b). The WE, RE, and CE were screen-printed on a flexible PET substrate and the Prussian-Blue carbon WE underwent chemical modification—crosslinking uricase and o-phenylenediamine (OPD). The PET-based sensor was integrated with an amperometric circuitry—miniaturized fabrication printed circuit board assembly—that consisted of a Bluetooth low energy communication System-on-Chip. The resulting wearable mouthguard biosensor showed high sensitivity to uric acid and had a wide linear range (covering both healthy and hyperuricemia people's salivary uric acid concentration levels) ($R^2 = 0.998$). Besides, saliva's suitability of non-invasive monitoring has been proven in previous literature, a good correlation between blood and saliva concentration of uric acid have been found [130,131]. This wearable salivary biosensing platform can be rapidly expanded to multiparameter detection of saliva analytes and would have various real applications in the future.

4.2. Optical Sensing

4.2.1. Fluorometry

Fluorometric sensing is a simple approach to circumvent the need for in situ processing and data transmission, and also the on-chip power supply. The fluorescent-based biosensor only serves as the sweat sampler and in situ bioreactor that generates measurable changes in fluorescent intensity. Although fluorometry relies on visual readout devices (e.g., mobile phone), this sensing modality not only reduces the complexity of the wearable biosensor, it also lowers the cost of mass-manufacture of the device due to the simple structure without embedded electronics. Besides, the intrinsic merits of fluorescent-based biosensors are high sensitivity, selectivity, and accuracy. Fluorometric sensing exploits the chemical probes which interact with the target analytes and emit measurable fluorescence. The fluorescence is excited by a given light source (e.g., light-emitting diode, LED) and quantified by a photodetector (e.g., smartphone camera). Sekine et al. [61] presented a ground-breaking microfluidic biosensor using fluorometric sensing to determine the concentrations of sweat chloride, sodium, and zinc (Figure 5). Sweat was routed from inlets to fill the microchambers with fluorometric reagents time-sequentially under the control of CBVs. After the mixture of sweat and fluorescent probes, the light-shielding film over the fluorometric microchambers was removed and the smartphone-based optical module was exploited to capture the fluorescent images of the microchambers. The optical module consisted of a smartphone and a dark box, which contained a blue excitation filter to delineate the range of wavelengths of the excitation light and a colored emission filter to enable detection of excited fluorescence intensities without interference from the excitation light. Meanwhile, the two-round microchambers, consisting of a mixture of an ionic liquid of known concentration and the fluorometric reagents, served as fluorescence reference makers. Finally, the fluorescent intensities of fluorometric microchambers measured by ImageJ were used to determine the concentrations of sweat chloride, sodium, and zinc. The results of human trials in field tests strongly correlated with those measured by conventional benchtop methods. Hence, fluorometric sensing avoids the interference of ambient light conditions, leading to higher accuracy, but has a minor drawback—increasing the reliance on the optical module (e.g., dark box).

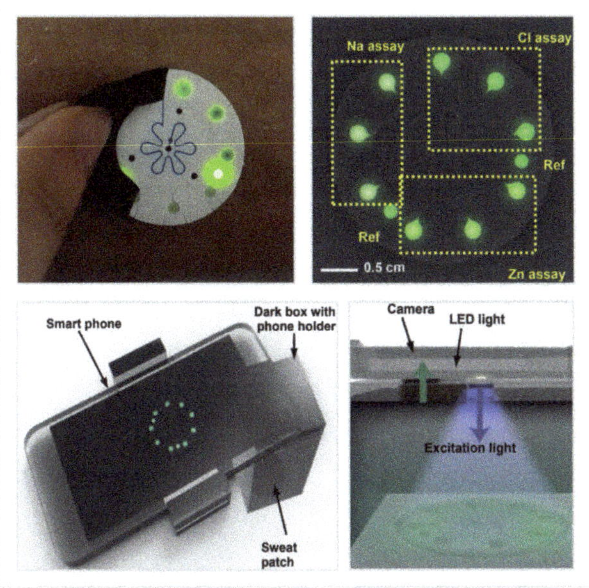

Figure 5. A fluorometric microfluidic biosensor for the detection of sweat biomarkers. Digital image of a fluorescent-based microfluidic device for in situ sensing the concentrations of sweat chloride, sodium, and zinc; fluorescent image illustrating the signals associated with targeted biomarkers; schematic illustrations of the smartphone-based optical module for fluorometric sensing [61]. Copyright 2018, Royal Society of Chemistry.

4.2.2. Bioluminescence

Bioluminescent (BL) sensing has been widely used for the detection of pollutants and toxicants, including organic chemicals (e.g., benzene, toluene) [132] and heavy metals (e.g., lead, zinc, cadmium, copper, and mercury) [132–134], in water, food, or the environment, especially for environmental toxicology and bioterrorism controls. Compared to fluorometry, which requires external excitation light [135], BL exploits genetically engineered luminous cells/bacteria to detect specific compounds by measuring the BL signals produced by these recombinant cells/bacteria in the presence of the analytes [132,134,136,137]. BL bacteria emit light in a hospitable environment but with decreasing BL levels in the presence of the increasing amount of the toxicants. The weaker the light emitted by BL bacteria, the higher the degree of toxicity [132]. However, bioluminescence has not yet been used for tracking analytes in biofluids due to the reliance on living cells/bacteria. But, when the storage and maintenance of BL cells/bacteria as well as the sensitivity of BL sensors, are greatly improved, in the future, BF sensors may be used for rapid in situ analysis of biomarkers by integrating wearable technology. Here, a portable BL biosensor will be discussed to further illustrate the principles and show the potential of bioluminescent sensing.

To facilitate on-field detection of toxicants, Cevenini et al. [138] developed a smartphone-based BL whole-cell toxicity biosensor for real-time, cost-effective, quantitative toxicity testing without trained personnel and laboratory instrumentation (Figure 6). This BL sensing platform consisted of two main components, including a 3D printed BL-cell-contained cartridge for quantification of the BL signals and an Android application, Tox-App, to capture, storage, and process the image, as well as provide real-time toxicity testing results. The human embryonic kidney cells (Hek293T) served as biosentinel cells that expressed a green-emitting luciferase. Dimethyl sulfoxide (DMSO) was used as a model toxic compound, and the percentage of BL signal normalized with respect to control (set as 100%) represented the cell viability value, which delimited the severity of toxicants: "Safe" (cell viability: 100–80%), "Harmful" (79–30%), and "Highly toxic" (<30%). This device provided real-time quantitative results about the toxicity of tested samples, and used BL cells as living biosentinels, enabling battery-free biosensing of toxicants and revealing the great potential of BL cell/bacteria sensor for portable quantitative biosensing.

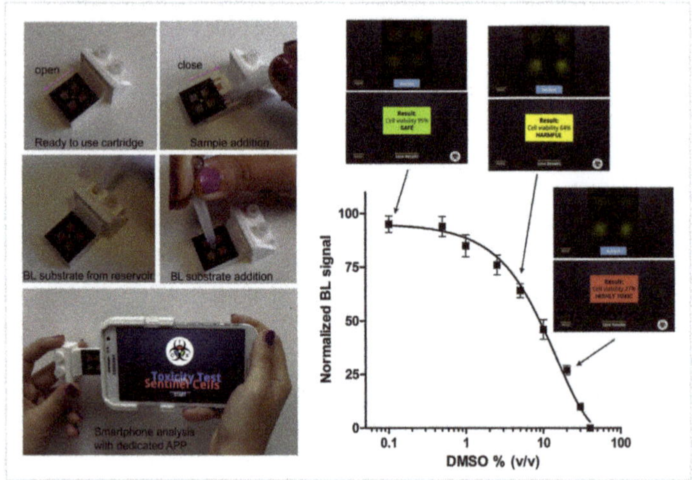

Figure 6. A smartphone-based bioluminescence (BL) whole-cell toxicity biosensor for on-field detection of toxicants; several easy steps required to carry out the toxicity test; dimethyl sulfoxide (DMSO) toxicity curve and the warming message (safe, harmful, or highly toxic) obtained with the smartphone-based biosensor and the Android application (Tox-App), set as "Safe" (cell viability: 100–80%), "Harmful" (79–30%), and "Highly toxic" (<30%) [138]. Copyright 2016, Elsevier.

4.2.3. Colorimetry

Colorimetric sensing modality using soft microfluidics is another way for non-invasive, in situ analysis of biofluids. The colorimetric sensing approach allows inexpensive, rapid, semiquantitative assessment of many sweat biomarkers, such as glucose, urea, lactate, and chloride [21,34]. The biochemical reactions between sweat biomarkers and reagents embedded in sensing platforms yield measurable changes in optical wavelength. Subsequently, capturing the optical information and converting it into quantitative data, such as pH and concentration of the analytes, were accomplished by high-quality digital image capture (e.g., digital camera, smartphone camera) and color extraction (RGB values). Finally, comparing the RGB values extracted from digital images with standard calibration curves measured in the laboratory beforehand can easily determine the concentration of the analytes. As the simplest optical sensing modalities, colorimetric sensing techniques offer multiple attractive features, such as ease of mass-manufacture, simultaneous monitoring of multiple analytes, and ultrathin, lightweight, flexible constructions [73,91,139]. These attributes make them extremely suitable for healthcare monitoring at home or sports analytics in the field. To further illustrate the capabilities of colorimetry, several representative examples will be hereby introduced.

The pioneering work was demonstrated by Koh et al. [21] in 2016, a wearable microfluidic device for colorimetric sensing of sweat. This device integrated four microchannels for guiding the sweat to fill four microreservoirs containing colorimetric reagents for analysis of biomarkers, including glucose, lactate, chloride, and pH. However, this microfluidic biosensor has limitations in accuracy. Recently, considerable efforts have been made to improve the performance of colorimetric sensors based on microfluidics and enzymatic reactions. Zhang et al. [34] established a microfluidic biosensor for colorimetric analysis of biomarkers relevant to kidney function. In Figure 7a, three separate sets of microreservoirs were introduced to achieve the time-sequential analysis of pH and concentrations of urea and creatinine. The printed color reference markers embedded in the capping layer facilitated the quantitative analysis of optical information gleaned by color extraction. This microfluidic biosensor not only improved the accuracy by providing time-sequential results of sweat biochemistry, but also circumvented the requirements for physical exertion by collecting sweat immediately after or during warm-water showering. Hence, this water-proof microfluidic device has vast potential for healthcare monitoring of vulnerable populations, such as infants and the elderly, as well as for sports analytics in aquatic settings [33].

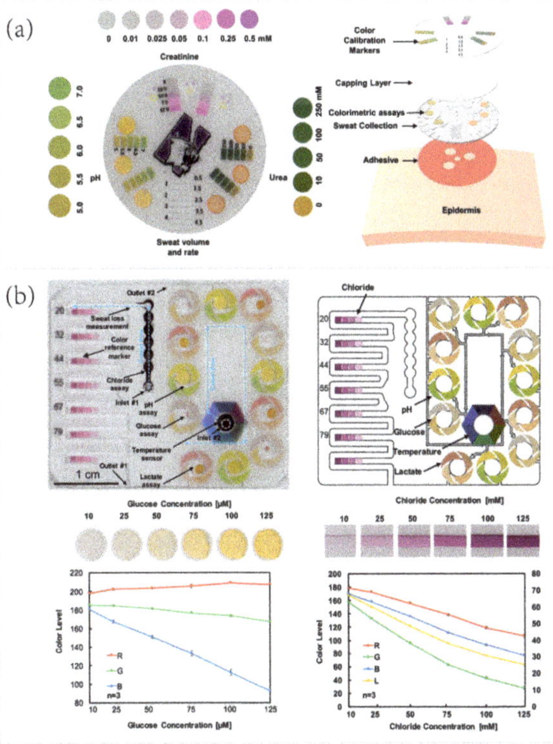

Figure 7. Microfluidic biosensor for colorimetric sensing biomarkers in sweat. (**a**) A soft microfluidic biosensor for colorimetric analysis of biomarkers relevant to kidney function. A top view of its layout and targeted biomarkers, and an exploded view of its different layers and components [34]. Copyright 2019, Royal Society of Chemistry. (**b**) A soft, skin-interfaced multifunctional microfluidic biosensor for accurate colorimetric sensing of sweat temperature, pH, chloride, glucose, and lactate. Two representative optical images of color development of assay microreservoirs as a function of sample concentrations and the corresponding RGB values of glucose and chloride [74]. Copyright 2019, American Chemical Society.

Another representative example of colorimetric sensing was a skin-interfaced multifunctional microfluidic system developed by Choi et al. [74] (Figure 7b). This device made good use of similar colorimetric modalities to monitoring pH, chloride, glucose, and lactate. But, importantly, this work further improved the accuracy of colorimetric sensing via its special layout. The color calibration markings laminated onto the top surface of the system, to surround each of the microreservoirs or each segment of the serpentine microchannel, as shown in Figure 7b. This layout facilitated the real-time quantitative analysis in various lighting conditions (such as under natural ambient conditions), which increased the accuracy of optical information gleaned from digital image analysis. The bottom two images showed in Figure 7b were the optical images of the color development of microreservoirs as a function of sample concentrations and the corresponding RGB values of glucose and chloride. The spectral information for color reference markers was obtained from a series of in vitro tests with standard samples of simulated sweat. Moreover, this platform also exploited thermochromic materials, a ternary cholesteric liquid crystalline mixture, for real-time temperature sensing (ranging from 31 to 37 °C) of sweat as it generated from the skin. Compared to electrochemical sensing modality, colorimetric sensing provides an easier way to monitoring sweat biochemistry and thermal conditions without the need for wearable power source, in situ data processing, and wired/wireless data transmission.

4.2.4. Optoelectronic Sensing

Advanced body-worn optoelectronic devices have been designed for quantitative, continuous sensing of vital signs. Recently, Kim et al. [26] reported an optoelectronic biosensor for pulse oximetry (Figure 8). Due to the high-power consumption of the LEDs, conventional pulse oximeters either rely on the bulky battery or hard-wired power supply [52]. To realize the on-field tests, wireless, battery-free, fingernail-mounted pulse oximetry was developed. This miniaturized oximeter exploited the bilayer loop antenna configured to maximize the efficiency of power harvesting and the distance for wireless data transmission. Wireless delivered power was used to drive a microcontroller and an operational amplifier. The infrared and red LEDs emitted light chronologically under the control of the microcontroller. The intermediate photodetector captured the backscattered light and an analog-digital converter amplified and digitized the resulting signals. Ultimately, the data of blood oxygenation (SpO$_2$), heart rate (HR), and heart rate variation (HRV) was wirelessly transmitted via NFC chipsets. This fingernail-mounted oximeter extends the mounting time up to 3 months and minimized the risk for irritation and discomfort. Long-term non-irritating integration with the human body allows for continuous monitoring of athletes' cardiovascular physiology, enabling easier management of daily athletic training and progress. Besides, "fingernail-mounted" modalities show several unnoticed but outstanding merits over "skin-interfaced" modalities. Compared to the skin, the fingernail can be exploited as a mounting location for long-term biosensing due to its optical transparency, mechanical rigidity, minimal interfacial water loss, and absence of nerve endings.

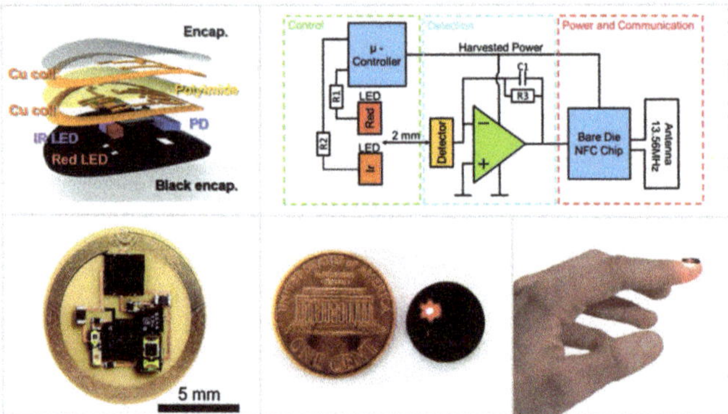

Figure 8. A novel optoelectronic wearable biosensor for wireless, battery-free pulse oximetry. A schematic illustration of the layout of the fingernail-mounted pulse oximetry, and a block diagram of the functional components; images of an unencapsulated device, a device next to a US one-cent coin, and a fingernail-mounted device during operation [26]. Copyright 2017, Wiley.

4.3. Other Sensing Modalities

Although the notable advances reported in the above-mentioned studies demonstrate the excellent performance of skin-interfaced wearable sensors for sweat biosensing, other sensing modalities still hold great potential. Over the past decade, novel sensing modalities emerge quickly, revealing promising solutions for long-term, non-irritating, and personalized therapeutics. Various mounting locations (e.g., tooth, fingernail) [26,52,53], integration methods (e.g., wearable, implantable) [140–143], and sensing modalities (e.g., volumetric, ultrasonic and optofluidic) [53,140,144] have opened up a spectrum of applications in wearable/ambulatory healthcare and sports analytics. Novel biosensing modalities for promoting these fields are urgently required, but herein we only present one novel sensing modality, volumetric sensing, to show the enormous potential of wearable biosensing.

Volumetry

Microfluidic systems demonstrate great potential for wearable biosensing. In the past five years, an innovative biosensing modality—volumetric biosensing, a quantitative method to ascertain sweat loss/rate for reflecting users' hydration status and providing suited rehydration strategies—quickly emerged. Herein, volumetric sensing of sweat will be introduced with two illustrative examples. Soft microfluidic biosensors make use of the pressure naturally-induced by the sweat glands to drive sweat through a series of microchannels to fill sweat collecting channels time-sequentially for real-time quantification of sweat loss/rate [63,145]. The microfluidic platforms, for volumetric sensing of sweat over local anatomical regions, usually consist of opening(s) for defining the sweat collecting area(s) where sweat can enter the platform, water-soluble dye(s) inserted adjacent to the inlet(s) to facilitate real-time visualization of sweat loss, and the serpentine microchannel(s) for sweat volumetry [21,34,43,74]. Finally, volumetric sensing of the colored sweat collected by the microchannel(s) is performed by digital image capture and analysis. Wearable biosensors for on-field tests usually exploit smartphone cameras for quantification of sweat loss/rate by image processing and real-time display of the data for end-users [21,27,73].

In recent years, various microfluidic platforms, characterized by different microchannel layouts and geometries, have been developed for volumetric sensing of sweat [43]. But, few of them can operate in aquatic settings (such as during showering, swimming, or diving). To fulfill the requirements of sweat volumetry for athletes participating in aquatic sports, Reeder et al. [33] developed a waterproof skin-interfaced microfluidic biosensor for sweat volumetry, biomarker analysis, and thermography (Figure 9a). Underwater sweat collection and volumetric sensing of sweat loss/rate were introduced. The special microfluidic inlet and outlet pores were essential for sweat loss visualization and device's waterproofing. As shown in Figure 9a, the water-soluble reagents, consisting of a food dye to facilitate the visual inspection of sweat collection, resided in the microchamber adjacent to the inlet pore. This food dye contained red/blue particles with different dissolution rates that generated a volume-dependent color gradient as the microchannel filled with sweat. The serpentine microchannel has 40 turns, each of which has a volume of 1.5 µL. Thus, the sweat loss/rate can be quantified by counting the sweat-filled turns. Besides, this device contained only one tiny outlet and a small amount of "dead volume" near the outlet pore, eliminating the interference from the outside aquatic environment, which was considered a highly attractive feature for aquatic sports analytics.

Another research presented an alternative method, instead of using colored dyes, to quantify sweat loss/ rate. Kim et al. [27] established a microfluidic/electronic biosensor for digital tracking of sweat loss/rate and electrolyte composition. As shown in Figure 9b, this platform also deployed a serpentine microchannel for volumetry of collected sweat, but a single pair of electrodes with five pairs of probing pads were used to trace the collected sweat. The total resistance (Rn) across the main electrodes depended on the number of pairs of probing pads that were bridged by sweat. The bottom two schematics of Figure 9b showed the correlations between the number of pairs of probing pads bridged by collected sweat and the amount of sweat collected by the microfluidic channel. This novel approach circumvents the need for digital image capture and analysis, which seemingly simplifies sweat volumetry. However, the volumes of microchannels between pairs are significantly different (from 2.6 µL to 27.2 µL). Despite the flaw mentioned here, future research efforts on the device design may provide a possible solution to equal the volumes of microchannels between different pairs of probing pads. Hence, optimizations of the geometry of the microfluidic channel and the corresponding layout of the electrodes present a promising direction for the improvement of digital sweat volumetry.

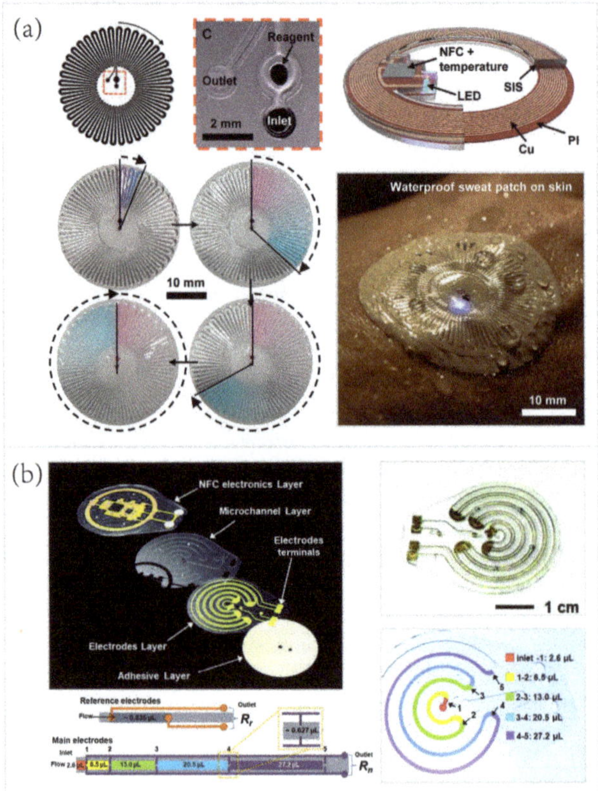

Figure 9. Wearable biosensors for sweat volumetry. (**a**) A waterproof skin-interfaced microfluidic biosensor for volumetric sensing, biomarker analysis, and thermography in aquatic settings; images showing the special inlet and outlet pores, as well as the volumetric sensing by using red/blue dyes to facilitate the visual inspection [33]. Copyright 2019, American Association for Advancement of Science. (**b**) A microfluidic/electronic biosensor for digital tracking of sweat loss/rate and electrolyte composition; illustrations showing the volumes of microfluidic channels that lay between different pairs of probing pads [27]. Copyright 2018, Wiley.

5. Key Analytes and Wearable Biosensing Platforms

5.1. Key Analytes in Wearable Biosensing

Biofluids are largely unexplored resources that contain a rich milieu of important biomarkers, from electrolytes [14,61,126], metabolites [21,34], heavy metals [31], cytokines [146,147], hormones [80,148,149] and amino acids [150] to exogenous drugs [77,83], each of which provides valuable insights into physiological status, health conditions, pharmacokinetics, and sports performance. Sweat and ISF are two representative biofluids that offer great potential for home-based healthcare monitoring or on-field sports performance assessment due to ease of non/minimally-invasive collection and analysis by using wearable biosensors [15,41,81]. Although traditional wearable platforms have demonstrated the effectiveness of measuring a few biophysical markers [25,49,50], skin-interfaced biosensors mainly exploit valuable resources—sweat and ISF biochemical markers for healthcare and sports monitoring as shown in Table 1. But, due to the space limitation, only the key sweat targets will be discussed herein, the key ISF analytes and a few wearable biosensors for ISF analysis will be presented in a limited manner in Section 5.3.

Electrolytes, the most abundant analytes in sweat, are widely tested in wearable biosensing [151]. Significant interest in Na^+, K^+, NH_4^+, and Ca^{2+} stems from their applications in disease diagnostics and sports analytics. For example, the breakdown of proteins can result in the presence of ammonium in blood before the liver converts them to urea [68]. There, high levels of NH_4^+ in excreted sweat can be used as indicators of hepatic diseases, such as hepatitis, cirrhosis, and related disease—hepatic encephalopathy (HE) [35,152]. Kim et al. [35] reported a wearable microfluidic biosensor with integrated enzymatic reactions for colorimetric sensing of the concentration of ammonia and ethanol in sweat. This sweat biosensor could serve as a diagnostic tool of HE. Besides, ionized calcium is another important marker of homeostasis. Excessive variations of Ca^{2+} levels in biofluids are detrimental to the human body, causing myeloma, renal failure, kidney stones, and hyperparathyroidism [32]. Moreover, electrolytes, especially Na^+ and K^+, play a critical role in maintaining the electrolyte balance and hydration status. Excessive loss of them could result in hyponatremia, hypokalemia, and muscle cramps, leading to inferior sports performance [14].

Metabolites are rich solutes in sweat which are highly attractive for diverse applications in wearable disease diagnostics and sports analytics. For instance, glucose and lactate attract most research efforts [45]. Real-time tracking of glucose concentration in biofluids not only has utility in the diagnosis of diabetes mellitus [153], but also can reflect energy availability and consumption for athletic monitoring [43]. Likewise, sweat lactate represents a key indicator of muscle fatigue, tissue hypoxia, and exercise intensity [21,75]. Thus, many wearable biosensors capable of simultaneously monitoring sweat glucose and lactate have been successfully developed [14,21,73–75], because of the importance of continuously gleaning metabolic information from sweat. Additionally, another three sweat metabolites, including urea, uric acid, and creatinine, create opportunities for diagnostics of kidney disorders. The levels of sweat urea and creatinine are typically higher in patients with kidney disorders than in healthy individuals, since the failure of glomerular function [79,154]. As described previously, Zhang et al. [34] reported a microfluidic biosensor for colorimetric analysis of urea and creatinine for POCT of kidney diseases. Similarly, newborns usually undergo sweat chloride tests for the diagnosis of cystic fibrosis (CF) [69,155].

Other key targets, such as hormones [80,148,149], cytokines [146], heavy metals [31,61,76], and exogenous drugs [83,156], have also attracted considerable attention in wearable biosensing for healthcare monitoring. As such, cortisol is a steroid hormone in sweat, and recent efforts have been made in wearable cortisol monitoring for stress level tracking and disease diagnostics, including Cushing's syndrome and Addison's disease [80,148,149]. Besides, cytokines and heavy metals have also been non-invasively analyzed by wearable sensors to gain insights into inflammatory response [146,147], heavy metal exposure and detoxication [31], respectively. Moreover, wearable biosensors have been exploited to monitoring exogenous drugs, such as Levodopa and caffeine [77,156]. Wearable biosensing has a considerable potential to transform the current state-of-the-art precision medicine and drug delivery research [83]. The concentration of drug in biofluids (especially sweat and ISF) gleaned from wearable monitoring can serve as feedback to improve personalized drug dosage, to establish closed-loop therapeutics, and to provide raw data for pharmacokinetics [77,153,157].

Table 1. Key sweat and ISF analytes in wearable healthcare and sports monitoring.

Key Targets	Analytes	References		Representative Applications
Electrolytes	[1] Na[+] [1,2] K[+] [1] NH$_4$[+] [1] Ca[2+]	[14,61,126] [158–160] [35,68,161] [32,161,162]	1. 2.	Disease diagnostics (Ca[2+] for kidney stones and hyperparathyroidism; NH$_4$[+] for [3] HE) Sports analytics (Na[+] and K[+] for electrolyte balance, muscle cramps, and de/rehydration)
Metabolites	[1,2] Glucose [1,2] Lactate [1] Creatinine [1] Chloride [1] Uric acid [1] Urea [2] Cholesterol	[37,75,153] [74,127,163] [21,34] [33,69,74,91] [150,164,165] [34,166] [167]	1. 2.	Disease diagnostics (glucose for diabetes, chloride for cystic fibrosis; urea, uric acid, and creatinine for kidney diseases, such as uremia; cholesterol for atherosclerosis) Sports analytics (lactate for muscle fatigue, tissue hypoxia, and exercise intensity; glucose for energy availability and consumption)
Heavy metals	[1] Zn [1] Cu,[1] Cd,[1] Pb,[1] Hg	[31,61,76] [31]	1. 2.	Monitoring of heavy metal intake and detoxification (excess or deficiency) Determination of heavy metal exposure
Cytokines	[1,3] IL-1β [1,3] CRP [1,2,3] TNF-α	[146] [146] [147,168]	1. 2.	Monitoring of inflammation (CRP, TNF-α) Precision medicine (IL-1β for [3] IBD treatment)
Hormones	[1] Cortisol	[80,148,149]	1. 2.	Disease diagnostics (Cushing's syndrome, Addison's disease) Mental health management (stress levels)
Amino acids	[1] Tyrosine [2] Glutamate	[150] [169]	1. 2.	Disease diagnostics (liver diseases, traumatic brain injury, Alzheimer's disease) Nutritional management (eating disorders)
Exogenous drugs	[2] β-lactam [2] Vancomycin [1] Levodopa [1] Caffeine	[83] [170] [77] [156]	1. 2.	Precision medicine (personalized drug dosage, closed-loop therapeutics) Drug delivery (pharmacokinetics, therapeutic drug monitoring for antibiotics)
Others	[1,2] Ethanol	[35,171,172]	1.	Alcohol consumption analysis
	[1,2] pH	[32,34,74,173]	1. 2.	Sports analytics (electrolyte balance) Disease diagnostics (metabolic alkalosis)
	[1] Sweat loss/rate	[21,33,34,73]	1. 2.	Sports analytics (de/rehydration) Disease diagnostics (autonomic nervous disorders; hyper/hypohidrosis)
	[2] Immunoglobulin	[174,175]	1. 2.	Early disease detection Investigation of vaccine effectiveness

Analytes sources: [1] sweat and [2] interstitial fluid (ISF); [3] Abbreviation: HE, hepatic encephalopathy; IL-1β, Interleukin-1β; CRR, C-reactive protein; TNF-α, Tumor necrosis factor-α; IBD, Inflammatory bowel disease.

5.2. Wearable Sweat Biosensing Platforms for Healthcare and Sports Monitoring

The rich composition of sweat analytes makes the wearable sweat biosensing platform applicable to healthcare and sports monitoring. Over the past decade, an increasingly large number of wearable biosensors have been developed, as we systematically reviewed in Section 4. However, the potential impact of wearable sweat biosensing technologies on sports analytics has not yet been fully recognized. To broaden the scope of wearable sweat sensing (like other analytes, integrated biosensors), several wearable platforms will be briefly introduced herein.

In one example, Reeder et al. [33] developed a waterproof microfluidic platform for sweat volumetry, biomarker analysis, and thermography in aquatic settings (Figure 10a). For biomarker analysis, colorimetric sensing was introduced: the reagent resided in the adjacent chamber reacted with sweat to create a colorimetric response corresponding to the concentration of chloride. A printed color reference dial at the center facilitated visual inspection. Besides, two representative biosensors exploited electrochemical sensing for in situ monitoring of heavy metals—Zn, Cd, Pb, Cu, and Hg [31] (Figure 10b), and an exogenous drug—caffeine [156] (Figure 10c). These systems based on electrochemical biosensing are characterized by high selectivity, sensitivity, and reproducibility. Tai et al. [156] reported a wearable biosensor with the capability of methylxanthine drug monitoring by electrochemical DPV, which is an important step to complement the traditional method of drug dosage tracking—blood sampling and analysis. Sports medicine personnel could be specifically interested in this work due to their ability to monitor the drug dosage during exercise intervention of the diseased. Another example of using wearable biosensors for drug monitoring was combined with microneedles technologies, and will be described in the following Section 5.3.1.

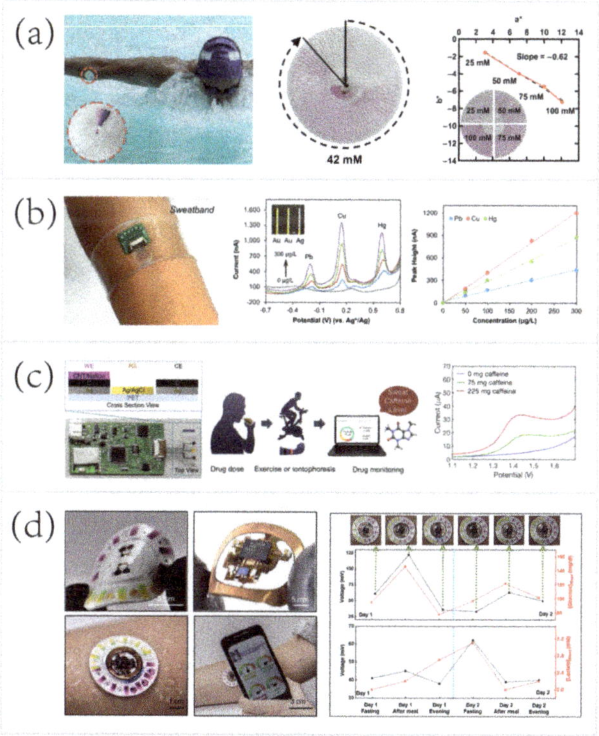

Figure 10. Wearable sweat biosensing platforms for healthcare and sports monitoring. (**a**) A waterproof skin-interfaced microfluidic biosensor for sweat chloride analysis in aquatic settings [33]. Copyright 2019, American Association for Advancement of Science. (**b**) A wearable biosensor for multiplexed monitoring of heavy metals, including Zn, Cd, Pb, Cu, and Hg [31].Copyright 2016, American Chemical Society (**c**) A wearable biosensor for methylxanthine drug (caffeine) monitoring [156]. Copyright 2018, Wiley. (**d**) A skin-mounted hybrid sensor for analysis of sweat glucose, lactate, pH, and chloride; correlations of data acquired from sweat biosensors (black line) with that acquired from blood glucose and lactate meters (red line), respectively [73].Copyright 2019, American Association for the Advancement of Science.

Recent progress has led to the advent of hybrid biosensors, which can monitor multiple parameters by simultaneously performing electrochemical, colorimetric, and volumetric sensing. To yield a comprehensive evaluation of the physiology and sports performance of end-users, Bandodkar et al. [73] developed a skin-mounted microfluidic/electronic biosensor for simultaneous monitoring of chloride, pH, lactate, glucose, and sweat loss (Figure 10d). The hybrid biosensor contained a reusable, thin NFC subsystem and a soft, disposable microfluidic subsystem. The NFC subsystem allowed for electrochemical sensing in a mode that targeted glucose and lactate, and spontaneously generated electrical signals proportional to their concentrations. The microfluidic subsystem embedded colorimetric reagents for tracking the pH and concentration of chloride, with an additional ratcheted microchannel for quantification of sweat rate and total sweat loss. As shown in Figure 10d (right), a field study aimed to suggest the potential for acquiring semi-quantitative data by analyzing sweat biochemistry. This study focused on comparing temporal variations in glucose and lactate levels in blood and sweat due to exercise engagement or food intake. The results showed a comparison between the data acquired from hybrid biosensors with the data acquired from blood glucose and lactate meters, respectively. Over two days, the blood glucose and lactate levels after each session followed trends that qualitatively similar to those of data measured in sweat. These correlations were confirmed by literature [69,153,176–178].

5.3. Microneedles Platforms

Several decades ago, microneedles (MNs) were first introduced to perform transdermal drug delivery [179]. MNs with small length (usually less than 1000 μm [180,181]) penetrate the stratum corneum and open windows for the ISF biosensing or form transient microchannels for drug delivery while preventing the nerves and blood vessels from stimulation and impairment. Currently, MNs have been utilized for applications in the following areas: ISF sampling [180,182], disease diagnostics [153,183,184], drug delivery [179,185,186], cosmetics [181,187–189], etc.

MNs have wide applications and show tremendous promise for applying biomedical advances into wearable healthcare and sports monitoring due to the three following intrinsic advantages.

1. Enhanced compliance. The patients and athletes will not suffer from pain, discomfort, and needle-phobia [179,180,190,191].
2. Easy-to-use. MNs are user-friendly biosensing and drug delivery devices that can be directly applied to the skin. Conventional methods, hypodermic injections, conversely, demand professional personnel who undergo rigorous medical training [179].
3. Real-time in situ biosensing and controllable/long-term drug delivery. MNs realize pain-free, in situ diagnostics even combine with feedback-based long-term drug delivery [160].

Recent advances in wearable MN patches capable of minimally-invasive, transdermal sensing biofluids [153,160,179,192,193] and pain-free drug delivery [184,194,195] provide an opportunity to perform sports monitoring, especially beneficial to sports nutrition and sports medicine [196]. These techniques combined with electrochemical/microfluidic biosensing modalities are expected to lead to several breakthroughs in wearable biosensing.

5.3.1. Microneedles for Transdermal Biosensing

In recent years, although the majority of MNs are still focused on painless drug delivery, MNs for transdermal biosensing are promising and receive considerable attention [190]. ISF contains rich physiological/pathological information than ever envisaged [197], in the meantime, the MN-based biosensor can serve as a portal for the pain-free acquisition of valuable information. Besides, recent research efforts have led to the emergence of advanced microfabrication and biophotonics that facilitate the manufacture of MNs [37,39,198]. More complicated MN platforms that integrate electrochemical biosensors [157,169,181] or biofuel cells [37] (harvesting energy from biofluids, such as sweat and ISF, to power wearable platforms) have been manufactured. These integrated biosensors demonstrate great promise for monitoring physiological status and athletic performance in field tests, as well as detecting doping in sports competitions.

MNs only penetrate the stratum corneum of the skin, therefore do not reach the capillary to sample blood for biosensing. Instead of hypodermic needles that sample blood for medical testing, the transdermal MN-based biosensors adopt a minimally-invasive way to sample the ISF for therapeutic drug monitoring (TDM). Using the TDM to optimize the drug dosage is important for improving therapeutic efficacy and predicting any adverse outcome, such as resistance and toxicity [199]. For example, Gowers et al. [83] developed a TDM system (Figure 11a), an MN-based biosensor for real-time continuous monitoring levels of β-lactam antibiotics in vivo. The MNs were coated with multiple layers, including a gold working electrode layer, a pH-sensitive iridium oxide layer, a β-lactamase hydrogel layer, and a poly(ethylenimine) layer. The pH-sensitive layer detected changes in local pH as a result of β-lactam hydrolysis with the presence of β-lactamase. Afterwards, the potentiometric modality was applied for establishing the relationship between variations in local pH and the potential measured. Such a biosensor is capable of real-time tracking penicillin concentrations, paving the way for personalized drug dosage that is a large step towards precision medicine. But, these MN-based biosensors are still at an early stage of development. A daunting challenge still exists, namely the need for a continuous, reliable power supply.

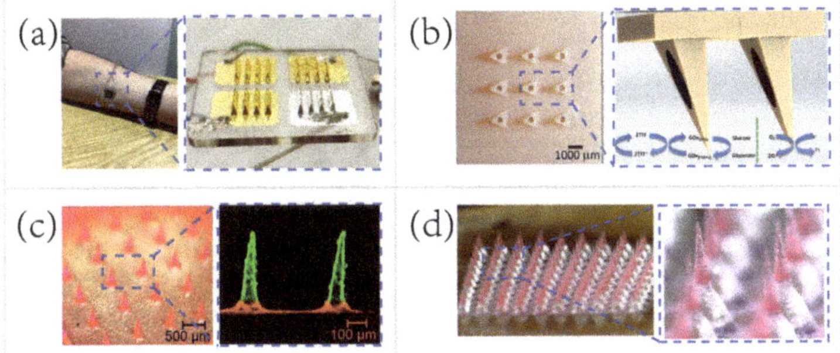

Figure 11. Representative applications of microneedles (MNs) for transdermal biosensing and pain-free drug delivery. (**a**) MN-based biosensor for continuous monitoring of β-lactam antibiotic concentration in vivo [83]. Copyright 2019, American Chemical Society. (**b**) MN-based self-powered glucose sensor and schematic reaction on two of the hollow MNs located within the MNs array [37]. Copyright 2014, Elsevier. (**c**) Multilayered pyramidal dissolving MNs, composed of silk fibroin tips supported on flexible pedestals, for improving drug delivery [38]. Copyright 2017, Elsevier. (**d**) MN patch with drug-loaded tips and effervescent backings for long-acting reversible contraception [40]. Copyright 2019, American Association for the Advancement of Science.

A majority of wearable biosensing devices rely on the external power supply [14,26,88,90,153], which causes numerous technical difficulties in the field tests. Thus, researchers begin to seek new methods to obviate the need for an external power supply. Bandodkar et al. [200] made a major thrust of biofuel-cell (BFC) wearable devices that scavenged energy from human sweat. The wearable energy harvester they developed converted readily available sweat lactate into electricity. Lightweight, ultrathin, stretchable, high power density were the attractive features of this wearable harvester. Nowadays most MN-based sensing patches are still relying on the external power supply [83,153,201]. To overcome this major drawback, Valdés-Ramírez et al. [37] developed an MN-based self-powered BFC glucose biosensor that harvested biochemical energy from the wearer's ISF (Figure 11b). The self-powered biosensor harnessed glucose as the fuel to eliminate the need for an external power supply, and also provided power signals proportional to the glucose concentration. Besides, this self-powered MN-based sensor displayed high selectivity and stability that showed considerable promise for battery-free transdermal glucose monitoring. In short, the combination of energy harvester, electrochemical sensor, and MN technology will address some key challenges, such as integrating monitoring multiple analytes in ISF and automatic feedback-based transdermal drug delivery using MN-based self-powered BFC biosensors.

5.3.2. Microneedles for Pain-free Drug Delivery

Microneedles for transdermal drug delivery reduce the systemic side effects and improve dosage efficacy and patient compliance [180] compared to hypodermic needles [38,180,202]. Typically, external small molecules (e.g., drug contents) can be minimally-invasive delivered percutaneously harnessing passive mass transfer. To facilitate the establishment of closed-loop diagnostic and therapeutic systems, MNs for drug delivery are also attractive, especially for controlled or long-term non-irritating drug release. In the past few years, great efforts have been made to find a way for controllable/long-term drug delivery using MNs, such as thermo-responsive MNs [153], stretch-triggered MNs [196], multilayered MNs [38], and effervescent MNs [40].

To deliver the drug contents in a controllable manner, Lau et al. [38] developed a multilayered pyramidal dissolving MN patch with flexible pedestals (Figure 11c). The drug release time and rate were regulated by the dissolution rate of diverse biomaterials in different layers of the MNs. Hence, the MNs were applicable to control the release rate of the anti-inflammatory drug to facilitate personalized sports medicine. For instance, the rapid dissolution of the tip's outermost layer of the multilayered pyramidal MNs can quickly control inflammation and continuously treat chronic inflammation via sustained dissolution of the inner layer.

Another group demonstrated the pursuit of long-term personalized drug delivery. Wei Li et al. [40] developed an effervescent system to increase access to long-acting contraception (Figure 11d). The MNs consisted of drug-loaded tips and effervescent backings containing polyvinylpyrrolidone (PVP), sodium bicarbonate, and citric acid. Once inserted in the skin, the ISF solubilized the effervescent backing and the CO_2 generated from the reaction of citric acid with sodium bicarbonate facilitated the separation of the interfaces between the MNs' tips and patch backings (detachment efficiency of 91.7 ± 2.4%). Long-acting contraception was achieved in vivo by the dissolution of MNs' tips to release levonorgestrel (LNG) for 2 months and maintained the LNG concentrations above the human contraceptive threshold level for >1 month (after rats being administered MNs).

In sum, MNs are not only appealing avenues to interrogate the ISF for biosensing, to carry out controllable/long-term transdermal drug delivery [179,185,186], but also can facilely incorporate with electrochemical biosensors [157,169,181] to realize closed-loop personalized diagnostics and therapeutics [153,193,203]. Moreover, recent advances in biofuel cells [89,200,204–206] could further address the key issue, lack of thin and reliable wearable power sources. These eminent features would enable a broad range of important applications in the sports field where real-time athletic performance monitoring and personalized sports injury treatments are urgently demanded [196]. We envisage that the advancement of MNs for biosensing and drug delivery will encourage the emergence of new generations of wearable athletic biosensors, that would revolutionize sports monitoring and sports medicine in the near future.

6. Unsolved Challenges for Future Research

Although considerable research efforts and remarkable progress have been made in wearable biosensing for healthcare monitoring and sports analytics, especially the field of microfluidics for sweat biosensing, several technical difficulties remain in aspects related to lack of well-established gold standard, as well as the selectivity, sensitivity, power supply, integration, cost of the device. First of all, unlike blood tests that have numerous well-established clinical gold standards, such as diabetes diagnosis (glucose concentration ≥ 7 mmol/L) [207], it is particularly difficult to establish gold standards for wearable biosensing of non-blood biofluids. Take sweat biosensing, for example. The sweat loss and the concentration of biomarkers show wide variations between different regions of the body. Besides, acquiring reliable concentrations of biomarkers (especially for scarce analytes) is challenging due to the relatively small amounts of collected sweat compared to total sweat loss. The calibration of biomarker concentrations is equally challenging since their strong relation to individual hydration status. Therefore, additional advances in wearable biosensors are needed, especially those related to diagnostic applications.

Furthermore, most wearable devices, especially electrochemical biosensors, rely on the continuous, reliable power supply, including miniaturized batteries, wireless power delivery devices, and biofuel-cell (BFC) energy harvesters. Miniaturized batteries, both coin cell and flexible batteries, have poor performance, requiring frequent recharging or large power source carried by wearers. The alternative, wireless power harvested from NFC chipsets, also has a tough problem that proximity (<1 m) is required for power delivery. Biofuel-cell energy harvesters scavenging energy from sweat or ISF metabolites (e.g., lactate, glucose) have some potential but power density still needs to elevate. Additionally, wearable biosensors, such as soft, skin-interfaced microfluidic/electrochemical biosensors, have to in situ monitor multiple analytes at extremely low concentrations without sample pretreatment. Thus, sensitivity and selectivity of biosensors are considered fundamental requirements for wearable healthcare monitoring and sports analytics. Besides, in situ monitoring biomarkers and real-time data acquisition demand the integration of monitoring subsystem, data processing subsystem, and wireless data transmission subsystem. This integration will yield significant additional expenditures and difficulties. More importantly, a majority of wearable biosensors meet difficulties to re-use since they do not discern the disposable and re-useable components of wearable biosensors. Nonetheless, future studies may focus efforts to design cheap, single-use plug-in parts of biosensors, which will explore possible avenues for the investigation and commercialization of this fascinating technology.

Unmet challenges also exist in the translation of biomedical advances into sports analytics. As mentioned previously, a vast majority of wearable technologies focus on healthcare monitoring. Firstly, well-established biomedical techniques have not yet been broadly applied in the sports field, since the enormous potential of the wearable biosensor for sports analytics has not been recognized by most researchers (e.g., biomedical engineers). Unfortunately, innumerable multidisciplinary studies concerning wearable technologies have been conducted within the medical and engineering fields instead of outside fields, such as sports analytics or sports engineering. Secondly, current existing sports analytical techniques, especially for chemical analysis of biofluids, are confined to the laboratory or hospital settings. Specifically for sports biochemists, difficulties largely follow from lack of engineering background and expertise, few opportunities for interdisciplinary collaborations. Ultimately, although several research groups at Northwestern University, University of California (Berkley) gain opportunities for in-depth interdisciplinary collaborations and develop numerous wearable biosensors for sports analytics, an obstacle, lack of practical data interpretation, in the path of the widespread application and commercialization of wearable biosensors still stubbornly persists. The wearable biosensors they developed have increasingly high performance related to sensitivity, selectivity, accuracy, and multifunctionality. Nevertheless, even if accurate data of sweat biomarkers gleaned by wearable biosensors can be facilely collected, the concentrations of sweat metabolites and electrolytes can be easily displayed on the smartphone, the end-users still do not know what these mean and how these data represent their health status and sports performance. Therefore, what next for wearable biosensors, from healthcare monitoring to sports analytics? It is perhaps a revolutionary technology for the future, although a great deal of work needs to be done before it achieves prosperity in academia and success in commercialization.

7. Conclusions and Perspectives

Recent advances in stretchable materials, microfluidics, optical/electrochemical sensors, image processing and analysis, NFC and wireless power supply, microneedles, as well as big data and cloud computing provide a robust foundation for wearable biosensing, especially the field of microfluidics for sweat and ISF biosensing. The emerging sampling and sensing modalities for obtaining sweat biochemical information offer profound insights into human physiology, metabolism, and sports performance, which complement traditional biophysical sensing modalities. Representative examples of wearable biosensors capable of in situ monitoring sweat biomarkers, from electrolytes, metabolites, heavy metals, cytokines [146], hormones [80,148,149], amino acids [150] to exogenous drugs [83,156], provide a window on the valuable biochemical resources that have a substantial impact on disease diagnostics, precision medicine, drug delivery, and sports analytics. However, the majority of advanced wearable sweat biosensors still focus on healthcare monitoring to improve home-based disease diagnosis. Although wearable biosensors for sports analytics are underdeveloped, the transition from focusing on healthcare monitoring to combining with sports analytics is an important step in a trend towards better preventive healthcare and sports & health science. It offers revolutionary capacities for future wearable technologies. It also in its infancy, and a great deal of work still needs to be done, such as integration of well-developed technologies (e.g., ultrasonics, implants, optofluidics, optoelectronics, microneedles, etc.), for building up a picture of future biosensing technologies. Despite the remarkable progress that several research groups achieved over the past five years, daunting challenges remain in the power supply, data interpretation, the cost of mass-fabrication, as well as in-depth interdisciplinary collaboration. With these challenges overcome, the commercialization and widespread adoption of wearable biosensors in sport-related fields will be fully realized, and state-of-the-art athletic monitoring and sports analytics will be transformed fundamentally.

Author Contributions: Writing—original draft preparation and editing, S.Y., S.F. and L.H.; supervision and funding acquisition, S.B. All authors have read and agreed to the published version of the manuscript.

Funding: This work was financially supported by the Beijing Sport University Scientific Research Program under Grant No. 2018qd01 and by the National Natural Science Foundation of China under Grant No. 52076074.

Conflicts of Interest: The authors declare no conflict of interest.

References

1. Figeys, D.; Pinto, D. Lab-on-a-chip: A revolution in biological and medical sciences. *Anal. Chem.* **2000**, *72*, 330A–335A. [CrossRef]
2. Manz, A.; Graber, N.; Widmer, H.M. Miniaturized total chemical analysis systems: A novel concept for chemical sensing. *Sens. Actuators B Chem.* **1990**, *1*, 244–248. [CrossRef]
3. Sackmann, E.K.; Fulton, A.L.; Beebe, D.J. The present and future role of microfluidics in biomedical research. *Nature* **2014**, *507*, 181–189. [CrossRef]
4. Mark, D.; Haeberle, S.; Roth, G.; von Stetten, F.; Zengerle, R. Microfluidic lab-on-a-chip platforms: Requirements, characteristics and applications. *Chem. Soc. Rev.* **2010**, *39*, 1153–1182. [CrossRef] [PubMed]
5. Lindström, S.; Andersson-Svahn, H. Miniaturization of biological assays—Overview on microwell devices for single-cell analyses. *Biochim. Biophys. Acta Gen. Subj.* **2011**, *1810*, 308–316. [CrossRef] [PubMed]
6. Kim, S.H.; Fujii, T. Efficient analysis of a small number of cancer cells at the single-cell level using an electroactive double-well array. *Lab Chip* **2016**, *16*, 2440–2449. [CrossRef] [PubMed]
7. Huang, L.; Bian, S.; Cheng, Y.; Shi, G.; Liu, P.; Ye, X.; Wang, W. Microfluidics cell sample preparation for analysis: Advances in efficient cell enrichment and precise single cell capture. *Biomicrofluidics* **2017**, *11*, 011501. [CrossRef] [PubMed]
8. Cheng, Y.-H.; Chen, Y.-C.; Brien, R.; Yoon, E. Scaling and automation of a high-throughput single-cell-derived tumor sphere assay chip. *Lab Chip* **2016**, *16*, 3708–3717. [CrossRef]
9. Bian, S.; Zhou, Y.; Hu, Y.; Cheng, J.; Chen, X.; Xu, Y.; Liu, P. High-throughput in situ cell electroporation microsystem for parallel delivery of single guide RNAs into mammalian cells. *Sci. Rep.* **2017**, *7*, 42512. [CrossRef]
10. Whitesides, G.M. The origins and the future of microfluidics. *Nature* **2006**, *442*, 368–373. [CrossRef]
11. Yılmaz, B.; Yılmaz, F. Chapter 8—Lab-on-a-chip technology and its applications. In *Omics Technologies and Bio-Engineering*; Barh, D., Azevedo, V., Eds.; Academic Press: Cambridge, MA, USA, 2018; pp. 145–153. [CrossRef]
12. Carlo, D.D.; Lee, L.P. Dynamic single-cell analysis for quantitative biology. *Anal. Chem.* **2006**, *78*, 7918–7925. [CrossRef] [PubMed]
13. Lin, C.-H.; Hsiao, Y.-H.; Chang, H.-C.; Yeh, C.-F.; He, C.-K.; Salm, E.M.; Chen, C.; Chiu, I.-M.; Hsu, C.-H. A microfluidic dual-well device for high-throughput single-cell capture and culture. *Lab Chip* **2015**, *15*, 2928–2938. [CrossRef] [PubMed]
14. Gao, W.; Emaminejad, S.; Nyein, H.Y.Y.; Challa, S.; Chen, K.; Peck, A.; Fahad, H.M.; Ota, H.; Shiraki, H.; Kiriya, D.; et al. Fully integrated wearable sensor arrays for multiplexed in situ perspiration analysis. *Nature* **2016**, *529*, 509. [CrossRef] [PubMed]
15. Kim, J.; Campbell, A.S.; de Ávila, B.E.-F.; Wang, J. Wearable biosensors for healthcare monitoring. *Nat. Biotechnol.* **2019**, *37*, 389–406. [CrossRef]

16. Tüdős, A.J.; Besselink, G.A.J.; Schasfoort, R.B.M. Trends in miniaturized total analysis systems for point-of-care testing in clinical chemistry. *Lab Chip* **2001**, *1*, 83–95. [CrossRef] [PubMed]

17. Dincer, C.; Kling, A.; Chatelle, C.; Armbrecht, L.; Kieninger, J.; Weber, W.; Urban, G.A. Designed miniaturization of microfluidic biosensor platforms using the stop-flow technique. *Analyst* **2016**, *141*, 6073–6079. [CrossRef]

18. Medina-Sánchez, M.; Miserere, S.; Merkoçi, A. Nanomaterials and lab-on-a-chip technologies. *Lab Chip* **2012**, *12*, 1932–1943. [CrossRef]

19. Zhang, M.; Wu, J.; Wang, L.; Xiao, K.; Wen, W. A simple method for fabricating multi-layer PDMS structures for 3D microfluidic chips. *Lab Chip* **2010**, *10*, 1199–1203. [CrossRef]

20. Walker, G.M.; Beebe, D.J. A passive pumping method for microfluidic devices. *Lab Chip* **2002**, *2*, 131–134. [CrossRef]

21. Koh, A.; Kang, D.; Xue, Y.; Lee, S.; Pielak, R.M.; Kim, J.; Hwang, T.; Min, S.; Banks, A.; Bastien, P.; et al. A soft, wearable microfluidic device for the capture, storage, and colorimetric sensing of sweat. *Sci. Transl. Med.* **2016**, *8*, 366ra165. [CrossRef]

22. Kaminski, T.S.; Scheler, O.; Garstecki, P. Droplet microfluidics for microbiology: Techniques, applications and challenges. *Lab Chip* **2016**, *16*, 2168–2187. [CrossRef] [PubMed]

23. Ray, T.; Choi, J.; Reeder, J.; Lee, S.P.; Aranyosi, A.J.; Ghaffari, R.; Rogers, J.A. Soft, skin-interfaced wearable systems for sports science and analytics. *Curr. Opin. Biomed. Eng.* **2019**, *9*, 47–56. [CrossRef]

24. Ying, M.; Bonifas, A.P.; Lu, N.; Su, Y.; Li, R.; Cheng, H.; Ameen, A.; Huang, Y.; Rogers, J.A. Silicon nanomembranes for fingertip electronics. *Nanotechnology* **2012**, *23*, 344004. [CrossRef] [PubMed]

25. He, S.; Feng, S.; Nag, A.; Afsarimanesh, N.; Han, T.; Mukhopadhyay, S.C. Recent progress in 3D printed mold-based sensors. *Sensors* **2020**, *20*, 703. [CrossRef] [PubMed]

26. Kim, J.; Gutruf, P.; Chiarelli, A.M.; Heo, S.Y.; Cho, K.; Xie, Z.; Banks, A.; Han, S.; Jang, K.-I.; Lee, J.W.; et al. Miniaturized battery-free wireless systems for wearable pulse oximetry. *Adv. Funct. Mater.* **2017**, *27*, 1604373. [CrossRef] [PubMed]

27. Kim, S.B.; Lee, K.; Raj, M.S.; Lee, B.; Reeder, J.T.; Koo, J.; Hourlier-Fargette, A.; Bandodkar, A.J.; Won, S.M.; Sekine, Y.; et al. Soft, skin-interfaced microfluidic systems with wireless, battery-free electronics for digital, real-time tracking of sweat loss and electrolyte composition. *Small* **2018**, *14*, 1802876. [CrossRef]

28. Krishnan, S.R.; Su, C.-J.; Xie, Z.; Patel, M.; Madhvapathy, S.R.; Xu, Y.; Freudman, J.; Ng, B.; Heo, S.Y.; Wang, H.; et al. Wireless, battery-free epidermal electronics for continuous, quantitative, multimodal thermal characterization of skin. *Small* **2018**, *14*, 1803192. [CrossRef]

29. Zhang, Y.; Castro, D.C.; Han, Y.; Wu, Y.; Guo, H.; Weng, Z.; Xue, Y.; Ausra, J.; Wang, X.; Li, R.; et al. Battery-free, lightweight, injectable microsystem for in vivo wireless pharmacology and optogenetics. *Proc. Natl. Acad. Sci. USA* **2019**, *116*, 21427–21437. [CrossRef]

30. Kim, J.; Banks, A.; Cheng, H.; Xie, Z.; Xu, S.; Jang, K.-I.; Lee, J.W.; Liu, Z.; Gutruf, P.; Huang, X.; et al. Epidermal electronics with advanced capabilities in near-field communication. *Small* **2015**, *11*, 906–912. [CrossRef]

31. Gao, W.; Nyein, H.Y.Y.; Shahpar, Z.; Fahad, H.M.; Chen, K.; Emaminejad, S.; Gao, Y.; Tai, L.-C.; Ota, H.; Wu, E.; et al. Wearable microsensor array for multiplexed heavy metal monitoring of body fluids. *ACS Sens.* **2016**, *1*, 866–874. [CrossRef]

32. Nyein, H.Y.Y.; Gao, W.; Shahpar, Z.; Emaminejad, S.; Challa, S.; Chen, K.; Fahad, H.M.; Tai, L.-C.; Ota, H.; Davis, R.W.; et al. A wearable electrochemical platform for noninvasive simultaneous monitoring of Ca^{2+} and pH. *ACS Nano* **2016**, *10*, 7216–7224. [CrossRef] [PubMed]

33. Reeder, J.T.; Choi, J.; Xue, Y.; Gutruf, P.; Hanson, J.; Liu, M.; Ray, T.; Bandodkar, A.J.; Avila, R.; Xia, W.; et al. Waterproof, electronics-enabled, epidermal microfluidic devices for sweat collection, biomarker analysis, and thermography in aquatic settings. *Sci. Adv.* **2019**, *5*, eaau6356. [CrossRef] [PubMed]

34. Zhang, Y.; Guo, H.; Kim, S.B.; Wu, Y.; Ostojich, D.; Park, S.H.; Wang, X.; Weng, Z.; Li, R.; Bandodkar, A.J.; et al. Passive sweat collection and colorimetric analysis of biomarkers relevant to kidney disorders using a soft microfluidic system. *Lab Chip* **2019**, *19*, 1545–1555. [CrossRef]

35. Kim, S.B.; Koo, J.; Yoon, J.; Hourlier-Fargette, A.; Lee, B.; Chen, S.; Jo, S.; Choi, J.; Oh, Y.S.; Lee, G.; et al. Soft, skin-interfaced microfluidic systems with integrated enzymatic assays for measuring the concentration of ammonia and ethanol in sweat. *Lab Chip* **2020**, *20*, 84–92. [CrossRef] [PubMed]

36. Kim, J.; Banks, A.; Xie, Z.; Heo, S.Y.; Gutruf, P.; Lee, J.W.; Xu, S.; Jang, K.-I.; Liu, F.; Brown, G.; et al. Miniaturized flexible electronic systems with wireless power and near-field communication capabilities. *Adv. Funct. Mater.* **2015**, *25*, 4761–4767. [CrossRef]

37. Valdés-Ramírez, G.; Li, Y.-C.; Kim, J.; Jia, W.; Bandodkar, A.J.; Nuñez-Flores, R.; Miller, P.R.; Wu, S.-Y.; Narayan, R.; Windmiller, J.R.; et al. Microneedle-based self-powered glucose sensor. *Electrochem. Commun.* **2014**, *47*, 58–62. [CrossRef]

38. Lau, S.; Fei, J.; Liu, H.; Chen, W.; Liu, R. Multilayered pyramidal dissolving microneedle patches with flexible pedestals for improving effective drug delivery. *J. Control. Release* **2017**, *265*, 113–119. [CrossRef] [PubMed]

39. Guo, S.; Lin, R.; Wang, L.; Lau, S.; Wang, Q.; Liu, R. Low melting point metal-based flexible 3D biomedical microelectrode array by phase transition method. *Mater. Sci. Eng. C* **2019**, *99*, 735–739. [CrossRef]

40. Li, W.; Tang, J.; Terry, R.N.; Li, S.; Brunie, A.; Callahan, R.L.; Noel, R.K.; Rodríguez, C.A.; Schwendeman, S.P.; Prausnitz, M.R. Long-acting reversible contraception by effervescent microneedle patch. *Sci. Adv.* **2019**, *5*, eaaw8145. [CrossRef]

41. Bariya, M.; Nyein, H.Y.Y.; Javey, A. Wearable sweat sensors. *Nat. Electron.* **2018**, *1*, 160–171. [CrossRef]

42. Yang, K.; Peretz-Soroka, H.; Liu, Y.; Lin, F. Novel developments in mobile sensing based on the integration of microfluidic devices and smartphones. *Lab Chip* **2016**, *16*, 943–958. [CrossRef] [PubMed]

43. Choi, J.; Ghaffari, R.; Baker, L.B.; Rogers, J.A. Skin-interfaced systems for sweat collection and analytics. *Sci. Adv.* **2018**, *4*, eaar3921. [CrossRef] [PubMed]

44. Heikenfeld, J.; Jajack, A.; Rogers, J.; Gutruf, P.; Tian, L.; Pan, T.; Li, R.; Khine, M.; Kim, J.; Wang, J.; et al. Wearable sensors: Modalities, challenges, and prospects. *Lab Chip* **2018**, *18*, 217–248. [CrossRef] [PubMed]

45. Bandodkar, A.J.; Jeang, W.J.; Ghaffari, R.; Rogers, J.A. Wearable sensors for biochemical sweat analysis. *Annu. Rev. Anal. Chem.* **2019**, *12*, 1–22. [CrossRef]

46. Farandos, N.M.; Yetisen, A.K.; Monteiro, M.J.; Lowe, C.R.; Yun, S.H. Contact lens sensors in ocular diagnostics. *Adv. Healthc. Mater.* **2015**, *4*, 792–810. [CrossRef]

47. Liu, Y.; Pharr, M.; Salvatore, G.A. Lab-on-skin: A review of flexible and stretchable electronics for wearable health monitoring. *ACS Nano* **2017**, *11*, 9614–9635. [CrossRef]

48. Yu, Y.; Nyein, H.Y.Y.; Gao, W.; Javey, A. Flexible electrochemical bioelectronics: The rise of in situ bioanalysis. *Adv. Mater.* **2019**, *32*, 1902083. [CrossRef]

49. Nag, A.; Afasrimanesh, N.; Feng, S.; Mukhopadhyay, S.C. Strain induced graphite/PDMS sensors for biomedical applications. *Sens. Actuator A Phys.* **2018**, *271*, 257–269. [CrossRef]

50. Nag, A.; Feng, S.; Mukhopadhyay, S.C.; Kosel, J.; Inglis, D. 3D printed mould-based graphite/PDMS sensor for low-force applications. *Sens. Actuator A Phys.* **2018**, *280*, 525–534. [CrossRef]

51. Lee, B.; Chen, S.; Sienko, K.H. A wearable device for real-time motion error detection and vibrotactile instructional cuing. *IEEE Trans. Neural Syst. Rehabil. Eng.* **2011**, *19*, 374–381. [CrossRef]

52. Haahr, R.G.; Duun, S.B.; Toft, M.H.; Belhage, B.; Larsen, J.; Birkelund, K.; Thomsen, E.V. An Electronic patch for wearable health monitoring by reflectance pulse oximetry. *IEEE Trans. Biomed. Circuits Syst.* **2012**, *6*, 45–53. [CrossRef] [PubMed]

53. Lochner, C.M.; Khan, Y.; Pierre, A.; Arias, A.C. All-organic optoelectronic sensor for pulse oximetry. *Nat. Commun.* **2014**, *5*, 5745. [CrossRef] [PubMed]

54. Feng, S.; Dong, T.; Yang, Z. Detection of urinary tract infections on lab-on-chip device by measuring photons emitted from ATP bioluminescence. In Proceedings of the 2014 36th Annual International Conference of the IEEE Engineering in Medicine and Biology Society, Chicago, IL, USA, 26–30 August 2014; pp. 3114–3117. [CrossRef]

55. Tseng, R.C.; Chen, C.-C.; Hsu, S.-M.; Chuang, H.-S. Contact-lens biosensors. *Sensors* **2018**, *18*, 2651. [CrossRef] [PubMed]

56. Kim, J.; Kim, M.; Lee, M.-S.; Kim, K.; Ji, S.; Kim, Y.-T.; Park, J.; Na, K.; Bae, K.-H.; Kim, H.K.; et al. Wearable smart sensor systems integrated on soft contact lenses for wireless ocular diagnostics. *Nat. Commun.* **2017**, *8*, 14997. [CrossRef]

57. Mannoor, M.S.; Tao, H.; Clayton, J.D.; Sengupta, A.; Kaplan, D.L.; Naik, R.R.; Verma, N.; Omenetto, F.G.; McAlpine, M.C. Graphene-based wireless bacteria detection on tooth enamel. *Nat. Commun.* **2012**, *3*, 763. [CrossRef]

58. Kim, J.; Imani, S.; de Araujo, W.R.; Warchall, J.; Valdés-Ramírez, G.; Paixão, T.R.L.C.; Mercier, P.P.; Wang, J. Wearable salivary uric acid mouthguard biosensor with integrated wireless electronics. *Biosens. Bioelectron.* **2015**, *74*, 1061–1068. [CrossRef]

59. Arakawa, T.; Kuroki, Y.; Nitta, H.; Chouhan, P.; Toma, K.; Sawada, S.-I.; Takeuchi, S.; Sekita, T.; Akiyoshi, K.; Minakuchi, S.; et al. Mouthguard biosensor with telemetry system for monitoring of saliva glucose: A novel cavitas sensor. *Biosens. Bioelectron.* **2016**, *84*, 106–111. [CrossRef]

60. Tseng, P.; Napier, B.; Garbarini, L.; Kaplan, D.L.; Omenetto, F.G. Functional, RF-trilayer sensors for tooth-mounted, wireless monitoring of the oral cavity and food consumption. *Adv. Mater.* **2018**, *30*, 1703257. [CrossRef]

61. Sekine, Y.; Kim, S.B.; Zhang, Y.; Bandodkar, A.J.; Xu, S.; Choi, J.; Irie, M.; Ray, T.R.; Kohli, P.; Kozai, N.; et al. A fluorometric skin-interfaced microfluidic device and smartphone imaging module for in situ quantitative analysis of sweat chemistry. *Lab Chip* **2018**, *18*, 2178–2186. [CrossRef]

62. Bandodkar, A.J.; Choi, J.; Lee, S.P.; Jeang, W.J.; Agyare, P.; Gutruf, P.; Wang, S.; Sponenburg, R.A.; Reeder, J.T.; Schon, S.; et al. Soft, skin-interfaced microfluidic systems with passive galvanic stopwatches for precise chronometric sampling of sweat. *Adv. Mater.* **2019**, *31*, 1902109. [CrossRef]

63. Choi, J.; Kang, D.; Han, S.; Kim, S.B.; Rogers, J.A. Thin, soft, skin-mounted microfluidic networks with capillary bursting valves for chrono-sampling of sweat. *Adv. Healthc. Mater.* **2017**, *6*, 1601355. [CrossRef] [PubMed]

64. Schazmann, B.; Morris, D.; Slater, C.; Beirne, S.; Fay, C.; Reuveny, R.; Moyna, N.; Diamond, D. A wearable electrochemical sensor for the real-time measurement of sweat sodium concentration. *Anal. Methods* **2010**, *2*, 342–348. [CrossRef]

65. Rose, D.P.; Ratterman, M.E.; Griffin, D.K.; Hou, L.; Kelley-Loughnane, N.; Naik, R.R.; Hagen, J.A.; Papautsky, I.; Heikenfeld, J.C. Adhesive RFID sensor patch for monitoring of sweat electrolytes. *IEEE Trans. Biomed. Eng.* **2015**, *62*, 1457–1465. [CrossRef]

66. Bandodkar, A.J.; Molinnus, D.; Mirza, O.; Guinovart, T.; Windmiller, J.R.; Valdés-Ramírez, G.; Andrade, F.J.; Schöning, M.J.; Wang, J. Epidermal tattoo potentiometric sodium sensors with wireless signal transduction for continuous non-invasive sweat monitoring. *Biosens. Bioelectron.* **2014**, *54*, 603–609. [CrossRef] [PubMed]

67. Anastasova, S.; Crewther, B.; Bembnowicz, P.; Curto, V.; Ip, H.M.D.; Rosa, B.; Yang, G.-Z. A wearable multisensing patch for continuous sweat monitoring. *Biosens. Bioelectron.* **2017**, *93*, 139–145. [CrossRef]

68. Guinovart, T.; Bandodkar, A.J.; Windmiller, J.R.; Andrade, F.J.; Wang, J. A potentiometric tattoo sensor for monitoring ammonium in sweat. *Analyst* **2013**, *138*, 7031–7038. [CrossRef]

69. Emaminejad, S.; Gao, W.; Wu, E.; Davies, Z.A.; Nyein, H.Y.; Challa, S.; Ryan, S.P.; Fahad, H.M.; Chen, K.; Shahpar, Z.; et al. Autonomous sweat extraction and analysis applied to cystic fibrosis and glucose monitoring using a fully integrated wearable platform. *Proc. Natl. Acad. Sci. USA* **2017**, *114*, 4625–4630. [CrossRef]

70. Cao, Q.; Liang, B.; Tu, T.; Wei, J.; Fang, L.; Ye, X. Three-dimensional paper-based microfluidic electrochemical integrated devices (3D-PMED) for wearable electrochemical glucose detection. *RSC Adv.* **2019**, *9*, 5674–5681. [CrossRef]

71. Lin, Y.; Bariya, M.; Nyein, H.Y.Y.; Kivimäki, L.; Uusitalo, S.; Jansson, E.; Ji, W.; Yuan, Z.; Happonen, T.; Liedert, C.; et al. Porous enzymatic membrane for nanotextured glucose sweat sensors with high stability toward reliable noninvasive health monitoring. *Adv. Funct. Mater.* **2019**, *29*, 1902521. [CrossRef]

72. Zhao, J.; Lin, Y.; Wu, J.; Nyein, H.Y.Y.; Bariya, M.; Tai, L.-C.; Chao, M.; Ji, W.; Zhang, G.; Fan, Z.; et al. A fully integrated and self-powered smartwatch for continuous sweat glucose monitoring. *ACS Sens.* **2019**, *4*, 1925–1933. [CrossRef]

73. Bandodkar, A.J.; Gutruf, P.; Choi, J.; Lee, K.; Sekine, Y.; Reeder, J.T.; Jeang, W.J.; Aranyosi, A.J.; Lee, S.P.; Model, J.B.; et al. Battery-free, skin-interfaced microfluidic/electronic systems for simultaneous electrochemical, colorimetric, and volumetric analysis of sweat. *Sci. Adv.* **2019**, *5*, eaav3294. [CrossRef] [PubMed]

74. Choi, J.; Bandodkar, A.J.; Reeder, J.T.; Ray, T.R.; Turnquist, A.; Kim, S.B.; Nyberg, N.; Hourlier-Fargette, A.; Model, J.B.; Aranyosi, A.J.; et al. Soft, skin-integrated multifunctional microfluidic systems for accurate colorimetric analysis of sweat biomarkers and temperature. *ACS Sens.* **2019**, *4*, 379–388. [CrossRef] [PubMed]

75. Zhang, Z.; Azizi, M.; Lee, M.; Davidowsky, P.; Lawrence, P.; Abbaspourrad, A. A versatile, cost-effective, and flexible wearable biosensor for in situ and ex situ sweat analysis, and personalized nutrition assessment. *Lab Chip* **2019**, *19*, 3448–3460. [CrossRef] [PubMed]

76. Kim, J.; de Araujo, W.R.; Samek, I.A.; Bandodkar, A.J.; Jia, W.; Brunetti, B.; Paixão, T.R.L.C.; Wang, J. Wearable temporary tattoo sensor for real-time trace metal monitoring in human sweat. *Electrochem. Commun.* **2015**, *51*, 41–45. [CrossRef]

77. Tai, L.-C.; Liaw, T.S.; Lin, Y.; Nyein, H.Y.Y.; Bariya, M.; Ji, W.; Hettick, M.; Zhao, C.; Zhao, J.; Hou, L.; et al. Wearable sweat band for noninvasive levodopa monitoring. *Nano Lett.* **2019**, *19*, 6346–6351. [CrossRef] [PubMed]

78. Buono, M.J.; Lee, N.V.L.; Miller, P.W. The relationship between exercise intensity and the sweat lactate excretion rate. *J. Physiol. Sci.* **2010**, *60*, 103–107. [CrossRef]

79. Tricoli, A.; Neri, G. Miniaturized bio-and chemical-sensors for point-of-care monitoring of chronic kidney diseases. *Sensors* **2018**, *18*, 942. [CrossRef]

80. Torrente-Rodríguez, R.M.; Tu, J.; Yang, Y.; Min, J.; Wang, M.; Song, Y.; Yu, Y.; Xu, C.; Ye, C.; IsHak, W.W.; et al. Investigation of cortisol dynamics in human sweat using a graphene-based wireless mHealth System. *Matter* **2020**, *2*, 921–937. [CrossRef]

81. Heikenfeld, J.; Jajack, A.; Feldman, B.; Granger, S.W.; Gaitonde, S.; Begtrup, G.; Katchman, B.A. Accessing analytes in biofluids for peripheral biochemical monitoring. *Nat. Biotechnol.* **2019**, *37*, 407–419. [CrossRef]

82. Ventrelli, L.; Strambini, L.M.; Barillaro, G. Microneedles for Transdermal Biosensing: Current Picture and Future Direction. *Adv. Healthc. Mater.* **2015**, *4*, 2606–2640. [CrossRef]

83. Gowers, S.A.N.; Freeman, D.M.E.; Rawson, T.M.; Rogers, M.L.; Wilson, R.C.; Holmes, A.H.; Cass, A.E.; O'Hare, D. Development of a Minimally Invasive Microneedle-Based Sensor for Continuous Monitoring of β-Lactam Antibiotic Concentrations in Vivo. *ACS Sens.* **2019**, *4*, 1072–1080. [CrossRef]

84. Kim, J.; Valdés-Ramírez, G.; Bandodkar, A.J.; Jia, W.; Martinez, A.G.; Ramírez, J.; Mercier, P.; Wang, J. Non-invasive mouthguard biosensor for continuous salivary monitoring of metabolites. *Analyst* **2014**, *139*, 1632–1636. [CrossRef]

85. Magiorkinis, E.; Diamantis, A. The fascinating story of urine examination: From uroscopy to the era of microscopy and beyond. *Diagn. Cytopathol.* **2015**, *43*, 1020–1036. [CrossRef] [PubMed]

86. Gao, L.; Xu, T.; Huang, G.; Jiang, S.; Gu, Y.; Chen, F. Oral microbiomes: More and more importance in oral cavity and whole body. *Protein Cell* **2018**, *9*, 488–500. [CrossRef] [PubMed]

87. Feng, S.; Caire, R.; Cortazar, B.; Turan, M.; Wong, A.; Ozcan, A. Immunochromatographic diagnostic test analysis using google glass. *ACS Nano* **2014**, *8*, 3069–3079. [CrossRef] [PubMed]

88. Chung, H.U.; Kim, B.H.; Lee, J.Y.; Lee, J.; Xie, Z.; Ibler, E.M.; Lee, K.; Banks, A.; Jeong, J.Y.; Kim, J.; et al. Binodal, wireless epidermal electronic systems with in-sensor analytics for neonatal intensive care. *Science* **2019**, *363*, eaau0780. [CrossRef]

89. Jeerapan, I.; Sempionatto, J.R.; Pavinatto, A.; You, J.-M.; Wang, J. Stretchable biofuel cells as wearable textile-based self-powered sensors. *J. Mater. Chem. A* **2016**, *4*, 18342–18353. [CrossRef]

90. Xu, B.; Akhtar, A.; Liu, Y.; Chen, H.; Yeo, W.-H.; Park, S.I.; Boyce, B.; Kim, H.; Yu, J.; Lai, H.-Y.; et al. An epidermal stimulation and sensing platform for sensorimotor prosthetic control, management of lower back exertion, and electrical muscle activation. *Adv. Mater.* **2016**, *28*, 4462–4471. [CrossRef]

91. Kim, S.B.; Zhang, Y.; Won, S.M.; Bandodkar, A.J.; Sekine, Y.; Xue, Y.; Koo, J.; Harshman, S.W.; Martin, J.A.; Park, J.M.; et al. Super-absorbent polymer valves and colorimetric chemistries for time-sequenced discrete sampling and chloride analysis of sweat via skin-mounted soft microfluidics. *Small* **2018**, *14*, 1703334. [CrossRef]

92. Lee, C.; Jug, L.; Meng, E. High strain biocompatible polydimethylsiloxane-based conductive graphene and multiwalled carbon nanotube nanocomposite strain sensors. *Appl. Phys. Lett.* **2013**, *102*, 183511. [CrossRef]

93. Nayak, L.; Mohanty, S.; Nayak, S.K.; Ramadoss, A. A review on inkjet printing of nanoparticle inks for flexible electronics. *J. Mater. Chem. C* **2019**, *7*, 8771–8795. [CrossRef]

94. Munje, R.D.; Muthukumar, S.; Prasad, S. Lancet-free and label-free diagnostics of glucose in sweat using zinc oxide based flexible bioelectronics. *Sens. Actuators B Chem.* **2017**, *238*, 482–490. [CrossRef]

95. Lee, C.-Y.; Lei, K.F.; Tsai, S.-W.; Tsang, N.-M. Development of graphene-based sensors on paper substrate for the measurement of pH value of analyte. *Biochip J.* **2016**, *10*, 182–188. [CrossRef]

96. Kanaparthi, S.; Badhulika, S. Low cost, flexible and biodegradable touch sensor fabricated by solvent-free processing of graphite on cellulose paper. *Sens. Actuators B Chem.* **2017**, *242*, 857–864. [CrossRef]

97. McNichol, L.; Lund, C.; Rosen, T.; Gray, M. Medical Adhesives and Patient Safety: State of the Science Consensus Statements for the Assessment, Prevention, and Treatment of Adhesive-Related Skin Injuries. *J. Wound Ostomy Cont. Nurs.* **2013**, *40*, 365–380. [CrossRef] [PubMed]

98. Rippon, M.; White, R.; Davies, P. Skin adhesives and their role in wound dressings. *Wounds UK* **2007**, *3*, 76.

99. Packham, D.E. *Handbook of Adhesion*, 2nd ed.; Wiley: West Sussex, UK, 2005.

100. Dykes, P.J.; Heggie, R.; Hill, S.A. Effects of adhesive dressings on the stratum corneum of the skin. *J. Wound Care* **2001**, *10*, 7–10. [CrossRef]

101. Wolff, H.-M.; Dodou, K. Investigations on the Viscoelastic Performance of Pressure Sensitive Adhesives in Drug-in-Adhesive Type Transdermal Films. *Pharm. Res.* **2014**, *31*, 2186–2202. [CrossRef]

102. Lee, B.K.; Ryu, J.H.; Baek, I.-B.; Kim, Y.; Jang, W.I.; Kim, S.-H.; Yoon, Y.S.; Kim, S.H.; Hong, S.-G.; Byun, S.; et al. Silicone-Based Adhesives with Highly Tunable Adhesion Force for Skin-Contact Applications. *Adv. Healthc. Mater.* **2017**, *6*, 1700621. [CrossRef]

103. Liao, X.; Liao, Q.; Yan, X.; Liang, Q.; Si, H.; Li, M.; Wu, H.; Cao, S.; Zhang, Y. Flexible and highly sensitive strain sensors fabricated by pencil drawn for wearable monitor. *Adv. Funct. Mater.* **2015**, *25*, 2395–2401. [CrossRef]

104. Lin, Z.; Young, S.; Chang, S. CO_2 gas sensors based on carbon nanotube thin films using a simple transfer method on flexible substrate. *IEEE Sens. J.* **2015**, *15*, 7017–7020. [CrossRef]

105. Zhang, P.; Chen, Y.; Li, Y.; Zhang, Y.; Zhang, J.; Huang, L. A flexible strain sensor based on the porous structure of a carbon black/carbon nanotube conducting network for human motion detection. *Sensors* **2020**, *20*, 1154. [CrossRef] [PubMed]

106. Yin, B.; Wen, Y.; Hong, T.; Xie, Z.; Yuan, G.; Ji, Q.; Jia, H. Highly stretchable, ultrasensitive, and wearable strain sensors based on facilely prepared reduced graphene oxide woven fabrics in an ethanol flame. *ACS Appl. Mater. Interfaces* **2017**, *9*, 32054–32064. [CrossRef]

107. Amjadi, M.; Pichitpajongkit, A.; Lee, S.; Ryu, S.; Park, I. Highly stretchable and sensitive strain sensor based on silver nanowire–elastomer nanocomposite. *ACS Nano* **2014**, *8*, 5154–5163. [CrossRef] [PubMed]

108. Amjadi, M.; Kyung, K.-U.; Park, I.; Sitti, M. Stretchable, skin-mountable, and wearable strain sensors and their potential applications: A review. *Adv. Funct. Mater.* **2016**, *26*, 1678–1698. [CrossRef]

109. Rogers, J.A.; Someya, T.; Huang, Y. Materials and mechanics for stretchable electronics. *Science* **2010**, *327*, 1603–1607. [CrossRef] [PubMed]

110. Sekitani, T.; Someya, T. Stretchable, large-area organic electronics. *Adv. Mater.* **2010**, *22*, 2228–2246. [CrossRef]

111. Han, Y.L.; Liu, H.; Ouyang, C.; Lu, T.J.; Xu, F. Liquid on paper: Rapid prototyping of soft functional components for paper electronics. *Sci. Rep.* **2015**, *5*, 11488. [CrossRef]

112. Li, G.; Wu, X.; Lee, D.-W. Selectively plated stretchable liquid metal wires for transparent electronics. *Sens. Actuators B Chem.* **2015**, *221*, 1114–1119. [CrossRef]

113. Eom, J.; Jaisutti, R.; Lee, H.; Lee, W.; Heo, J.-S.; Lee, J.-Y.; Park, S.K.; Kim, Y.-H. Highly sensitive textile strain sensors and wireless user-interface devices using all-polymeric conducting fibers. *ACS Appl. Mater. Interfaces* **2017**, *9*, 10190–10197. [CrossRef]

114. Shin, K.-Y.; Lee, J.S.; Jang, J. Highly sensitive, wearable and wireless pressure sensor using free-standing ZnO nanoneedle/PVDF hybrid thin film for heart rate monitoring. *Nano Energy* **2016**, *22*, 95–104. [CrossRef]

115. Luheng, W.; Tianhuai, D.; Peng, W. Influence of carbon black concentration on piezoresistivity for carbon-black-filled silicone rubber composite. *Carbon* **2009**, *47*, 3151–3157. [CrossRef]

116. Zheng, Y.; Li, Y.; Li, Z.; Wang, Y.; Dai, K.; Zheng, G.; Liu, C.; Shen, C. The effect of filler dimensionality on the electromechanical performance of polydimethylsiloxane based conductive nanocomposites for flexible strain sensors. *Compos. Sci. Technol.* **2017**, *139*, 64–73. [CrossRef]

117. Tian, L.; Li, Y.; Webb, R.C.; Krishnan, S.; Bian, Z.; Song, J.; Ning, X.; Crawford, K.; Kurniawan, J.; Bonifas, A.; et al. Flexible and stretchable 3ω sensors for thermal characterization of human skin. *Adv. Funct. Mater.* **2017**, *27*, 1701282. [CrossRef]

118. Kim, D.-H.; Lu, N.; Ma, R.; Kim, Y.-S.; Kim, R.-H.; Wang, S.; Wu, J.; Won, S.M.; Tao, H.; Islam, A.; et al. Epidermal electronics. *Science* **2011**, *333*, 838–843. [CrossRef]

119. Chen, S.; Wei, Y.; Wei, S.; Lin, Y.; Liu, L. Ultrasensitive cracking-assisted strain sensors based on silver nanowires/graphene hybrid particles. *ACS Appl. Mater. Interfaces* **2016**, *8*, 25563–25570. [CrossRef]

120. Zhao, Y.; Huang, X. Mechanisms and materials of flexible and stretchable skin sensors. *Micromachines* **2017**, *8*, 69. [CrossRef]

121. Li, Z.; Liu, H.; Ouyang, C.; Wee, W.H.; Cui, X.; Lu, T.J.; Pingguan-Murphy, B.; Li, F.; Xu, F. Recent advances in pen-based writing electronics and their emerging applications. *Adv. Funct. Mater.* **2016**, *26*, 165–180. [CrossRef]

122. Yin, Z.; Huang, Y.; Bu, N.; Wang, X.; Xiong, Y. Inkjet printing for flexible electronics: Materials, processes and equipments. *Chin. Sci. Bull.* **2010**, *55*, 3383–3407. [CrossRef]

123. Angeli, M.A.C.; Caronna, F.; Cramer, T.; Gastaldi, D.; Magagnin, L.; Fraboni, B.; Vena, P. Strain mapping inkjet-printed resistive sensors array. *IEEE Sens. J.* **2020**, *20*, 4087–4095. [CrossRef]

124. Karim, N.; Afroj, S.; Malandraki, A.; Butterworth, S.; Beach, C.; Rigout, M.; Novoselov, K.S.; Casson, A.J.; Yeates, S.G. All inkjet-printed graphene-based conductive patterns for wearable e-textile applications. *J. Mater. Chem. C* **2017**, *5*, 11640–11648. [CrossRef]

125. Neophytou, M.; Cambarau, W.; Hermerschmidt, F.; Waldauf, C.; Christodoulou, C.; Pacios, R.; Choulis, S.A. Inkjet-printed polymer–fullerene blends for organic electronic applications. *Microelectron. Eng.* **2012**, *95*, 102–106. [CrossRef]

126. Oncescu, V.; O'Dell, D.; Erickson, D. Smartphone based health accessory for colorimetric detection of biomarkers in sweat and saliva. *Lab Chip* **2013**, *13*, 3232–3238. [CrossRef] [PubMed]

127. Zhang, Q.; Jiang, D.; Xu, C.; Yuancai, G.; Liu, X.; Wei, Q.; Huang, L.; Ren, X.; Wang, C.; Wang, Y. Wearable electrochemical biosensor based on molecularly imprinted Ag nanowires for noninvasive monitoring lactate in human sweat. *Sens. Actuators B Chem.* **2020**, *320*, 128325. [CrossRef]

128. Zhang, Y.; Chen, Y.; Huang, J.; Liu, Y.; Peng, J.; Chen, S.; Song, K.; Ouyang, X.; Cheng, H.; Wang, X. Skin-interfaced microfluidic devices with one-opening chambers and hydrophobic valves for sweat collection and analysis. *Lab Chip* **2020**, 2635–2645. [CrossRef]

129. Ogasawara, K.; Tsuru, T.; Mitsubayashi, K.; Karube, I. Electrical conductivity of tear fluid in healthy persons and keratoconjunctivitis sicca patients measured by a flexible conductimetric sensor. *Graefes Arch. Clin. Exp. Ophthalmol.* **1996**, *234*, 542–546. [CrossRef]

130. Shibasaki, K.; Kimura, M.; Ikarashi, R.; Yamaguchi, A.; Watanabe, T. Uric acid concentration in saliva and its changes with the patients receiving treatment for hyperuricemia. *Metabolomics* **2012**, *8*, 484–491. [CrossRef]

131. Soukup, M.; Biesiada, I.; Henderson, A.; Idowu, B.; Rodeback, D.; Ridpath, L.; Bridges, E.G.; Nazar, A.M.; Bridges, K.G. Salivary uric acid as a noninvasive biomarker of metabolic syndrome. *Diab. Metab. Syndr.* **2012**, *4*, 14. [CrossRef]

132. Girotti, S.; Bolelli, L.; Roda, A.; Gentilomi, G.; Musiani, M. Improved detection of toxic chemicals using bioluminescent bacteria. *Anal. Chim. Acta* **2002**, *471*, 113–120. [CrossRef]

133. Petänen, T.; Romantschuk, M. Use of bioluminescent bacterial sensors as an alternative method for measuring heavy metals in soil extracts. *Anal. Chim. Acta* **2002**, *456*, 55–61. [CrossRef]

134. Jouanneau, S.; Durand, M.-J.; Courcoux, P.; Blusseau, T.; Thouand, G. Improvement of the Identification of Four Heavy Metals in Environmental Samples by Using Predictive Decision Tree Models Coupled with a Set of Five Bioluminescent Bacteria. *Environ. Sci. Technol.* **2011**, *45*, 2925–2931. [CrossRef] [PubMed]

135. Yeh, H.-W.; Ai, H.-W. Development and Applications of Bioluminescent and Chemiluminescent Reporters and Biosensors. *Annu. Rev. Anal. Chem.* **2019**, *12*, 129–150. [CrossRef] [PubMed]

136. Roda, A.; Cevenini, L.; Michelini, E.; Branchini, B.R. A portable bioluminescence engineered cell-based biosensor for on-site applications. *Biosens. Bioelectron.* **2011**, *26*, 3647–3653. [CrossRef] [PubMed]

137. Burlage, R.S.; Sayler, G.S.; Larimer, F. Monitoring of naphthalene catabolism by bioluminescence with nah-lux transcriptional fusions. *J. Bacteriol.* **1990**, *172*, 4749–4757. [CrossRef]

138. Cevenini, L.; Calabretta, M.M.; Tarantino, G.; Michelini, E.; Roda, A. Smartphone-interfaced 3D printed toxicity biosensor integrating bioluminescent "sentinel cells". *Sens. Actuators B Chem.* **2016**, *225*, 249–257. [CrossRef]

139. Araki, H.; Kim, J.; Zhang, S.; Banks, A.; Crawford, K.E.; Sheng, X.; Gutruf, P.; Shi, Y.; Pielak, R.M.; Rogers, J.A. Materials and device designs for an epidermal UV colorimetric dosimeter with near field communication capabilities. *Adv. Funct. Mater.* **2017**, *27*, 1604465. [CrossRef]

140. Wang, C.; Li, X.; Hu, H.; Zhang, L.; Huang, Z.; Lin, M.; Zhang, Z.; Yin, Z.; Huang, B.; Gong, H.; et al. Monitoring of the central blood pressure waveform via a conformal ultrasonic device. *Nat. Biomed. Eng.* **2018**, *2*, 687–695. [CrossRef]

141. Boutry, C.M.; Beker, L.; Kaizawa, Y.; Vassos, C.; Tran, H.; Hinckley, A.C.; Pfattner, R.; Niu, S.; Li, J.; Claverie, J.; et al. Biodegradable and flexible arterial-pulse sensor for the wireless monitoring of blood flow. *Nat. Biomed. Eng.* **2019**, *3*, 47–57. [CrossRef]

142. Tian, L.; Zimmerman, B.; Akhtar, A.; Yu, K.J.; Moore, M.; Wu, J.; Larsen, R.J.; Lee, J.W.; Li, J.; Liu, Y.; et al. Large-area MRI-compatible epidermal electronic interfaces for prosthetic control and cognitive monitoring. *Nat. Biomed. Eng.* **2019**, *3*, 194–205. [CrossRef]

143. Qi, M.; Huang, J.; Wei, H.; Cao, C.; Feng, S.; Guo, Q.; Goldys, E.M.; Li, R.; Liu, G. Graphene oxide thin film with dual function integrated into a nanosandwich device for in vivo monitoring of interleukin-6. *ACS Appl. Mater. Interfaces* **2017**, *9*, 41659–41668. [CrossRef]

144. Zhang, Y.; Mickle, A.D.; Gutruf, P.; McIlvried, L.A.; Guo, H.; Wu, Y.; Golden, J.P.; Xue, Y.; Grajales-Reyes, J.G.; Wang, X.; et al. Battery-free, fully implantable optofluidic cuff system for wireless optogenetic and pharmacological neuromodulation of peripheral nerves. *Sci. Adv.* **2019**, *5*, eaaw5296. [CrossRef] [PubMed]

145. Choi, J.; Xue, Y.; Xia, W.; Ray, T.R.; Reeder, J.T.; Bandodkar, A.J.; Kang, D.; Xu, S.; Huang, Y.; Rogers, J.A. Soft, skin-mounted microfluidic systems for measuring secretory fluidic pressures generated at the surface of the skin by eccrine sweat glands. *Lab Chip* **2017**, *17*, 2572–2580. [CrossRef] [PubMed]

146. Jagannath, B.; Lin, K.-C.; Pali, M.; Sankhala, D.; Muthukumar, S.; Prasad, S. A sweat-based wearable enabling technology for real-time monitoring of IL-1β and CRP as potential markers for inflammatory bowel disease. *Inflamm. Bowel Dis.* **2020**, *26*, 1533–1542. [CrossRef] [PubMed]

147. Hao, Z.; Wang, Z.; Li, Y.; Zhu, Y.; Wang, X.; De Moraes, C.G.; Pan, Y.; Zhao, X.; Lin, Q. Measurement of cytokine biomarkers using an aptamer-based affinity graphene nanosensor on a flexible substrate toward wearable applications. *Nanoscale* **2018**, *10*, 21681–21688. [CrossRef]

148. Parlak, O.; Keene, S.T.; Marais, A.; Curto, V.F.; Salleo, A. Molecularly selective nanoporous membrane-based wearable organic electrochemical device for noninvasive cortisol sensing. *Sci. Adv.* **2018**, *4*, eaar2904. [CrossRef] [PubMed]

149. Kinnamon, D.; Ghanta, R.; Lin, K.-C.; Muthukumar, S.; Prasad, S. Portable biosensor for monitoring cortisol in low-volume perspired human sweat. *Sci. Rep.* **2017**, *7*, 13312. [CrossRef]

150. Yang, Y.; Song, Y.; Bo, X.; Min, J.; Pak, O.S.; Zhu, L.; Wang, M.; Tu, J.; Kogan, A.; Zhang, H.; et al. A laser-engraved wearable sensor for sensitive detection of uric acid and tyrosine in sweat. *Nat. Biotechnol.* **2019**, 217–224. [CrossRef]

151. Qiao, L.; Benzigar, M.R.; Subramony, J.A.; Lovell, N.H.; Liu, G. Advances in sweat wearables: Sample extraction, real-time biosensing, and flexible platforms. *ACS Appl. Mater. Interfaces* **2020**, *12*, 34337–34361. [CrossRef]

152. Shawcross, D.L.; Shabbir, S.S.; Taylor, N.J.; Hughes, R.D. Ammonia and the neutrophil in the pathogenesis of hepatic encephalopathy in cirrhosis. *Hepatology* **2010**, *51*, 1062–1069. [CrossRef]

153. Lee, H.; Choi, T.K.; Lee, Y.B.; Cho, H.R.; Ghaffari, R.; Wang, L.; Choi, H.J.; Chung, T.D.; Lu, N.; Hyeon, T.; et al. A graphene-based electrochemical device with thermoresponsive microneedles for diabetes monitoring and therapy. *Nat. Nanotechnol.* **2016**, *11*, 566. [CrossRef]

154. Huang, C.-T.; Chen, M.-L.; Huang, L.-L.; Mao, I.F. Uric acid and urea in human sweat. *Chin. J. Physiol.* **2002**, *45*, 109–115. [CrossRef] [PubMed]

155. Farrell, P.M.; Rosenstein, B.J.; White, T.B.; Accurso, F.J.; Castellani, C.; Cutting, G.R.; Durie, P.R.; LeGrys, V.A.; Massie, J.; Parad, R.B.; et al. Guidelines for diagnosis of cystic fibrosis in newborns through older adults: Cystic fibrosis foundation consensus report. *J. Pediatr.* **2008**, *153*, S4–S14. [CrossRef] [PubMed]

156. Tai, L.-C.; Gao, W.; Chao, M.; Bariya, M.; Ngo, Q.P.; Shahpar, Z.; Nyein, H.Y.Y.; Park, H.; Sun, J.; Jung, Y.; et al. Methylxanthine drug monitoring with wearable sweat sensors. *Adv. Mater.* **2018**, *30*, 1707442. [CrossRef] [PubMed]

157. Rawson, T.M.; Gowers, S.A.N.; Freeman, D.M.E.; Wilson, R.C.; Sharma, S.; Gilchrist, M.; MacGowan, A.; Lovering, A.; Bayliss, M.; Kyriakides, M.; et al. Microneedle biosensors for real-time, minimally invasive drug monitoring of phenoxymethylpenicillin: A first-in-human evaluation in healthy volunteers. *Lancet Digit. Health* **2019**, *1*, e335–e343. [CrossRef]

158. Nyein, H.Y.Y.; Tai, L.-C.; Ngo, Q.P.; Chao, M.; Zhang, G.B.; Gao, W.; Bariya, M.; Bullock, J.; Kim, H.; Fahad, H.M.; et al. A wearable microfluidic sensing patch for dynamic sweat secretion analysis. *ACS Sens.* **2018**, *3*, 944–952. [CrossRef]

159. Liang, B.; Cao, Q.; Mao, X.; Pan, W.; Tu, T.; Fang, L.; Ye, X. An integrated paper-based microfluidic device for real-time sweat potassium monitoring. *IEEE Sens. J.* **2020**. [CrossRef]

160. Miller, P.R.; Xiao, X.; Brener, I.; Burckel, D.B.; Narayan, R.; Polsky, R. Microneedle-based transdermal sensor for on-chip potentiometric determination of k+. *Adv. Healthc. Mater.* **2014**, *3*, 876–881. [CrossRef]

161. Keene, S.T.; Fogarty, D.; Cooke, R.; Casadevall, C.D.; Salleo, A.; Parlak, O. Wearable organic electrochemical transistor patch for multiplexed sensing of calcium and ammonium ions from human perspiration. *Adv. Healthc. Mater.* **2019**, *8*, 1901321. [CrossRef]

162. Xu, G.; Cheng, C.; Yuan, W.; Liu, Z.; Zhu, L.; Li, X.; Lu, Y.; Chen, Z.; Liu, J.; Cui, Z.; et al. Smartphone-based battery-free and flexible electrochemical patch for calcium and chloride ions detections in biofluids. *Sens. Actuators B Chem.* **2019**, *297*, 126743. [CrossRef]

163. Bollella, P.; Sharma, S.; Cass, A.E.G.; Antiochia, R. Microneedle-based biosensor for minimally-invasive lactate detection. *Biosens. Bioelectron.* **2019**, *123*, 152–159. [CrossRef]

164. Windmiller, J.R.; Bandodkar, A.J.; Valdés-Ramírez, G.; Parkhomovsky, S.; Martinez, A.G.; Wang, J. Electrochemical sensing based on printable temporary transfer tattoos. *Chem. Commun.* **2012**, *48*, 6794–6796. [CrossRef] [PubMed]

165. Windmiller, J.R.; Bandodkar, A.J.; Parkhomovsky, S.; Wang, J. Stamp transfer electrodes for electrochemical sensing on non-planar and oversized surfaces. *Analyst* **2012**, *137*, 1570–1575. [CrossRef] [PubMed]

166. Liu, Y.-L.; Liu, R.; Qin, Y.; Qiu, Q.-F.; Chen, Z.; Cheng, S.-B.; Huang, W.-H. Flexible electrochemical urea sensor based on surface molecularly imprinted nanotubes for detection of human sweat. *Anal. Chem.* **2018**, *90*, 13081–13087. [CrossRef] [PubMed]

167. Chang, H.; Zheng, M.; Yu, X.; Than, A.; Seeni, R.Z.; Kang, R.; Tian, J.; Khanh, D.P.; Liu, L.; Chen, P.; et al. A swellable microneedle patch to rapidly extract skin interstitial fluid for timely metabolic analysis. *Adv. Mater.* **2017**, *29*, 1702243. [CrossRef]

168. Ng, K.W.; Lau, W.M.; Williams, A.C. Towards pain-free diagnosis of skin diseases through multiplexed microneedles: Biomarker extraction and detection using a highly sensitive blotting method. *Drug Deliv. Transl. Res.* **2015**, *5*, 387–396. [CrossRef]

169. Windmiller, J.R.; Valdés-Ramírez, G.; Zhou, N.; Zhou, M.; Miller, P.R.; Jin, C.; Brozik, S.M.; Polsky, R.; Katz, E.; Narayan, R.; et al. Bicomponent microneedle array biosensor for minimally-invasive glutamate monitoring. *Electroanalysis* **2011**, *23*, 2302–2309. [CrossRef]

170. Ranamukhaarachchi, S.A.; Padeste, C.; Dübner, M.; Häfeli, U.O.; Stoeber, B.; Cadarso, V.J. Integrated hollow microneedle-optofluidic biosensor for therapeutic drug monitoring in sub-nanoliter volumes. *Sci. Rep.* **2016**, *6*, 29075. [CrossRef]

171. Kim, J.; Jeerapan, I.; Imani, S.; Cho, T.N.; Bandodkar, A.; Cinti, S.; Mercier, P.P.; Wang, J. Noninvasive alcohol monitoring using a wearable tattoo-based iontophoretic-biosensing system. *ACS Sens.* **2016**, *1*, 1011–1019. [CrossRef]

172. Mohan, A.M.V.; Windmiller, J.R.; Mishra, R.K.; Wang, J. Continuous minimally-invasive alcohol monitoring using microneedle sensor arrays. *Biosens. Bioelectron.* **2017**, *91*, 574–579. [CrossRef]

173. Miller, P.R.; Skoog, S.A.; Edwards, T.L.; Lopez, D.M.; Wheeler, D.R.; Arango, D.C.; Xiao, X.; Brozik, S.M.; Wang, J.; Polsky, R.; et al. Multiplexed microneedle-based biosensor array for characterization of metabolic acidosis. *Talanta* **2012**, *88*, 739–742. [CrossRef]

174. Corrie, S.R.; Fernando, G.J.P.; Crichton, M.L.; Brunck, M.E.G.; Anderson, C.D.; Kendall, M.A.F. Surface-modified microprojection arrays for intradermal biomarker capture, with low non-specific protein binding. *Lab Chip* **2010**, *10*, 2655–2658. [CrossRef] [PubMed]

175. Coffey, J.W.; Corrie, S.R.; Kendall, M.A.F. Early circulating biomarker detection using a wearable microprojection array skin patch. *Biomaterials* **2013**, *34*, 9572–9583. [CrossRef] [PubMed]

176. Biagi, S.; Ghimenti, S.; Onor, M.; Bramanti, E. Simultaneous determination of lactate and pyruvate in human sweat using reversed-phase high-performance liquid chromatography: A noninvasive approach. *Biomed. Chromatogr.* **2012**, *26*, 1408–1415. [CrossRef] [PubMed]

177. Kim, J.; Campbell, A.S.; Wang, J. Wearable non-invasive epidermal glucose sensors: A review. *Talanta* **2018**, *177*, 163–170. [CrossRef]

178. Moyer, J.; Wilson, D.; Finkelshtein, I.; Wong, B.; Potts, R. Correlation between sweat glucose and blood glucose in subjects with diabetes. *Diabetes Technol. Ther.* **2012**, *14*, 398–402. [CrossRef]

179. Kim, Y.-C.; Park, J.-H.; Prausnitz, M.R. Microneedles for drug and vaccine delivery. *Adv. Drug Deliv. Rev.* **2012**, *64*, 1547–1568. [CrossRef]

180. Wang, M.; Hu, L.; Xu, C. Recent advances in the design of polymeric microneedles for transdermal drug delivery and biosensing. *Lab Chip* **2017**, *17*, 1373–1387. [CrossRef]

181. Tasca, F.; Tortolini, C.; Bollella, P.; Antiochia, R. Microneedle-based electrochemical devices for transdermal biosensing: A review. *Curr. Opin. Electrochem.* **2019**, *16*, 42–49. [CrossRef]

182. Wang, P.M.; Cornwell, M.; Prausnitz, M.R. Minimally invasive extraction of dermal interstitial fluid for glucose monitoring using microneedles. *Diabetes Technol. Ther.* **2005**, *7*, 131–141. [CrossRef]

183. Bariya, S.H.; Gohel, M.C.; Mehta, T.A.; Sharma, O.P. Microneedles: An emerging transdermal drug delivery system. *J. Pharm. Pharmacol.* **2012**, *64*, 11–29. [CrossRef]

184. Ye, Y.; Wang, J.; Hu, Q.; Hochu, G.M.; Xin, H.; Wang, C.; Gu, Z. Synergistic transcutaneous immunotherapy enhances antitumor immune responses through delivery of checkpoint inhibitors. *ACS Nano* **2016**, *10*, 8956–8963. [CrossRef] [PubMed]

185. Zhang, Y.; Brown, K.; Siebenaler, K.; Determan, A.; Dohmeier, D.; Hansen, K. Development of lidocaine-coated microneedle product for rapid, safe, and prolonged local analgesic action. *Pharm. Res.* **2012**, *29*, 170–177. [CrossRef] [PubMed]

186. Pere, C.P.P.; Economidou, S.N.; Lall, G.; Ziraud, C.; Boateng, J.S.; Alexander, B.D.; Lamprou, D.A.; Douroumis, D. 3D printed microneedles for insulin skin delivery. *Int. J. Pharm.* **2018**, *544*, 425–432. [CrossRef] [PubMed]

187. Aust, M.C.; Fernandes, D.; Kolokythas, P.; Kaplan, H.M.; Vogt, P.M. Percutaneous collagen induction therapy: An alternative treatment for scars, wrinkles, and skin laxity. *Plast. Reconstr. Surg.* **2008**, *121*, 1421–1429. [CrossRef]

188. Fabbrocini, G.; Fardella, N.; Monfrecola, A.; Proietti, I.; Innocenzi, D. Acne scarring treatment using skin needling. *Clin. Exp. Dermatol.* **2009**, *34*, 874–879. [CrossRef]

189. Kim, S.T.; Lee, K.H.; Sim, H.J.; Suh, K.S.; Jang, M.S. Treatment of acne vulgaris with fractional radiofrequency microneedling. *J. Dermatol.* **2014**, *41*, 586–591. [CrossRef]

190. Babity, S.; Roohnikan, M.; Brambilla, D. Advances in the design of transdermal microneedles for diagnostic and monitoring applications. *Small* **2018**, *14*, 1803186. [CrossRef]

191. Nir, Y.; Paz, A.; Sabo, E.; Potasman, I. Fear of injections in young adults: Prevalence and associations. *Am. J. Trop. Med. Hyg.* **2003**, *68*, 341–344. [CrossRef]

192. Yang, J.; Liu, X.; Fu, Y.; Song, Y. Recent advances of microneedles for biomedical applications: Drug delivery and beyond. *Acta Pharm. Sin. B* **2019**, *9*, 469–483. [CrossRef]

193. Yu, J.; Zhang, Y.; Ye, Y.; DiSanto, R.; Sun, W.; Ranson, D.; Ligler, F.S.; Buse, J.B.; Gu, Z. Microneedle-array patches loaded with hypoxia-sensitive vesicles provide fast glucose-responsive insulin delivery. *Proc. Natl. Acad. Sci. USA* **2015**, *112*, 8260–8265. [CrossRef]

194. Chen, W.; Tian, R.; Xu, C.; Yung, B.C.; Wang, G.; Liu, Y.; Ni, Q.; Zhang, F.; Zhou, Z.; Wang, J.; et al. Microneedle-array patches loaded with dual mineralized protein/peptide particles for type 2 diabetes therapy. *Nat. Commun.* **2017**, *8*, 1777. [CrossRef]

195. Than, A.; Liang, K.; Xu, S.; Sun, L.; Duan, H.; Xi, F.; Xu, C.; Chen, P. Transdermal delivery of anti-obesity compounds to subcutaneous adipose tissue with polymeric microneedle patches. *Small Methods* **2017**, *1*, 1700269. [CrossRef]

196. Di, J.; Yao, S.; Ye, Y.; Cui, Z.; Yu, J.; Ghosh, T.K.; Zhu, Y.; Gu, Z. Stretch-triggered drug eelivery from wearable elastomer films containing therapeutic depots. *ACS Nano* **2015**, *9*, 9407–9415. [CrossRef] [PubMed]

197. Dardano, P.; Rea, I.; De Stefano, L. Microneedles-based electrochemical sensors: New tools for advanced biosensing. *Curr. Opin. Electrochem.* **2019**, *17*, 121–127. [CrossRef]

198. Liu, R.; Zhang, M.; Jin, C. In vivo and in situ imaging of controlled-release dissolving silk microneedles into the skin by optical coherence tomography. *J. Biophotonics* **2017**, *10*, 870–877. [CrossRef]

199. Mouton, J.W.; Ambrose, P.G.; Canton, R.; Drusano, G.L.; Harbarth, S.; MacGowan, A.; Theuretzbacher, U.; Turnidge, J. Conserving antibiotics for the future: New ways to use old and new drugs from a pharmacokinetic and pharmacodynamic perspective. *Drug Resist. Updat.* **2011**, *14*, 107–117. [CrossRef] [PubMed]

200. Bandodkar, A.J.; You, J.-M.; Kim, N.-H.; Gu, Y.; Kumar, R.; Mohan, A.M.V.; Kurniawan, J.; Imani, S.; Nakagawa, T.; Parish, B.; et al. Soft, stretchable, high power density electronic skin-based biofuel cells for scavenging energy from human sweat. *Energy Environ. Sci.* **2017**, *10*, 1581–1589. [CrossRef]

201. Rawson, T.M.; Sharma, S.; Georgiou, P.; Holmes, A.; Cass, A.; O'Hare, D. Towards a minimally invasive device for beta-lactam monitoring in humans. *Electrochem. Commun.* **2017**, *82*, 1–5. [CrossRef]

202. Amjadi, M.; Sheykhansari, S.; Nelson, B.J.; Sitti, M. Recent advances in wearable transdermal delivery systems. *Adv. Mater.* **2018**, *30*, 1704530. [CrossRef]

203. Lee, H.; Song, C.; Hong, Y.S.; Kim, M.S.; Cho, H.R.; Kang, T.; Shin, K.; Choi, S.H.; Hyeon, T.; Kim, D.-H. Wearable/disposable sweat-based glucose monitoring device with multistage transdermal drug delivery module. *Sci. Adv.* **2017**, *3*, e1601314. [CrossRef]

204. Jia, W.; Valdés-Ramírez, G.; Bandodkar, A.J.; Windmiller, J.R.; Wang, J. Epidermal biofuel cells: Energy harvesting from human perspiration. *Angew. Chem. Int. Ed.* **2013**, *52*, 7233–7236. [CrossRef] [PubMed]

205. Katz, E.; MacVittie, K. Implanted biofuel cells operating in vivo–methods, applications and perspectives–feature article. *Energy Environ. Sci.* **2013**, *6*, 2791–2803. [CrossRef]

206. Zhao, C.-E.; Gai, P.; Song, R.; Chen, Y.; Zhang, J.; Zhu, J.-J. Nanostructured material-based biofuel cells: Recent advances and future prospects. *Chem. Soc. Rev.* **2017**, *46*, 1545–1564. [CrossRef] [PubMed]

207. Giugliano, D.; Ceriello, A.; Esposito, K. Glucose metabolism and hyperglycemia. *Am. J. Clin. Nutr.* **2008**, *87*, 217S–222S. [CrossRef]

Publisher's Note: MDPI stays neutral with regard to jurisdictional claims in published maps and institutional affiliations.

Review

Modular and Integrated Systems for Nanoparticle and Microparticle Synthesis—A Review

Hongda Lu [1], Shi-Yang Tang [2,*], Guolin Yun [1], Haiyue Li [3], Yuxin Zhang [2], Ruirui Qiao [4,*] and Weihua Li [2,*]

1   School of Mechanical, Materials, Mechatronic and Biomedical Engineering, University of Wollongong, Wollongong, NSW 2522, Australia; hl108@uowmail.edu.au (H.L.); gy417@uowmail.edu.au (G.Y.)
2   Department of Electronic, Electrical and Systems Engineering, University of Birmingham, Edgbaston, Birmingham B15 2TT, UK; YXZ048@student.bham.ac.uk
3   Department of Chemistry and Biochemistry, University of California, San Diego, CA 92093, USA; hal412@ucsd.edu
4   ARC Centre of Excellence in Convergent Bio-Nano Science and Technology and Australian Institute for Bioengineering and Nanotechnology, The University of Queensland, Brisbane, QLD 4072, Australia
*   Correspondence: S.Tang@bham.ac.uk (S.-Y.T.); r.qiao@uq.edu.au (R.Q.); weihuali@uow.edu.au (W.L.)

Received: 1 October 2020; Accepted: 29 October 2020; Published: 3 November 2020

Abstract: Nanoparticles (NPs) and microparticles (MPs) have been widely used in different areas of research such as materials science, energy, and biotechnology. On-demand synthesis of NPs and MPs with desired chemical and physical properties is essential for different applications. However, most of the conventional methods for producing NPs/MPs require bulky and expensive equipment, which occupies large space and generally need complex operation with dedicated expertise and labour. These limitations hinder inexperienced researchers to harness the advantages of NPs and MPs in their fields of research. When problems individual researchers accumulate, the overall interdisciplinary innovations for unleashing a wider range of directions are undermined. In recent years, modular and integrated systems are developed for resolving the ongoing dilemma. In this review, we focus on the development of modular and integrated systems that assist the production of NPs and MPs. We categorise these systems into two major groups: systems for the synthesis of (1) NPs and (2) MPs; systems for producing NPs are further divided into two sections based on top-down and bottom-up approaches. The mechanisms of each synthesis method are explained, and the properties of produced NPs/MPs are compared. Finally, we discuss existing challenges and outline the potentials for the development of modular and integrated systems.

Keywords: synthesis; nanoparticles; microparticles; microfluidics; integrated systems; modularisation

---

1. Introduction

Particles, generally referred to as small pieces of material with sizes ranging from a few nanometres to hundreds of micrometres, possess many unique chemical and physical properties that are different from those of bulk materials. At large, particles can be categorised into two groups: nanoparticles (NPs) and microparticles (MPs) based on sizes. Owing to their large surface areas in relation to the mass, NPs and MPs are applied in areas such as catalysis [1–3], sensing [4,5], and imaging [6,7]. Moreover, materials in such small scale can be easily handled, transported, and processed compared with the same materials in bulk.

In general, NPs (with sizes range from a few to hundreds of nanometres) have various shapes and structures such as sphere, rod, wire, tube, cube, cone, and spiral. Besides, NPs can be crystalline or amorphous. Inorganic NPs such as nanocrystals made from noble metals or rare-earth elements can be used in drug delivery [8] and biological sensing [9]. In amorphous cases, liquid metal (LM)

NPs made from gallium (Ga)-based alloys are gradually attracting attentions in fields of drug delivery, tumour therapy, and soft electronics [10–12]. Besides, organic polymeric nanocomposites with good biocompatibility show potentials in drug delivery vehicles [13]. The applications for various NPs are outlined and summarised in Table 1.

Table 1. Overview of applications for different types of nanoparticles (NPs).

Type to NPs	Applications	Reference
Liquid metal (EGaIn, Galinstan, other Ga-based alloys)	Soft/flexible/wearable electronics Biomedical applications (e.g., drug delivery, medical imaging, therapeutics, and antimicrobial activities)	[12,14–16]
Nobel metal (Au, Ag, platinum group of metals)	Antimicrobial activities Optoelectronics Catalysis Biomedical applications (e.g., drug delivery, medical imaging, and photothermal therapeutics)	[17–20]
Transition metal (Cu, Ni, etc.)	Wastewater treatment Antimicrobial and anticancer activities Catalysis Biomedicine Energy storage	[21–24]
Oxides of metals (Fe_2O_3, SnO_2, Al_2O_3, etc.)	Anti-infective applications Electrochemical sensing and biosensing Catalysis Optoelectronics Medical imaging	[25–27]
Semiconductor quantum dots (CdSe, CdTe, CdSeTe, etc.)	Catalysis Solar concentrators Medical imaging Cellular imaging	[28–31]
Carbon-based materials (CNTs, graphene, CB, etc.)	Electrochemical sensing Energy storage Catalysis Cellular imaging Biomedical applications (e.g., bioimaging, biosensors, drug delivery, theranostics, and tissue engineering)	[4,32–35]
Organic polymer (PLGA, PLGA@HF, PEG, etc.)	Mostly biomedical applications (e.g., drug delivery, tissue regeneration, molecular imaging, and cancer phototherapy)	[36–38]

A variety of synthesis approaches for NPs have been developed, and they can be categorised as either the top-down or bottom-up method, as shown in Figure 1. For the top-down approach, methods such as ultrasonication, laser ablation, and thermal decomposition have been introduced to break bulk materials into nanoscales [39,40]. In such approaches, high energy or pressure is required, and it is hard to produce NPs with sizes below 100 nm [41]. This process is also not suitable to produce uniform NPs and complex structured nanocomposites due to the uncontrollable disruptive forces. As for the bottom-up approach, NPs are synthesized by the built-up of materials from atoms to clusters to nanoscale structures. Methods such as photochemical reduction, chemical precipitation, microemulsion, microbial reduction, and hydrothermal methods are commonly used in synthesising NPs [39,41].

Both organic and inorganic MPs with sizes from a few to hundreds of micrometres have been frequently used as vehicles to carry functional molecules, such as proteins and drugs [42]. Complex MPs with customised sizes, morphologies, and compositions have been widely investigated by researchers in recent years. Preparing such MPs using conventional emulsification methods is challenging because the interfacial tension drives the automatic morphoring of emulsions into spherical shape [43]. In the past two decades, microfluidics has been adopted as a novel synthesis technique for producing MPs with customised properties [44,45]. Using monodispersed microdroplets produced by microfluidics as templates, MPs can be formed after solidification [46]. For instance, microspheres can be produced from single emulsion, while core/shell, Janus, and other irregular shapes are available from transformation of double or multiemulsions. In addition to microfluidics, millifluidic reactor (reactor with channel diameter in millimetre scales) systems have also been implemented for MPs synthesis to increase the productivity. These techniques can easily connect with typical laboratory instruments [47]. In addition, systems using acoustic [48], centrifugal [49], and jetting mechanisms [50] are developed to synthesis functional MPs. Such systems can be established easily without the need of complex microfabrication processes.

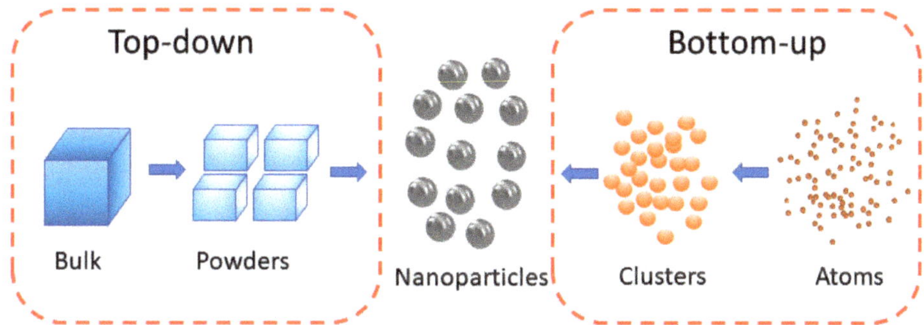

Figure 1. Scheme of top-down and bottom-up synthesis of nanoparticles (NPs).

Optimising synthesis methods is beneficial in the production of quality NPs and MPs to further stimulate the development of electronic, medical, and sensing research. Despite this, most of the conventional methods require bulky and expensive equipment, leading to complex, labour-consuming operations. These limitations impede inexperienced researchers with different backgrounds to synthesise customised NPs/MPs. Additionally, without automatic control systems, most of the conventional methods follow specific protocols that require extensive experience. This lack of flexibility and controllability would require enormous and unnecessary efforts for researchers to synthesise NPs/MPs with repeatable quality and desired properties, causing inadequate utilisation of NPs/MPs in their fields of research. For example, conventional bottom-up methods produce NPs in small batches through a series of chemical reactions; the discrepancy in manual operation may cause variation of size and properties of the produced NPs from batch to batch. To address these limitations, a feasible solution is the development of "modular" and "integrated" systems for assisting the synthesis of NPs/MPs in a repeatable manner. In this solution, complex production systems are divided into various manageable modules, and the individual modules can be readily assembled and dissembled on a need base. After assembling into an integrated system, each module interacts with others and exhibits different functions. Based on the requirements of researchers, suitable modules can be chosen and assembled to produce customised NPs/MPs.

This review will discuss current modular and integrated systems for the synthesis of NPs/MPs. We categorize the systems into two main groups: systems for producing (1) NPs and (2) MPs. Systems for synthesising NPs are elucidated by top-down and bottom-up approaches. These versatile systems utilise various mechanisms such as sonication, laser ablation, microfluidics, flame synthesis, centrifugal, and spinning force for the on-demand production of NPs/MPs. Several representative examples are provided to illustrate the mechanisms and production efficiency of such systems. Finally, we envision the challenges and opportunities for the development of future modular and integrated systems.

2. Integrated Systems for Synthesising NPs

Numerous integrated systems have been developed to facilitate the production of NPs for research in laboratories or application in the industry without complicated operations. Here, we categorise the systems based on top-down and bottom-up methods. We first elucidate top-down production methods enabled by (1) sonication, (2) laser ablation, (3) microfluidics and millifluidics, and (4) flame synthesis. The bottom-up methods would be explained in later text.

2.1. Systems Based on Top-Down Methods

Synthesis mechanisms like sonication and laser ablation are suitable to produce metallic NPs for top-down methods. Bulk materials can be comminuted into nanoscales, and the existence of ligands or the rapid formation of oxide layer prevents the products from assembling back into large scales [51,52].

2.1.1. Sonication Systems

Sonication is applied as sound energy to agitate bulk materials, as shown in Figure 2A. The sonication systems such as probe sonicator and ultrasonic bath have attracted attention for the production of LM NPs. The spontaneous formation of an oxide skin on the surface prevents LM NPs from assembling back into bulk materials. For the top-down production of LM NPs, sonication probe systems are commonly used, which can provide larger power density and yield NPs more efficiently than that of ultrasonic bath. However, the rapid heating induced by the high-power intensity of sonication probe drives undesired oxidation, dealloying and phase transitions of LM NPs [53]. To eliminate the undesired effects and efficiently produce LM NPs, a dynamic temperature control system was developed [54]. Figure 2B displays the exploded schematic of the system, combining sonication probe, cooling modules, control module, and power supply. The system adjusted the supplying power to Peltier cool pads at the same time by detecting the real-time temperature of the vial through the thermocouple to keep the vial in target temperature. This system is capable of producing NPs smaller than 50 nm, which is beneficial for applications in nanomedicine.

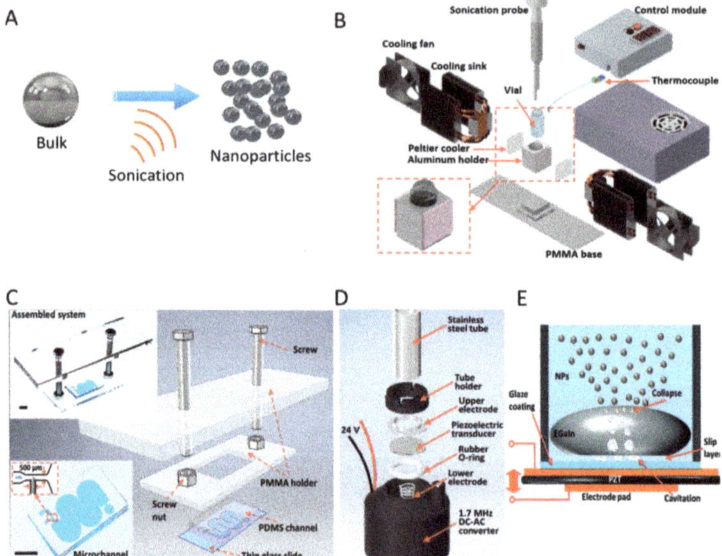

Figure 2. Schematics of ultrasonic mechanism and representative platforms. (**A**) Schematic of sonication mechanism. (**B**) Exploded schematic of the liquid metal (LM) NP production platform with a dynamic temperature control system. Reprinted with permission from ref [54]. Copyright (2020) American Chemical Society. (**C**) Experimental setup of the on-chip LM NP production platform. Reprinted with permission from ref [51]. (**D**) Exploded schematic of a liquid-based nebulization system. Reprinted with permission from ref [55]. (**E**) The schematic of the mechanism for producing EGaIn LM NPs.

Despite the mentioned advantages of probe sonicator, the bulky and expensive nature limit its use. As a substitute, a microfluidic LM NP production platform using ultrasonic bath was created, as shown in Figure 2C [51]. This platform utilises microfluidic chips to generate microdroplets of LM and further breaks them into NPs in ultrasonic bath. By adjusting the dimensions of microchannels and the applied centrifugal force on the suspension, NPs with different size distributions of NPs can be collected. This simple and low-cost platform can be applied in any laboratories with inexpensive ultrasonic bath. In addition, Tang et al. designed a liquid-based nebulization system for producing LM NPs, as shown in Figure 2D. This cost-effective platform can achieve one-step production of stable and functional LM NPs [55]. The lack of severe turbulence after activating the transducer indicates that the production process is different from the case of using sonication probe systems, and the production mechanism can be explained using the cavitation hypothesis, as shown in Figure 2E. The generated vapor cavities of water within the slip layer between LM droplet and the surface of transducer collapse at the EGaIn–solution interface, which liberates EGaIn NPs into the surrounding suspending medium. Additionally, the thickness of oxide skin on the NPs can be controlled using an integrated electrochemistry system.

2.1.2. Laser Ablation Systems

The core of laser ablation is to ablate the surface of the solid target material by a femtosecond–nanosecond pulsed laser to synthesize NPs. It constitutes a green technique that does not need any metal precursors and reductants to stabilise the colloidal dispersions and the highly pure colloids are produced without any by-products [56]. In addition, the production process is conducted under ambient conditions without the need of extreme temperature and pressure. The typical setup of liquid-phase laser ablation is schematically presented in Figure 3A. This technique can cover a wide range of suitable target materials and solvents. By tuning laser parameters and other assisting factors, the size and shape of NPs can be controlled. Due to the formation of new phase in laser ablation involves both liquid and solid, researchers are able to select and combine various solid targets and liquids to synthesize NPs for fundamental research and industrial applications [57]. The exact mechanism is complicated and still under investigation. Generally, this method involves photon-induced material ionization, plasma phase formation, and the nucleation of NPs [58]. Due to the simplicity of processing, the production of metal NPs such as gold (Au), silver (Ag), and copper (Cu) by laser ablation is of a great scientific interest. The lack of grafting agents in general laser ablation systems causes aggregation of NPs in the solution. To fix this issue, Hu et al. established an ultrasonic-assisted pulse laser ablation (PLA) system by applying ultrasonic bath to assist the laser ablation process, resulting in a decrease in particle size and an enhanced fabrication rate, as shown in Figure 3B [59]. This system used a laser with a wavelength of 532 nm to produce Au and Ag NPs, with smaller average sizes compared with the system without an ultrasonic bath.

To efficiently produce NPs with controllable sizes, Yu et al. established a multiple-pulse picosecond fibre laser system with a 3D stage that can be adjusted to change target position while producing NPs [62]. By altering the subpulse number in an envelope, NPs with sizes ranging from 4 to 120 nm (standard error of 5%) were produced. Moreover, Mahdieh et al. utilised a Nd:YAG laser system with a Joulemeter measuring pulse energy to investigate the effects of water depth and laser pulse numbers on the size of the produced colloidal NPs including aluminium (Al) and titanium (Ti) [63]. At a certain water depth, the increase in laser pulses numbers can lead to smaller mean NP sizes. Kőrösi et al. also established a picosecond laser system for producing ligand-free size-controllable Ag NPs by applying different laser wavelengths, which can maintain stable without any additives and be applied for antimicrobial activities [64]. Herbani et al. produced Au, Ag, and Cu NPs with average sizes of 20–40 nm by a pulse laser ablation system [65]. The effective laser wavelengths for correlated NPs production are found, e.g., the effective laser wavelength for Au and Ag targets is 532 nm and that for Cu target is 1064 nm. Apart from the laser wavelength, the effects of liquid medium were also investigated using a femtosecond laser ablation-assisted system. It was determined that the deionised water medium have positive impacts on the properties of the produced Ag NPs [66].

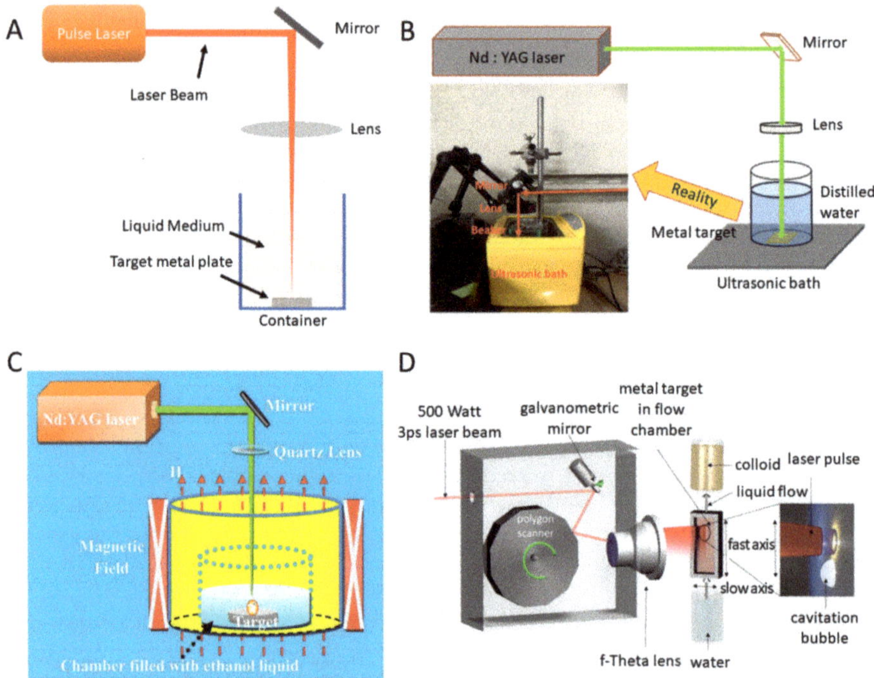

Figure 3. Schematic of laser ablation systems for producing NPs. (**A**) Schematic representation of a typical setup for liquid phase laser ablation. (**B**) An ultrasonic-assisted pulse laser ablation (PLA) system. Reprinted with permission from ref [59]. (**C**) Illustration of magnetic field assisted laser ablation system. Reprinted with permission from ref [60]. (**D**) Schematic showing the laser ablation in liquid setup. A laser beam is deflected by two scanning systems: a polygon scanner for the vertical axis and a galvanometric mirror for the horizontal axis. Reprinted with permission from ref [61] © The Optical Society.

In addition to laser and liquid medium parameters that affect the synthesis of NPs, external environment also influences laser ablation [40]. Various field-assisted laser ablation systems have been established. These fields include electric [67], magnetic [60,68,69], and electrochemical ones [70–72]. The germanium dioxide (GeO_2) NPs with a high refractive index can be produced by applying different electric fields assisting laser ablation. Shapes of NPs, such as cubes or spindles, are obtained with the assistance of electric fields, in whose absence would result in spherical particles [67]. Apart from using the magnetic field to assist laser ablation, one step fabrication of one-dimensional (1D, which means a very large aspect ratio) magnetic NPs (such as cobalt carbide spheres) can be achieved, which effectively simplifies the NPs synthesis process [60]. Figure 3C illustrated the setup of the laser ablation system. During the period when laser beam is focused on the target sample, a strong and uniform magnetic field in vertical direction is applied to achieve the synthesis of NPs and one-step fabrication of 1D chains. Moreover, external magnetic field enables the acceleration of Au NPs formation [68] and can increase the aspect ratio (up to 17–18) of Au NPs [69]. Additionally, electrochemistry-assisted laser ablation can change the morphology and composition of produced NPs. The products can possess excellent optical multiabsorption [72] and magnetic [70] properties. To meet the demand of industrial production, Streubel et al. created a novel laser system that achieved a continuous ablation rate up to

4 g/h to produce a variety of metal NPs, such as Au, Ag, and Pt [61,73]. This platform consisted of a 3 ps laser system (*Amphos 500flex*), two scanning systems (a polygon scanner for the vertical (fast) axis, a galvanometric mirror for the horizontal (slow) axis), a water-filled chamber, and a liquid flow system, as shown in Figure 3D. The fast axis is controlled by the polygon scanner, which reaches speeds up to 500 m/s, while the slow axis utilises the galvanometer mirror with speeds up to 10 m/s. The liquid flow system can pump the liquid continuously to take away the generated NPs that remain in the path and shield the laser pulses. The platform adapted a repetition rate of 10 MHz to bypass the cavitation bubbles, minimizing the plasma-induced cavitation bubbles that limits the productivity of NPs.

To make it convenient for researchers to distinguish and pick up feasible systems for their researches, the integrated systems developed in recent years based on these two mechanisms are outlined and summarised in Table 2. As the costs may impede researchers conducting their academic experiments, we mark the cost of each system by asterisks to discriminate different systems. The systems including bulk expensive equipment such as laser generators, syringe pumps, high temperature, and pressure reactors are regarded as high costs and marked with high number asterisks; the systems incorporating commercially available components or modules that are easy to get are regarded as low cost and gain low number of asterisks.

Table 2. Overview of integrated systems for the top-down production of nanoparticles (NPs).

NPs Type	Enabling Technologies/Modules	Crucial Parameters	NP Size (nm)	Costs [1]	Year	Reference
EGaIn	Microfluidics Ultrasonic bath	Dimension of microchannel s Centrifugal force	200–700 (peak)	★★	2018	Tang [51]
EGaIn	Liquid-based nebulization	Input voltage	~160–200	★	2019	Tang [55]
EGaIn additive	Ultrasonic bath Cooling water machine		286 ± 21	★★★	2020	Guo [74]
Au	Laser ablation	Subpulse number in an envelope	~4–120	★★★	2017	Yu [62]
Ag	Laser ablation	Liquid medium	3.4–15.4	★★★	2020	Menazea [66]
Au, Ag	Laser ablation Ultrasonic bath	Ultrasonic field	5.4–7.8 (Au)/ 7.9–12.1 (Ag)	★★★★	2020	Hu [59]
Au	Laser ablation	pH	13 ± 3	★★★	2017	Palazzo [75]
Au	Laser ablation		14 ± 2.1	★★★	2015	Affandi [76]
Au	Laser ablation Magnetic field	Field tensity	~3–8	★★★★	2016	Serkov [68]
Au	Laser ablation Magnetic field	Residence time in the external magnetic field	~20	★★★★	2019	Shafeev [69]
Ag	Laser ablation	Laser pulse energy	~10	★★★	2015	Valverde-Alva [77]
Au	Laser ablation	Laser fluence Liquid media	~3.16 (average)	★★★	2015	Tomko [78]
Ag	Laser ablation	Laser wavelength	3 and 20	★★★	2016	Kőrösi [64]
Ag, Cu, Ag-Cu alloy	Femtosecond laser ablation Laser irradiation		~33.4(Ag)/~22.7(Cu)/ ~23.8(Ag-Cu alloy)	★★★	2019	Bharati [79]
Copper (I and II) oxide	Continuous flow Laser ablation		~14	★★★	2019	Al-Antaki [80]
Pt, Au, CuO	High-speed pulsed laser ablation	Laser fluences Repetition rates Ablation time	4–7	★★★	2016	Streubel [73]
Al, Ti	Laser ablation	Laser pulse number Water depth	19–38 (Ti)/29–41 (Al)	★★★	2015	Mahdieh [63]
Pt, Au, Ag, Al, Cu, Ti	Laser ablation Two scanning systems	Repetition rate of laser	7	★★★★	2016	Streubel [61]
CuO	Laser ablation in liquid	Laser energy	3–40	★★★	2016	Khashan [81]
$Cu_3Mo_2O_9$ nanorods	Laser ablation Electrochemistry		~100 (diameter) ~3 μm (length)	★★★	2011	Liu [70]
CdO	Pulsed laser ablation		~47	★★★	2017	Mostafa [82]

Table 2. *Cont.*

NPs Type	Enabling Technologies/Modules	Crucial Parameters	NP Size (nm)	Costs [1]	Year	Reference
Au@CdO	Pulsed laser ablation		~11.35	★★★	2017	Mostafa [83]
Transition metal vanadates nanostructures	Laser ablation Electrochemistry	Applied voltage	~300 (diameter) ~100–140 (thickness)	★★★	2012	Liang [72]
Cobalt oxide/hydroxide	Laser ablation	Laser wavelength Laser fluence	~10–22 (average)	★★★	2014	Hu [84]
CeO$_2$/Pd	Pulse laser ablation		~20(CeO$_2$)/~9(Pd)	★★★	2015	Ma [85]
GeO2 nanotubes/spindles	Laser ablation Electrical field Ultrasonic vibrator	Applied electrical field	~200–500 (nanotube) ~200–400 (spindle)	★★★★	2008	Liu [67]
FePO$_4$	Ultrasonic intensification Impinging jet reactor	Ultrasonic power	107–191	★★★★	2019	Guo [86]
α-Fe$_2$O$_3$	laser ablation	Laser fluencies	50–110	★★★	2015	Ismail [87]
Fe$_2$O$_3$	Laser ablation/fragmentation technique	Liquid media	50–200	★★★	2014	Pandey [88]
Magnetic NPs	Laser ablation Magnetic field		~200–500	★★★★	2014	Liang [60]
Carbon nanotube	Laser ablation	Laser wavelength	1.3	★★★	2015	Chrzanowska [89]
Carbon	Pulsed laser ablation in vacuum		~33	★★★	2017	Kazemizadeh [90]

[1] The number of asterisks (★) represents the cost of synthesis system; 1 means relatively low cost, while 5 means expensive.

2.2. Systems Based on Bottom-Up Methods

Because the synthesis of monodispersed NPs is challenging by top-down methods, microfluidic and flame synthesis systems have been developed based on bottom-up mechanisms. By precisely controlling the crucial reaction parameters, micro-/millifluidic systems enable the production of a wide variety of organic and inorganic NPs [44], and some scalable architectures that combine individual units together in parallel. This technique promotes the potential in industrial production level [91]. In addition, without complex steps, flame synthesis systems are promising and capable of producing high-purity products like Ag [92] and carbon [93]. These two methods will be elaborated in detail in the following sections.

2.2.1. Microfluidic and Millifluidic Systems

The concept of microfluidics emerged in the beginning of the 1980s. This technology can precisely control the fluid in channels on a scale of tens to hundreds of micrometres and with the features of small volumes, small size, low energy consumption, and microdomain effects. It has been widely used in engineering, physics, chemistry, and biotechnology. Regarding using microfluidic chips as microreactors, the systems can realize the synthesis of NPs, nanoframeworks, and nanocrystals. Additionally, millifluidic systems have also been implemented for NPs synthesis, achieving high production rate.

Semiconductor NPs such as cadmium selenide (CdSe) quantum dots (QDs) with highly crystalline and narrow size distributions have been applied in biological imaging. To synthesise CdSe QDs with high quality in good control, Marre et al. designed a microfluidic system to achieve supercritical continuous-microflow synthesis process, as shown in Figure 4A [94]. This system consists of a high-pressure, high-temperature microreactor, a compression-cooling aluminium part, a high-pressure syringe pump, a 5-way high-pressure valve, and a high-pressure reservoir containing four vials. To improve the syntheses process in microreactors, high pressure is applied to the whole system. The microreactor was pressurised first from inlet to outlet by pressurised gas cylinder. Then, the two high-pressure syringe pumps assured the flow rate of precursor is controllable. The precursor in the high pressure microreactor remained liquid or become supercritical, leading to fast mixing and narrow residence-time distribution. With this system, quantum dots with different size distributions and nuclei concentrations were synthesized by adjusting the precursors initial concentration, temperature, and residence time in the microreactor. This system verified the possibility of materials synthesis in supercritical media in a continuous style at the microscale. Additionally, Toyota et al. presented a combinatorial synthesis system that contains several microreactors with high reproducibility to efficiently obtain high-performance CdSe NPs, as shown in Figure 4B [95]. This system has four sections: syringe pump, mixing section, heated reaction section, and spectroscopic measurement section. Applying the parallel operation of microreactors, the optimum sets of five reaction parameters, including the temperature, linear velocity, capillary length, reaction time, and reaction additive concentration, can be obtained to produce NPs with desired properties. The whole processing time for synthesising one batch of NPs, including raw material solutions preparation, system setup, reaction, production measurement, analysis, and instruments clean-up, is just around 20 min.

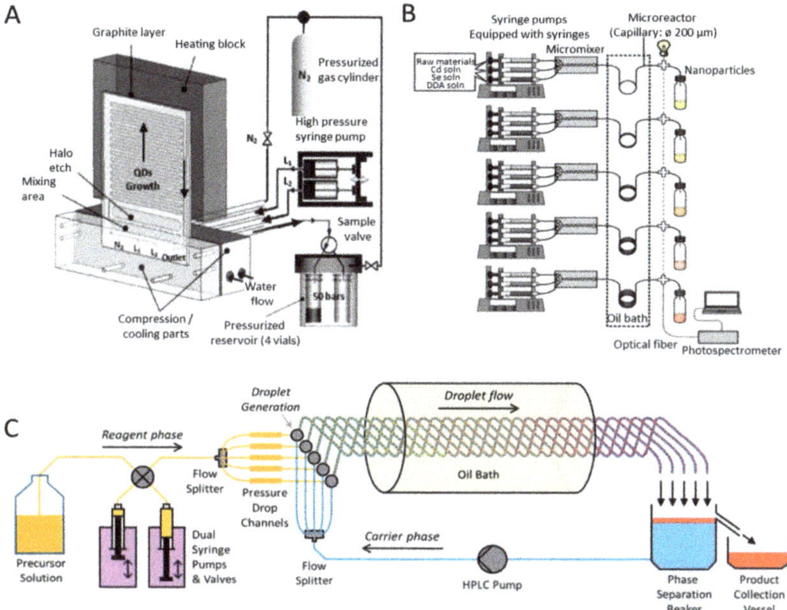

Figure 4. Microfluidic/millifluidic systems for producing NPs. (**A**) microfluidic system for producing CdSe quantum dots (QDs). Reprinted with permission from ref [94]. (**B**) A combinatorial synthesis system contains several microreactors for CdSe NPs production [95]. (**C**) A multichannel droplet microfluidic reactor. Reprinted with permission from ref [91].

Scale-up synthesis of CdSe nanocrystals, which can meet the demand of industry production, is admirable. However, it is still challenging to develop scalable architecture that allows multiple channels to be supplied from a small number of feed reservoirs. Crucially, individual channels must be stable for operation and keep operations over extended periods of time. Nightingale et al. reported a multichannel droplet microfluidic reactor for achieving large-scale synthesis of nanocrystals [91]. A common set of feed reservoirs, two pumping systems, and five-way passive flow-drivers were contained in this platform, as shown in Figure 4C. A high-pressure liquid chromatography pump that withdraws carrier fluid can maintain a continuous flow. By adjusting back-pressure and pressure drop channels to regulate the reagent phase, the system enables the stable balanced flow distribution, providing the possibility of high yield synthesis of NPs. This system can yield 1.5 g of CdSe in the first hour, while cadmium telluride (CdTe) and cadmium selenide tellurium (CdSeTe) can be synthesized with the yields of 3.7 and 2.1 g, respectively. Moreover, the system can work a maximum of 9 h continuously to produce 145 g of CdTe and showed no signs of degeneration of performance during the running time.

However, commercial applications with Cd-based NPs are limited because of its high toxicity. Cd-free alternatives, such as indium phosphide (InP) QDs, have been of booming interest for research and industry areas. A millifluidic reactor system was designed for multistep and continuous synthesis of InP/stilleite (ZnSeS) NPs by Vikram et al. [96]. The reason of choosing millifluidic reactors instead of microfluidic reactors is the ability of producing high-quality NPs at high production rate for large-scale synthesis. The modular system included two or more distinct reactors, for core formation, shell growth, and cooling modules, and one or more inline mixing units, for adequate and fast mixing the precursors outside the reactors. A static mixer was incorporated inside the reactor modules to minimize the residence time distribution of the reactants in the reactor as it broke up the parabolic flow profile. This system yields NPs with narrow size distribution and desired optical properties by precise control over rapid and uniform heating, cooling, and mixing of flows at a high production rate. It can be further optimised by adding other modules for crystal growth kinetics studies. Additionally, the real-time observation of nucleation and the growth of lead sulphide (PbS) QDs at high temperature was first achieved in microfluidic systems by Lignos et al. [97]. The droplet-based microfluidic platform was integrated with online absorbance and fluorescence detection (collecting the real-time data during the synthesis process) for kinetic analysis of PbS QDs synthesis. More importantly, this system can be used to study the fast kinetics of nanomaterial syntheses and ion exchange reactions (e.g., the time of the nucleation of PbS occurred is <1 s).

Apart from semiconductor NPs, noble metals including Au, Ag, and platinum (Pt) group metals with different sizes and shapes in nanoscale exhibit different optical properties, which have shown potential in applications of sensing and catalysis [98]. Duraiswamy et al. presented the synthesis of anisotropic Au nanocrystal dispersion with a droplet-based microfluidic method with T-junction [99]. However, this system failed to enable a fully continuous process and limit the scaling up synthesis, as it lacked real-time monitoring for the synthesis of Au NP seeds and had limited residence time of growing nanocrystals on-chip. Lohse et al. developed a simple millifluidic benchtop reactor system for high-throughput synthesis to obtain Au NPs with different sizes and shapes [100]. This system included multiple commercially available modular components and peristaltic pumps, which can be used in almost any chemistry research laboratory. Flow rates between 35.0 and 60 mL/min for the system can produce monodisperse Au NPs. The reason for using millifluidic reactors is that they have the advantages of easy fabrication, good resistance to fouling, and the possibility for large-scale synthesis. To achieve scaling up synthesis of Au NPs, a new approach based on millilitre-sized droplet reactors was displayed by Zhang et al. [101]. Commercially available components such as syringe pump, T-connectors, and polytetrafluoroethylene (PTFE) tubes were assembled into the device. Au, palladium (Pd), and Pd-M (M = Au, Pt, and Ag) nanocrystals with controllable sizes, shapes, compositions, and structures on a yield of 1–10 g per hour have been synthesized successfully without losing uniformity, providing exciting opportunities to biomedical and petroleum industries. Cattaneo et al. utilised different setup of millifluidic reactors to synthesize monometallic Au, bimetallic AuPd, and Au@Pd core–shell NPs in continuous flow style [102]. They produced titanium dioxide (TiO2)-supported bimetallic AuPd with a mean particle size of ~2 nm, and the size distribution is narrower compared to the distribution of conventional batch methods.

For synthesising Ag NPs, Matthias et al. described a microfluidic system by combining a split-and-recombine-mixer, a T-shaped mixer and a Dean-flow-mixer, to synthesize uniform seed particles for forming anisotropic Ag triangles [103]. In comparison with batch methods, the size, edge length and thickness of the seed particles can be adjusted precisely. Likewise, Cheol et al. developed a customised microfluidic reactor by assembling single functional microfluidic units, which overcame the limitations of the device reconfiguration in conventional polydimethylsiloxane (PDMS)-based microfluidic platforms [104]. The sizes of the Ag NPs can be adjusted from 4.3 to 11.45 nm with different flow rates. Since the reactor was easy to assemble and the size of the Ag NPs was controlled precisely by simply changing the injection flow rate of the precursor, the reactor can be suitable for the synthesis of NPs requiring precise control of temperature and chemical concentrations. To synthesize various noble metal NPs, Niu et al. designed a droplet-reactor system with the capabilities of both water and oil separation and the product purification, as shown in Figure 5A [105]. By Niu's method, noble metal NPs with uniform sizes and well-controlled shapes can be easily obtained through assembling four different modules: reaction, cooling, water/oil separation, and purification. Such a droplet-reactor system is possible for automated operation by assembling the modules of both water and oil separation and product purification into a millifluidic system to further achieve continuous and scalable production of noble metal NPs in industrial scale.

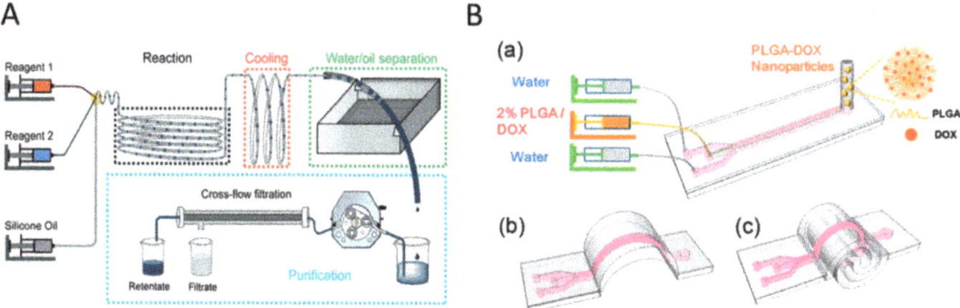

Figure 5. Schematic of modular microfluidic systems for producing NPs. (**A**) A droplet-reactor system with the potential for automation. Reprinted with permission from ref [105]. (**B**) A microfluidic origami chip with different configurations for enhancing mixing. Reprinted with permission from ref [106].

Besides noble metal NPs, polymeric NPs with different functionalities provide the potential in biomedical applications. Poly(lactic-co-glycolic acid) (PLGA) NPs with great biocompatibility and biodegradability are applicable for making drug delivery vehicles, which can reduce cell toxicity and improve therapeutic efficacy [106]. To realize the best cellular uptake and anticancer efficacy, NPs with a diameter of ~100 nm are desirable. Jiashu et al. presented a microfluidic origami chip for the synthesis of PLGA NPs [106]. The novel 3D origami structure can avoid rapid velocity changes (Figure 5B), shorten the mixing distance, and reduce the mixing time, as it has smooth turns and two counter-rotating cortices perpendicular to the flow direction. Applying the novel chip, the doxorubicin-loaded PLGA NPs, which were around 100 nm with a monodisperse size distribution, were synthesized efficiently. The productivity of NPs can reach 1200 mg per day. After this work, Liu et al. designed a versatile and robust microfluidic platform with glass capillaries, which took advantage of chemical resistance [107]. The platform adopted a coflow capillary setup in which a tapered glass capillary was inserted into another bigger cylindrical one. This platform enables high linear velocity ratio between the inner and outer fluids, resulting in the fast mixing and uniform self-assembly of NP precursors with controlled microvortices and unsteady jetting. This platform can synthesize homogeneous NPs with a productivity of 242.8 g per day, a much higher than that of the origami one. Regardless of the type of organic solvents or stabilizers used, NPs with a narrow size distribution can be obtained. To further scale up the production rate and achieve precise control over the NP structure, a superfast sequential nanoprecipitation microfluidic platform was designed [108]. The platform consisted of three nested cylindrical glass capillary tubes assembled in the coaxial arrangement. High-throughput production of core/shell nanocomposites was demonstrated. By sequentially mixing nanocomposites in superfast time intervals (in microseconds range), the nanoprecipitation processes can produce stable nanocomposite cores without any stabilizers. When operated in the continuous mode, this platform can offer a high production rate of drug nanocrystal encapsulated nanocomposites up to 700 g per day, meeting the standard of clinical studies and industry production.

2.2.2. Flame Synthesis Systems

Combustion is of great interest for material synthesis like carbon and Ag NPs for its scalability. Such a technique does not need tedious steps and can easily form high-purity products with metastable compositions [109]. The typical schematic of flame aerosol synthesis is outlined in Figure 6A. The energy from the flame flows to precursor materials to cause evaporation and drives the reaction of nucleation, agglomeration, and growth, and then, NPs formed are collected by a holder. In terms of synthesising inorganic nonmetallic NPs, Mädler et al. designed a spray apparatus consisting of an external-mixing gas-assisted nozzle, syringe pump, a glass microfiber filter, a water-cooled holder, and a vacuum pump [110]. By creating gas flow using vacuum pumps, the particles can deposit on the filter during flame spray. This system synthesized nanostructured silica (SiO_2) particles with specific surface area using different oxidant flow rates and precursor/fuel compositions. Moreover, SiO_2 production rate can reach up to 9 g/h (diameter of 7–40 nm) using a liquid mixture of hexamethyldisiloxane (HMDSO)/ethanol at a molar ratio of 1:10.

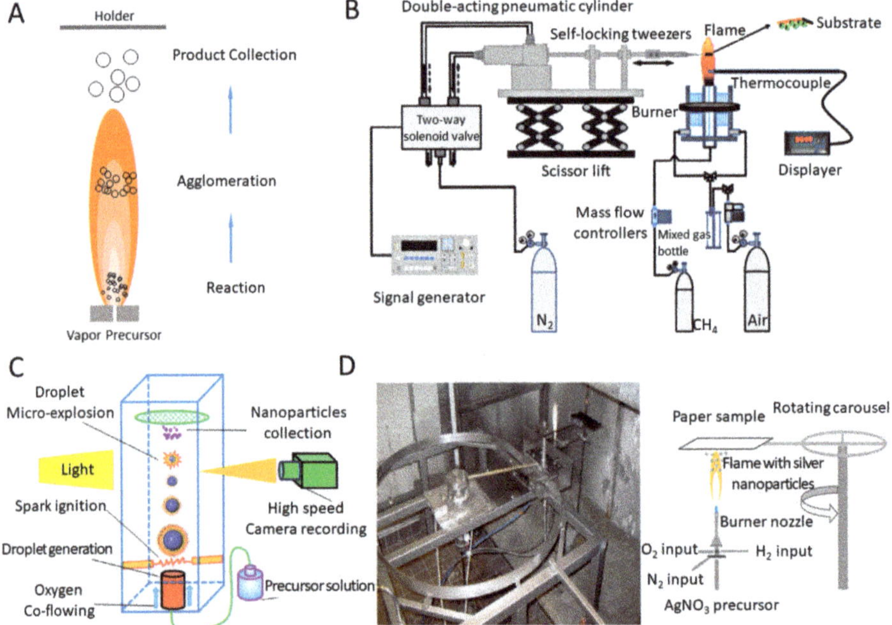

Figure 6. Flame synthesis systems for producing NPs. (**A**) Schematic representation of a typical flame synthesis system. (**B**) Schematic of methane coaxial jet diffusion flam system. Reprinted with permission from ref [111]. (**C**) Experimental setup for single isolated droplet combustion. Reprinted with permission from ref [112]. (**D**) The setup and schematic of the liquid flame spray (LFS) system. Reprinted with permission from ref [92].

Mohapatra et al. was able to simplify the process of the flame synthesis and introduced a facile wick-and-oil flame system to synthesize high-quality hydrophilic onion-like carbon NPs [93]. Without any complex instruments and catalyst, the NPs produced by such system have high purity and narrow size distribution for 25 nm with a standard derivation of 5 nm. Besides, Esmeryan et al. designed a novel system that contains conical chimney with adjustable air-inlet opening [113]. They obtained amorphous, graphitic-like and diamond-like phases of carbon coatings by changing the size of opening. To dig into the effect of sampling substrate, time, and height, a methane coaxial jet diffusion flame system combines a carbon nanomaterial collection system, a burner, a flow control device, and a gas distribution system together [111]. The collection system includes a scissor lift to adjust the sampling height and a double-acting pneumatic cylinder operated by signal generator. The system can be adjusted to change the sampling time by controlling the motion of substrate in horizon direction and the tip of self-locking tweezers attached to replaceable substrate, as shown in Figure 6B. Compared with substrates of copper and nickel-chromium, carbon nanotubes collected at nickel foam substrate are produced with highest yields.

Apart from inorganic nonmetallic NPs, the synthesis of metal NPs also get benefits from the flame spray pyrolysis (FSP) system. Various integrated synthesis systems have been established on demand. For example, Li et al. established a single droplet combustion system for synthesizing tin dioxide (SnO_2) NPs in order to provide fundamental information for the process of FSP [112]. This system combined a droplet-on demand generator, a spark ignition module, and carbon-coated Cu grids. Droplets ejected from the generator are ignited by spark from two electrodes and the formed NPs are collected on carbon-coated Cu grids, as shown in Figure 6C. Two paths of NPs formation were found, in which large SnO_2 NPs are formed during the period for microexplosions of droplet surface, while the small ones are obtained during the steady combustion. Additionally, Brobbey et al. utilised a one-step flame synthesis method with liquid flame spray (LFS) system to produce homogeneous monolayer of Ag NPs on a paper surface [92]. Figure 6D illustrates the process by which a burner nozzle generates flame with Ag NPs, which are deposited on a paper with a rotating carousel that adjusted the passing times and speed. This environment-friendly method can deposit 30 nm NPs without effluents, and the amount of the produced NPs can be effectively adjusted by changing the passing times, which highlights potential in industry production by roll-to-roll processing.

Additionally, alumina (Al_2O_3) NPs can be synthesized by in-flight oxidation of flame synthesis [114]. This system adopted flames, which flow in the horizontal direction, with different ratios of oxygen and acetyl. Micro Al powders as precursors were dropped from the top, generating various sizes of NPs collected in the powder collector. Moreover, Fe ions have impacts on the synthesis and characterizations of Al_2O_3 NPs in the FSP system [115]. In this system, the reaction occurred in the combustion chamber and the final NPs can be collected in the microfiber filters via a vacuum pump. Different from commercial Al powders as precursors, this system used Al acetylacetonate and Ferrocene. Researchers found that two types of Al_2O_3 (θ- and η-) were obtained and the ability of fluoride removal is most efficient in the absence of iron, and the ferrocene added was capable of suppressing the formation of θ-Al_2O_3. Highly crystalline hexagonal caesium tungsten bronze ($Cs_{0.32}WO_3$) NPs are available by FSP followed by annealing [116]. The system integrated an ultrasonic nebulizer, to form droplets and avoid precipitation; a glass flame reactor; and a bag filter to collect NPs. The particles gained from the system were then annealed inside a tubular furnace and high-purity $Cs_{0.32}WO_3$ NPs were obtained after cooling naturally. Compared with conventional methods, such a system requires shorter reaction time and has high energy efficiency.

The representative integrated systems using milli-/microfluidics and flame synthesis systems are summarised in Table 3.

Table 3. Overview of integrated systems for bottom-up methods.

NPs Type	Enabling Technologies/Modules	Crucial Parameters	NP Size (nm)	Costs [1]	Year	Reference
CdSe	Continuous-microflow synthesis High-pressure microreactor	Solvent phase Concentration Temperature Residence time	~3–6	★★★★	2008	Marre [94]
CdSe	Combinational microreactors	Temperature Reaction time Reaction additive concentration	~2–4.5	★★★★	2010	Toyota [95]

Table 3. *Cont.*

NPs Type	Enabling Technologies/Modules	Crucial Parameters	NP Size (nm)	Costs [1]	Year	Reference
CdTe, CdSe, alloy CdSeTe	Multichannel droplet reactor			★★★★	2013	Nightingale [91]
InP/ZnSeS	Millifluidic reactor system	Flow rate Reactor temperature Shell precursor concentration	5.9 ± 1.2	★★★	2018	Vikram [96]
PbS	Droplet-based microfluidic	Temperature Flow rate	2–6	★★	2015	Lignos [97]
Au	Millifluidic benchtop reactor system "Y" mixer Flow synthesis	Concentration	~2–40	★★★	2013	Lohse [100]
Au	Zigzag micromixer	Seeds volume Residence channel length	75 ± 6	★★★	2017	Thiele [117]
Au, bimetallic AuPd	Millifluidics Continuous flow	Flow rate Reactor geometry	6.4 ± 1.5 (I-shape connection)/6.3 ± 1.3 (helical reactor)	★★	2019	Cattaneo [102]
Ag	Droplet-based microfluidic reactor	Static mixing Temperature Flow rate	4.37–11.45	★★★	2018	Kwak [104]
Ag	Drop-based microfluidics	Concentration ratios Flow rates	4.9 ± 1.2	★★	2016	Xu [118]
Nobel metal	Millilitre-sized droplet reactors	Capping agent Reductant Reaction temperature	~9–50	★★★	2014	Zhang [101]
Nobel metal	Multichannel droplet reactor	Capping agent Reductant Reaction temperature	~2.5	★★★	2018	Niu [105]
Pd-Pt, Pd@Au (core@shell)	Duo-microreactor	Concentration	18.0 ± 2.7	★★★★	2019	Santana [119]
BaSO$_4$, Au, CaCO$_3$	Segmented flow microchannel Passive picoinjection	Injection volume	30–40 (BaSO$_4$) 32–91 (Au)	★★★★	2018	Du [120]
Superparamagnetic iron oxide	Micellar electrospray		36 ± 6	★★★★	2014	Duong [121]
Ni	Continuous flow microreactor	Flow rates	~6.43–8.76	★★★	2015	Xu [122]
Fe$_3$O$_4$	Flow synthesis "T" mixer	Linear velocity Residence time Reactor dimension	4.9 ± 0.7	★★	2015	Jiao [123]
PLGA@HF, PLGA@AcDX	Multiplex microfluidics	Flow rates	~60–550	★★	2017	Liu [108]
PLGA	Microfluidic origami chip	Flow rates	~100	★★	2013	Sun [106]
PLGA, hydrophobic chitosan, acetylated dextran	3D coaxial flows Glass capillaries		~100–400	★★★	2015	Liu [107]
Metal-organic frameworks (MIL-88B)	Nanolitre continuous reactor Segmented flow	Residence time Temperature Volume slug	90–900	★★★	2013	Paseta [124]
Silica, polymersomes, niosomes	Microreaction technology	Flow rates	238–361 (silica)/275–75 (niosomes)	★★★	2019	Bomhard [125]
Ag	Liquid flame spray	Passing times	~10–100	★★★	2017	Brobbey [92]
SnO$_2$	Single droplet combustion Flame spray pyrolysis	Metal-precursor concentration	~3–39	★★★★	2020	Li [112]

Table 3. *Cont.*

NPs Type	Enabling Technologies/Modules	Crucial Parameters	NP Size (nm)	Costs [1]	Year	Reference
α-Al$_2$O$_3$	Flame synthesis	Ratios of oxygen and acetyl	50–150	★★★	2014	Kathirvel [114]
Cs$_{0.32}$WO$_3$	Flame-assisted spray pyrolysis	Flame temperature Flow rate	~6–300	★★★	2018	Hirano [116]
Fe/Al$_2$O$_3$	Flame spray pyrolysis	Precursor molar ratio Multicomponent structures	183–187	★★★	2016	Hafshejani [115]
Carbon	Flame synthesis Conical chimney	Combustion regime	~200	★★★	2016	Esmeryan [113]
Carbon nanotube	Flame synthesis Methane diffusion flames	Sampling time Sampling height Sampling substrate	30–110	★★★	2018	Chu [111]
Onion-like carbon	"Wick-and-oil" flame synthesis		~25 ± 5	★★★	2016	Mohapatra [93]

[1] The number of asterisks (★) represents the cost of synthesis system; 1 means relatively low cost, while 5 means expensive.

3. Integrated Systems for MPs Production

MPs with specific size and composition exhibit a promising potential in numerous areas of applications such as adductive manufacturing [126], eco-friendly electronic systems [127], and drug delivery systems [128]. Apart from traditional synthesis methods of MPs, which require redundant process and complex operation, microfluidics has been adopted as a novel synthesis technique due to its advantages of low volume and high controllability [43]. Besides, off-chip generation platforms are desirable because they are easy to operate, which facilitate inexperienced researchers [129]. Certain systems show unique advantages as they can produce MPs with a certain size range or reduce the cost of producing a certain type of MPs. For example, magnetic droplets with a diameter of ~100 μm can be simply produced by magnetically driven step emulsion device [130]. Off-the-shelf self-setting rubber can be introduced in a flow-focusing microfluidic system for the efficient and low-cost synthesis of W/O emulsions with a large size range (from 100 to 500 μm) by simply adjusting the diameter of the rubber nozzle [131]. Furthermore, some systems can achieve the versatile production of various MPs, such as LM, hydrogel, and double emulsion [129]. Based on the different production mechanisms, this section will discuss integrated systems utilising techniques including: (1) microfluidics, (2) acoustics, (3) centrifugal and spinning force, and (4) jetting.

3.1. Microfluidic Systems

Microfluidics is capable of producing microdroplets with specific properties by controlling and manipulating the microscale fluid in a small chip [44]. Microfluidic platforms that consists of a microfluidic chip and syringe pump can produce a variety of microspheres in one step, such as poly(3,4-ethylenedioxythiophene) (PEDOT)-based microspheres [127], enzyme-immobilised reusable polymerised microcapsules [132], and biodegradable chitosan microspheres [133]. The microfluidic chips fabrication usually requires cleanroom facilities and photolithography, impeding research carried out in laboratories without such conditions. To fabricate easy-to-make fluidic devices for producing MPs, Lapierre et al. designed a pipette tip based on microfluidic system using commercially available self-setting rubber and conventional components [131]. Such a microfluidic device can be fabricated by combining a micropipette tip with tubing in the appropriate materials and size. The entire assembly process takes less than 10 min, without any microfabrication expertise requirement. The droplet size is determined by nozzle diameter and the flow rate of each phase. Monodisperse emulsion with the size ranging from 100 to 500 μm can be generated by simply using nozzles with different inner diameters, while double water-in-oil-in-water (W/O/W) emulsions can be generated by regulating the flow rates of the three phases. The produced emulsions can be used in tissue engineering, 3D cell culture, biocatalysis, and hydrogen storage materials. In addition, Kahkeshani et al. developed a novel step emulsification device, which is driven by magnets to emulsify ferrofluids [130]. Without using pumps, the system only needs magnetic field to control the emulsification of ferrofluid-containing solutions. The device consists of a sample reservoir, connecting channels, terrace, step, continuous phase reservoir, and magnet, as shown in Figure 7A. Detailed configurations of the platform is given in Figure 7B. The magnetic field breaks the bulk ferrofluid within the emulsification device and the droplet size depends primarily on the channel geometry instead of the flow rates of the ferrofluid, interfacial forces, or magnetic force. Moreover, the droplet generation rate can be adjusted by changing the movement frequency of the magnet and the number of connecting channels between the ferrofluid reservoir and continuous-phase reservoir. This pumpless, novel method could be beneficial for point-of-care assays and cell sorting in droplets.

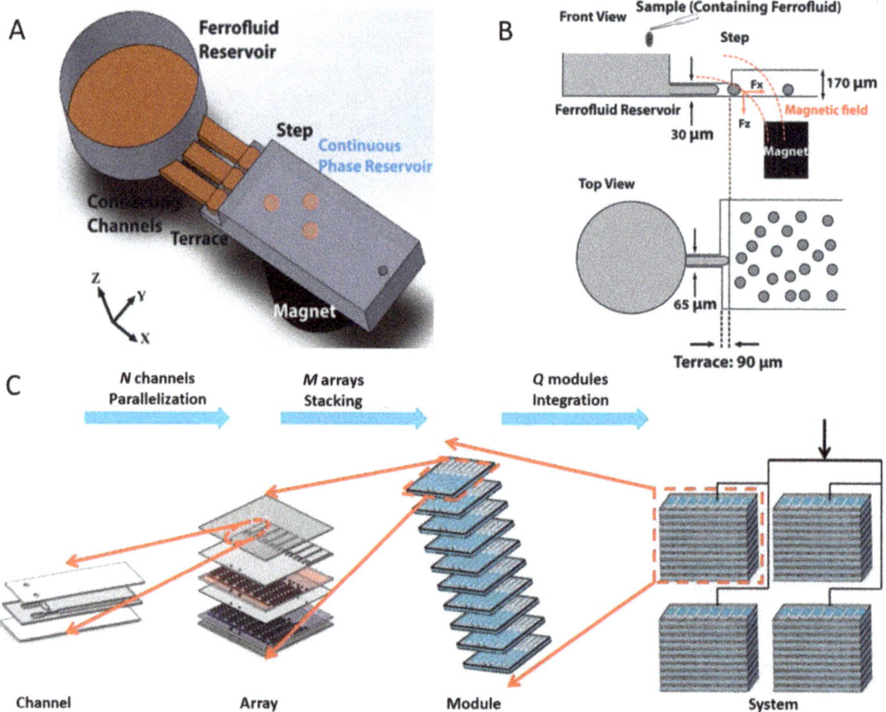

Figure 7. Microfluidic systems for producing microparticles (MPs). (**A**) Schematic view of a magnetically driven microfluidic droplet generation technique using ferrofluids (without any pumps). (**B**) Top and front views of the microfluidic device with dimensions. Reprinted with permission from ref [130]. (**C**) Overview of multidimensional scale-up strategy (not to scale). Reprinted with permission from ref [134].

Apart from one-step fabrication methods with microfluidic flow-focusing systems, high throughput and scaling up productions are also widely investigated in recent years. Amstad et al. presented a microfluidic device named "millipede device." The core of this new design is to increase the number of individual nozzles in parallel [135]. The device can produce monodisperse droplets with a diameter ranging from 20 to 160 µm and a coefficient of variations (CV) of 3%. With 550 individual nozzles in parallel, the throughput reached up to 150 mL per hour. If the millipede device is packed into array, 800 L of 160 μm droplets can be produced in the dripping regime, which provides potential in industrial applications. Kim et al. [133] presented a 512-microchannel geometrical passive breakup device was presented. Microspheres with diameters of 40.0 ± 2.2 μm can be produced (3 mL/h). The same group further improved this design by adding 256 T-junctions in the last branch of the device [136], and the micro water droplets produced can have a smaller size of 35.29 μm (CV of 8.8%) at a higher production rate of 10 mL/h. To enhance the reusability, a glass microfluidic device was designed to achieve high-throughput step emulsification with 364 linearly parallelised droplet makers and the throughput can reach up to 25 mL/h [137].

Additionally, integrated systems that combines the microchannel and different modules are developed to generate NPs with higher yield. Han et al. proposed the concept of factory-on-chip [134]. They established a large-scale 400-microchannel high-throughput system by assembling a 3D microfluidic network from channel to system. In this type of device, multiple microchannels are lined in parallel on a two-dimensional array, and multiple arrays are stacked in a reconfigurable manner in the third dimension as a module. The integration of Q modules could achieve a production rate proportional to three scale-up numbers N, M, and Q, as shown in Figure 7C. The configuration of circular array causes smaller pressure imbalance among the channels than the parallel array under the same operation conditions. Gravity and flow resistance lead to the stacking effect (i.e., the produced NPs diameters vary among different array layers) as the arrays are stacked in the vertical direction. Applying the synthesis system based on the scale-up strategy mentioned above, chitosan/TiO$_2$ functional MPs with a diameter of 539.65 μm (CV of 3.59%) were synthesized in a large throughput (80 mL/min). Mohamed et al. developed a continuous and integrated microfluidic platform to produce monodisperse cell-laden microgel droplets [138]. Three main functional modules including a flow-focusing junction, an on-chip photo cross-linking chamber, and a hydrodynamic filter unit were assembled into the system. With these modules, droplets generation, separation, and other required processes can be successfully achieved. Goff et al. designed a bench-top contact flow lithography system that consists of an illumination unit, a stage unit, and an imaging unit [139]. By applying a multichannel microfluidic chip to contact flow-lithography, the platform dramatically increased the synthesis rate for chemically homogenous particles (>10^6 particles per hour). Likewise, Jeyhani et al. developed a microfluidic flow focusing device with microneedles, which are commercially available [128]. The microneedle inserted into the microchannel formed a 3D focusing configuration that can remove the wetting effects of the disperse phase. The 3D flow focusing system enabled the control production of microdroplets with diameters ranging from 5 to 65 μm. The generation of water-in-water microdroplets can reach a production rate up to 850 Hz. To produce MPs containing radiotracers, Jia et al. for the first time designed an integrated system containing an automated radionuclide concentrator and an automated microdroplet synthesis platform. This system made possible the synthesis of MPS for patient doses of [^{18}F] fluoride [140].

Furthermore, control modules that aims at fabricating monodisperse or specific-shaped MPs have been introduced to some microfluidic systems. For microdroplets produced by the conventional T-junction microfluidic channels, Zeng et al. compared two different driven methods using syringe pump and pressure pump [141]. Pressure-driven method exhibited lower polydispersity than the other one, providing theoretical support for the design of complex flow-control microfluidic systems. Furthermore, the mathematical model relating pressure fluctuations was established, which indicates that an increase in microchannel height and a decrease in width can reduce the effect of pressure fluctuations on MPs production when using T-junction microfluidic channels [142]. Crawford et al. combined imaging-based feedback and pressure-driven pumping to develop an imaged-based closed-loop feedback system for producing microdroplets with specific sizes [143]. This was achieved by acquiring volume parameters using a high-speed camera and by utilising the proportional-integral-derivative (PID) algorithm to control both the volume and frequency of droplet production. Other than imaging-based methods, Fu et al. presented an electrical-detection droplet microfluidic closed-loop control system, which could increase the efficiency and accuracy of the measurement of the droplet size [144]. Furthermore, artificial neural network (ANN) displays the potential for microdroplet prediction, and Mahdi et al. presented an ANN-based microfluidic system [145]. Two dimensionless numbers (Reynolds number and capillary number in continuous and dispersed phase, respectively; total four parameters) are picked as four neurons for input layer. The neural network architecture has 10 neurons in the hidden layer, and the dimensionless length of the drops is selected as the neuron for the output layer. Overall, 742 experiments were conducted to build up a database for the network, and the platform was capable of predicting droplet size in a high precision (mean square error is ~1.4×10^{-6}).

Moreover, 3D-printing has been regarded as a novel technology being low-cost and convenient process and can, therefore, be applied to fabricate microfluidic devices. After assembling the device together, different types of droplets with desirable sizes can be obtained. Zhang et al. reported a 3D-printed "plug-and-play" device consisting of a generator, a commercial tubing and a fingertight fitting for generating microdroplets with a size of ~50 μm [146]. Figure 8A illustrates a 3D-printed screw-and-nut droplet generator [147]. The nut was attached vertically to T-junction droplet channels, while the screw with an external thread (total 75 teeth) was used to adjust the height of the channel. By simply rotating the screw along the axial, the gap height is adjusted between 0 and 75 μm, precisely controlling droplets with various diameters generation. The size of the produced droplets ranged from 39.0 ± 2.6 to 1404.3 ± 23.3 μm. The water-in-oil (W/O) and oil-in-water (O/W) emulsions provide the application in chemical reactions, drug testing, chemical and biological analyses, and medical engineering. To scale up production, Hwang et al. developed a 3D-printed chimney-shaped millifluidic device for producing droplets with diameters ranging from 36 to 616 μm (CV < 4%) [148]. As shown in Figure 8B, two inlets in the opposite direction at the low position of the device side allow the fluid to flow into the tapered void chamber and form droplets in a parallelised manner, which reduces the overlapping of flow channels. One outlet at the top of the chamber is responsible for transferring out the produced droplets. The control of the droplet size was achieved by adjusting the flow rate ratio between the dispersed aqueous phase and the continuous oil phase, as well as the apex angle. Moreover, the possibility of scaling up production was investigated by fabricating four chimneys with the same dimensions and incorporating them into an integrated device (4-PC); a distributer connected to the device was built to ensure the uniform supply and distribution of both the dispersed and continuous phase into the device, as shown in Figure 8C. The result showed that uniform droplets (CV less than 5%) were generated smoothly, indicating that the device can be further modularised to meet different demands and achieve scaling-up production. Furthermore, magnetic LM droplets with controllable sizes can be generated by a 3D-printed coaxial microfluidic device [149]. The device integrated three modules to produce magnetic LM droplets with diameters ranging from 650 to 1900 μm (CV < 5%) by changing the flow rate ratio of the two phases of magnetic liquid metal and poly(ethylene glycol) (PEG) solution.

3.2. Acoustic Systems

Acoustic waves can break large droplets into microdroplets. Piezoelectric elements are often used for making low-cost acoustic wave generators for microdroplets production. Using piezoelectric transducers, Kishi et al. designed a droplet-generating system [150]. The system consisted of a compressor, a regulator, a constant pressure pump, a voltage amplifier, a high-speed camera, and an ultrasonic torsional transducer, as shown in Figure 9A. The constant pressure pump continuously pumped liquid into the device and the piezoelectric transducer generated acoustic waves at a regular interval to produce microdroplets with a uniform size. The droplet diameter is determined by the applied flow rate, the ultrasound frequency, and the radius of the nozzle. Besides, the produced droplets can be applied in noncontact deposition for liquid crystal displays, drug delivery systems, and biomedical research. Similarly, Fujimoro et al. developed a droplet production system with a torsional bolt-clamped Langevin-type transducer [151]. The two constant pressure pumps ensure that the phases of water and oil are steady and adjustable; through the transducer, droplets were generated in ambient liquid; the inset illustrated the structure of the designed transducer incorporated with a micropore plate and piezoelectric elements. This platform can generate droplets continuously as long as the pressure pump keeps injecting liquid through the transducer. The diameter of the generated water-in-oil (W/O) and silicone emulsions was 62.5 ± 2.6 μm, while the driving frequency and the vibrational velocity of micropore kept constant at 37.0 kHz and 96.5 mm/s.

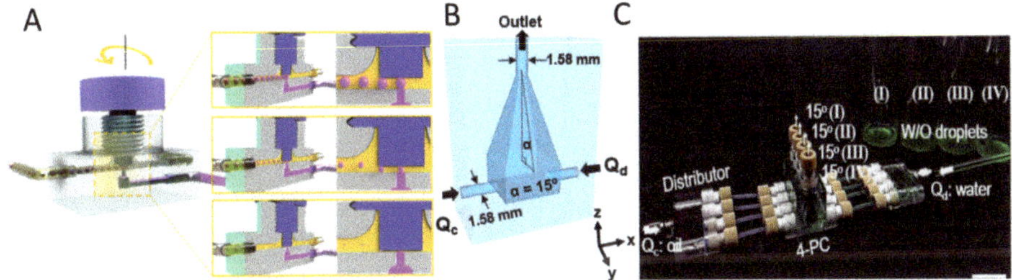

Figure 8. Microfluidic systems applying 3D printing technology. (**A**) Setup of a 3D-printed screw-and-nut droplet generator, the schematic also illustrates the control of the droplet size. Reprinted with permission from ref [147]. (**B**) Virtual object photographs of the 3D-printed millifluidic device with two inlets for continuous and dispersed phase and one outlet. Reprinted with permission from ref [148]. (**C**) Real image of the four-parallelised-chimneys device with the same apex angle. Scale bar is 2 cm.

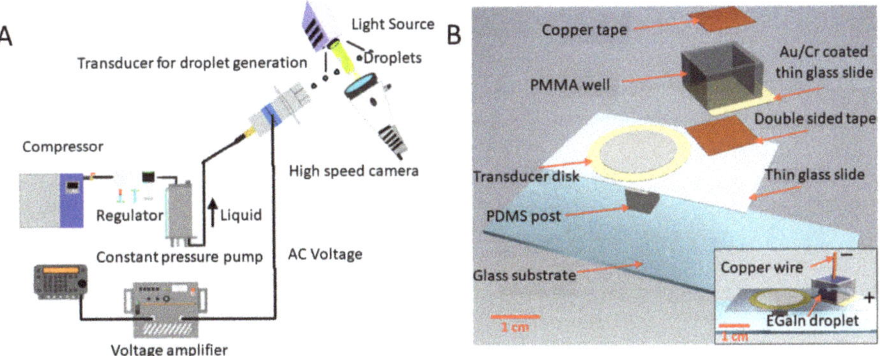

Figure 9. Schematics of acoustic systems for MPs production. (**A**) Schematic of a droplet generation system incorporating an ultrasonic torsional transducer and a micropore plate. Reprinted with permission from ref [150]. (**B**) Exploded schematic of the acoustic-based LM microdroplet production platform. The inset is the assembled view. Reprinted with permission from ref [48].

Apart from aqueous microdroplets, Figure 9B illustrates an acoustic-based miniaturised system for producing LM MPs [48]. The piezoelectric transducer is stuck to the thin glass slide that adheres to the PDMS post to maximize the vibration; the Cu tape is used as electrode. The polymethylmethacrylate (PMMA) wall is attached to the thin glass slide with the 45° angle, leading to the most efficient production of EGaIn microdroplets. The acoustic waves generated from the piezoelectric transducer disk can transfer destructive forces to the droplets to compete with cohesive forces on the liquid interface, breaking it into microscale (40–50 μm) drops. This system enables the production of microdroplets with controllable size by tuning applied voltage to change the surface tension of the liquid metal and effectively avoids the use of cumbersome and expensive sonication bath or probe. Besides, such miniaturised system shows the versatility that can also be used as electrochemical sensor for heavy metal ion detection. For example, the EGaIn microdroplets coated WO_3 NPs fabricated by the platform can detect the Pb^{2+} with the sensitivity minimizing to 500 ppb.

3.3. Centrifugal and Spinning Systems

Centrifugal forces can also be harnessed to break aqueous solutions into microdroplets. Centrifugal forces are generated by the rotation of liquid, and the microdroplets are pinched off from the orifice (no limitation of the number of capillary orifices) when the centrifugal force surpasses the interfacial tension exerted by the capillary orifice. Gelation of the produced sodium alginate microdroplets can be induced using calcium chloride ($CaCl_2$) solution [152]. Furthermore, the shapes and morphologies of MPs can be tuned effectively by the partial dissolution of the products and manipulating the 3D microflows to deform the precursor microdroplets, realizing complex-shaped 3D multicompartmental MPs [153]. This type of devices has numerous advantages including easy fabrication, low cost, small volume required, and less sensitivity to fluid's properties. A centrifuge-based axisymmetric coflowing microfluidic device can be prepared without the need of any photolithography process. It consists of two round capillaries (inner and outer capillaries), a capillary holder, and a sampling microtube, which can be used for producing microdroplets, as shown in Figure 10A [154]. The screw is introduced to adjust the distance between lower and upper part of the holders. By rotating the microtube, W/O microdroplets were pinched off from inner capillary; the sizes of monodisperse water microdroplets increase as the inner capillary diameters enlarge. Additionally, Shin et al. developed a centrifuge-based step emulsification device for producing microdroplets with diameters ranging from of 18 to 90 μm (CV < 2%) [155]. The device includes a reservoir part for storage of the dispersed phase, a triangular microchannel part and a step at the channel end for droplet formation, as shown in Figure 10B. Such device is put into a commercial centrifuge with a fixed angle of 45° to provide excessive pressure on the dispersed phase to infuse the dispersed phase into a microchannel. Then, the microdroplets are pinched off and stored into the microtube. With a fixed channel height, monodisperse droplets are generated under high levels of the oil phase; on the contrary, polydisperse droplets are produced under low levels of the oil phase. The high monodisperse droplets can be generated in a high centrifugal force (~>1500× g), which is yet to be observed in the conventional step emulsification. Moreover, under the constant aspect ratio of the microchannel (width/height) and centrifugal force, the higher the height of the microchannel is, the larger droplets can be obtained. Such droplets highlight the potential applications in biochemical reactors and food, cosmetics, and medical industries.

Other than rotating the microtube, Chen et al. designed a liquid emulsion generator to produce W/O microdroplets by moving a micropipette in a revolving manner [49]. The system consisted of a hydrophobic-coated glass micropipette, a servo motor, an eccentric wheel, a syringe, and a syringe pump (Figure 10C). The water phase flows through the micropipette under the pressure of syringe pump while it is spun by the servo controller and then breaks into uniform droplets. Due to the higher density of the droplets compared with the mineral oil, the products can smoothly sink down the bottom of the microtube. Microdroplets with diameters ranging from 25 to 230 μm (CV < 5%), which are suitable for single cell analysis, were produced by tuning the flow rate of the continuous phase and the motion velocity of the micropipette tip. However, when the platform works under the high flow rate of water phase and the high motion velocity of the micropipette, droplets with a high polydispersity will be generated. These satellite droplets are the result of high dispensing rate of water phase, and droplets near the rotation region of the tip of micropipette are smashed by the moving micropipette with a high speed. A centrifugal microchannel array droplet generating system was established for digital polymerase chain reaction (PCR) [156]. This system not only eliminates the usage of complex microfluidic devices and control systems but also greatly suppresses the loss of materials and cross-contamination. The diameter of microdroplets depends on the size of the microchannel and the centrifugal force.

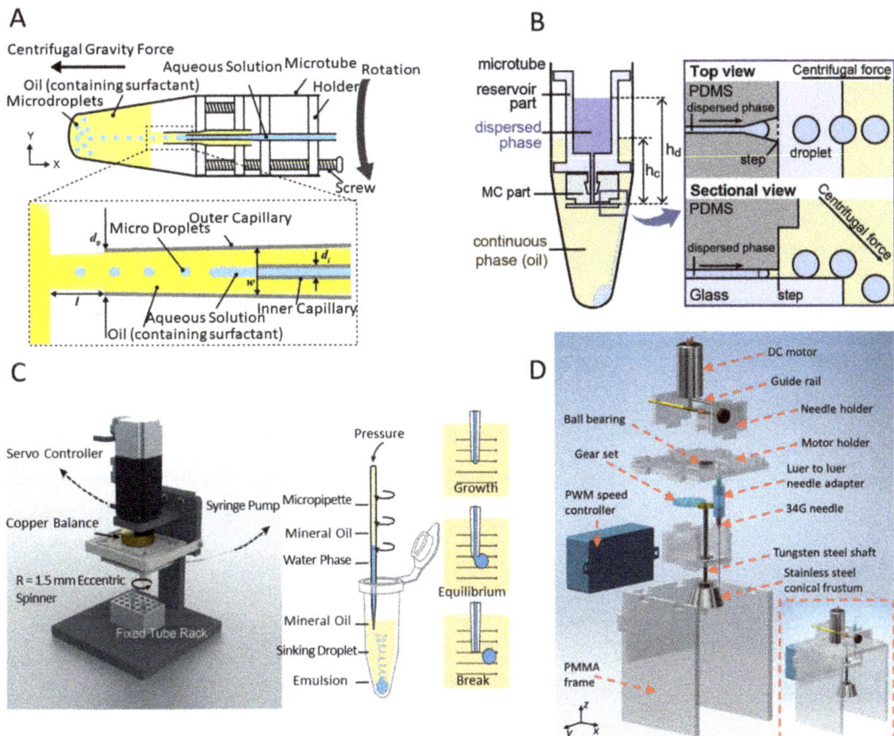

Figure 10. Schematics of centrifugal and spinning systems. (**A**) Illustration of the centrifuge-based axisymmetric coflowing microfluidic device. Reprinted with permission from ref [154]. (**B**) Components and schematic of the centrifuge-based step emulsion device. Reprinted with permission from ref [155]. (**C**) Schematic illustration of the liquid emulsion generator and process of droplet formation. Reprinted with permission from ref [49]. (**D**) Schematic representation of the off-chip spinning microdroplet generator. Reprinted with permission from ref [129].

To further improve the throughput of the microdroplets generation, an off-chip microdroplet generator using a spinning conical frustum was developed, as shown in Figure 10D [129]. The direct-current motor controlled by a PWM motor controller drives the tungsten-steel shaft via a gear set, and a stainless-steel conical frustrum is fixed at the end of the shaft. The diameters of microdroplets can be tuned by rotation speed, flow rate of the continuous phase, the gap between the needle tip and the surface of the frustrum, and the application of an electric field. The higher the speed the conical frustum spins at, the smaller the diameters of microdroplets obtained. After applying the electric field between the needle and the frustum, the droplet size can further decrease due to the reduction in the interfacial tension between water and oil. Apart from water droplets, LM droplets that are relatively difficult to produce in microfluidic platforms due to the high density, viscosity, and surface tension can also be generated within the corn syrup–water mixture. The LM MPs can be adopted in microelectromechanical systems, electrochemical sensors, and constructing 3D structures. Similarly, an off-chip spinning disk microdroplet production platform using submerged electrodispersion was designed to simplify the process of the production of LM microdroplets [157]. The platform consisted of a Cu electrode, a Teflon sleeve, a 34G needle, a tungsten steel shaft, and a polymethylmethacrylate (PMMA) disk. LM microdroplets with uniform sizes (<5%) were produced using the submerged electrodispersion technique. The spinning disk provided the dragging force to take the produced microdroplets away from the ground electrode and therefore avoided the undesired coalescence of droplets.

3.4. Jetting Systems

Inkjet printing technology can rapidly create and release liquid droplets for precise deposition. In brief, jetting is induced by driving an actuator (typically pneumatic or piezoelectric) at the print head to propel the fluid, then microdroplets are expelled from the nozzle to the solution or a target surface, as shown in Figure 11A. Inkjet printing has become a standard production process in the industry and can accommodate a wide range of fluids. Numerous integrated jetting systems have been developed to address a wide variety of applications. Gao et al. presented a drop-on-demand jetting system by integrating a lift platform, a pressure control system, a syringe filter, and an imaging system [158]. With this integrated system, microdroplets with a diameter of 100 μm were made with a jetting velocity at 2.42 m/s. For the formation of microdroplets of liquid with a high viscosity, a pneumatically driven inkjet printing system was designed [50]. By assembling a pneumatic printing system for generating droplets by alternatively applying negative and positive air pressure to the precursor, a monitoring system for detecting the droplet formation process, and a measuring system for analysing the printing volume together, viscous microdroplets (1-384.5 Cp) were formed (CV of the diameter ≤1.07%).

Apart from pneumatic-driven systems, the alternating viscous and inertial forcing jetting (AVIFJ) mechanism was successfully applied in the jetting system [159]. The AVIFJ-based microdroplet deposition system contains modules of data processing, microdroplet dispensing, materials delivery, 3D motion, and observation. The piezoceramic is driven by a periodic voltage signal and moves in a reciprocating motion; the glass nozzle is fastened to the holder and the holder is connected to the piezoceramic. In the first half of the reciprocating micromotion, the fluid in the nozzle is driven by the viscous force between the fluid and the nozzle and moves downward. In the second half of the reciprocating micromotion, the fluid keeps moving downward by inertia, while the nozzle moves upward. Consequently, production of microdroplets can be achieved using the reciprocating micromotion of the nozzle. The system can form microdroplets (diameter of 52–72 μm, with a nozzle diameter of 45 μm and jetting speed of 0.4–2.0 m/s), which can be applied in addictive manufacturing and cell printing.

In addition to water, solder droplets that are suitable for metal additive manufacturing can be formed using this jetting technology. Ming et al. presented a piezoelectric membrane-piston-based jetting technology (PMPJT) system, as shown in Figure 11B [160]. A computer numerical control system is introduced to produce an electrical pulse signal that excites the piezoelectric ceramic, forcing the vibration bar to move. Figure 11C simply illustrates the working principle of PMPJT. When the electrical pulse signal applies, the vibration bar moves downward to stretch the metal membranes and some molten metal flows through the orifice of the nozzle to form a stream. After turning off the pulse signal, the membranes begin to restore and induce the shrinking of the stream at the orifice. Due to gravity and inertia of the stream, molten metal is finally separated from the main body. The spherical microdroplets are finally formed because of the surface tension. By adjusting the voltage and length of the electrical pulses and temperature of the metal, the diameter of microdroplets formed in this system ranged from around 85 to 121 μm. The system can operate at a working temperature higher than 180 °C to generate solder MPs. Similarly, StarJet technology was used to generate Al alloy MPs [126]. The assembly of the StarJet printhead includes the inserted Macor reservoir, the cap, and the body (made from Inconel 718 alloy). The reservoir outlet tube with diameter of 400–600 μm prevents the melting metal from flowing out of the printhead by counteracting capillary forces, and the nozzle chip is directly mounted under the reservoir outlet tube. The melting metal is pneumatically pushed towards the nozzle chip and forms microdroplets. Such a prototype of printhead in drop-on-demand mode enables the direct printing MPs of Al alloys, which extends to the application of Al alloy MPs in additive manufacturing. Moreover, using the principle of laser-induced forward transfer, Zenou et al. utilised a laser system to induce Al microdroplets jetting from donor to accepter, as shown in Figure 11E [161]. By measuring the electrical signal obtained once the Al microdroplet touches the accepter, the velocity of Al microdroplets can be estimated. Such a jetting process demonstrates the effective energy transfer from the laser to the Al microdroplets. Above-introduced integrated systems for synthesising MPs are summarised in Table 4.

Table 4. Overview of integrated systems for synthesis of microparticles (MPs.)

MPs Type	Enabling Technologies/Modules	Crucial Parameters	MP Size (μm)	Costs [1]	Year	Reference
PEDOT/PSS-agarose hybrid MPs	Microfluidic droplet generator	Continuous oil flow rate	20–80	★★	2016	Lee [127]
Solid core enzyme-immobilised microcapsules	Flow focusing		580 ± 10	★★	2019	Varshney [132]
Magnetic droplets	Step emulsion device Magnetically driven microfluidic droplet generation technique	Dimensions of channels	85–125	★★	2016	Kahkeshani [130]
W/O emulsions W/O/W emulsions	Flow focusing Droplet-based microfluidics Commercially available self-setting rubber	Flow rate Nozzle diameter	100–500	★	2015	Lapierre [131]
Chitosan microspheres	512-microchannel geometrical passive breakup device T-junction	Flow rate	40.0 ± 2.2	★★	2019	Kim [133]
PLGA microspheres	512-channel geometric droplet-splitting microfluidic device 256 T-junction		6.56	★★	2020	Kim [136]
Cell-laden microgel	Flow-focusing platform On-chip	Cell concentration	~240–300	★★	2019	Mohamed [138]
Drops	Parallelised microfluidic device Millipede device	Device geometry	20–160	★★	2016	Amstad [135]
Free-floating polymer (PEGDA)	Contact flow lithography system	Microchannel dimensions	20–150	★★★	2015	Goff [139]
W/O and O/W emulsions	Glass microfluidic device Step emulsification		80.9 (CV = 2.8%)	★★	2017	Ofner [137]
Chitosan/TiO$_2$ composite	Factory-on-chip Modularised microfluidic reactors		539.65	★★★	2017	Han [134]
Water-in-water (W/W) emulsions	Microneedle-assistance Microfluidics Flow focusing	Column pressure	5–65	★★★	2019	Jeyhani [128]
W/O emulsions	Electrical detection Microfluidics Closed-loop control	Flow rate	200	★★★★	2017	Fu [144]
Liquid metal	Microfluidic flow-focusing device	Electrical potential Flow rate	~80–160	★★★	2016	Tang [48]
W/O and oil-in water (O/W) emulsions	3D-printed droplet generator Plug-and-play	Liquid flow rate ratio Viscosity of the dispersed phase	~50	★★	2016	Zhang [146]
PEGDA	3D-printed generator Screw-and-nut	T-junction gap height Flow rates	34–1404	★★	2019	Nguyen [147]
W/O droplets	3D-printing technology Millifluidics Chimney-shaped void geometry	Flow rates Apex angle	36–616	★★	2019	Hwang [148]

Table 4. *Cont.*

MPs Type	Enabling Technologies/Modules	Crucial Parameters	MP Size (μm)	Costs [1]	Year	Reference
Magnetic liquid metal	3D-printed coaxial microfluidic device	Orifice diameter Flow rate ratio	650–1900	★★★	2020	He [149]
EGaIn	Acoustic waves Electrochemistry Electrocapillary	Oxidative/reducing voltages Activating frequency	10–80	★★★	2016	Tang [48]
Water-in-oil (W/O) emulsions	Ultrasonic transducer	Vibrational velocity Pressure	62.5 ± 2.6	★★★	2018	Fujimoro [151]
Pure water, silicone oils	Ultrasonic torsional transducer	Pressure Resonance frequency Diameter of liquid column	~80–120	★★★	2015	Kishi [150]
W/O microdroplets	Glass-capillary-based microfluidic device Tabletop minicentrifuge	Diameter of inner and outer capillary orifice	~6.6–13.8	★★	2014	Yamashita [154]
W/O emulsions	Spinning micropipette liquid emulsion generator	Flow rate Motion velocity of the micropipette	25–230	★★	2016	Chen [49]
W/O emulsion	Centrifugal microchannel	Size of microchannels Centrifugal force	~52.5	★★	2017	Chen [156]
Calcium alginate	Centrifugal microfluidic technique	Centrifugal force Circumference of the channel outlet	~109–269	★★★	2015	Liu [162]
W/O picolitre droplets	Centrifuge-based step emulsification device	Level of oil phase Centrifugal force Height of microchannel	18–90	★★★	2019	Shin [155]
Gallium-based liquid metal	Submerged electrodispersion technique Spinning disk	Electric field Flow rate Rotation speed of the disk	~10–800	★★★	2019	Zhang [157]
Water, liquid metal, hydrogel, double emulsions	Spinning conical frustum	Rotational speed Applied voltage Flow rate	~200–550	★	2019	Tang [129]
Sodium alginate multicompartmental particles	Centrifuge-based droplet shooting device	Barrel configuration Diameter of capillary orifice	99 and 16	★★★	2012	Maeda [152]
Sodium alginate with complex shape	Centrifuge 3D nonequilibrium-induced microflows	Diffusional flow Marangoni microflows	~112.4–135.1 (various shapes)	★★★	2016	Hayakawa [153]
Janus MPs	Centrifugal gravity UV irradiation		282 (mean)	★★★	2020	Tsuchiya [163]
Solder (Sn63Pb37)	Piezoelectric membrane-piston-based jetting technology	Pulse length Voltage value Temperature	~85	★★★★	2019	Ma [160]
PDMS, UV-curing optical glue (high viscosity >2000 cps)	Tip-assisted electric field intensity enhancement effect High-resolution capability of EHD printing	Applied voltage Gap distance Nozzle inner diameter Deposition time	>2.3	★★★★	2019	Zou [164]
Al	Pneumatic drop-on-demand technology	The aspect ratio of the nozzle hole The distance between inlet hole and nozzle hole	359.9	★★★★	2017	Zhong [165]
Ink drops	Pneumatic valve Feedback control Ejection technology Machine vision	Solenoid valve "ON" time		★★★★★	2018	Wang [166]

Table 4. *Cont.*

MPs Type	Enabling Technologies/Modules	Crucial Parameters	MP Size (μm)	Costs [1]	Year	Reference
Al alloys (AlSi12)	StarJet technology	Applied pressures	235 ± 15	★★★★	2017	Gerdes [126]
Alginate	Drop-on-demand jetting Piezoelectric print-head	Voltage waveform Microdroplet velocity Concentration of CaCl₂ solution	~80–110	★★★★	2016	Gao [158]
Water drops	Piezo-actuated microdroplet generator Drop on demand	Deflection voltage Suction and compression time Nozzle diameter	450–1000	★★★★	2014	Sadeghian [167]
Chitosan aerogel	Jet cutting Supercritical drying of gel	Nozzle diameter Cutting disc velocity Number of wires of the cutting disc	700–900	★★★	2020	López-Iglesias [168]
Sodium alginate	Alternating viscous and inertial force jetting mechanism	Applied voltage Nozzle diameter Fluid viscosity	~30–80	★★★★	2017	Zhao [169]
Sodium alginate	Alternating viscous and inertial force jetting mechanism	Actuation signal waveforms Nozzle dimensional features Solution velocity	53–72	★★★★	2015	Zhao [159]
Al	Supersonic laser-induced jetting	Incubation time Droplet velocity	~3.9	★★★★	2015	Zenou [161]
High viscous microdroplets	Pneumatically driven inkjet printing system	Droplet volume Standoff distance frequency	~143–247 (12.2–63.5 nL)	★★★★	2016	Choi [50]

[1] The number of asterisks (★) represents the cost of synthesis system; 1 means relatively low cost, while 5 means expensive.

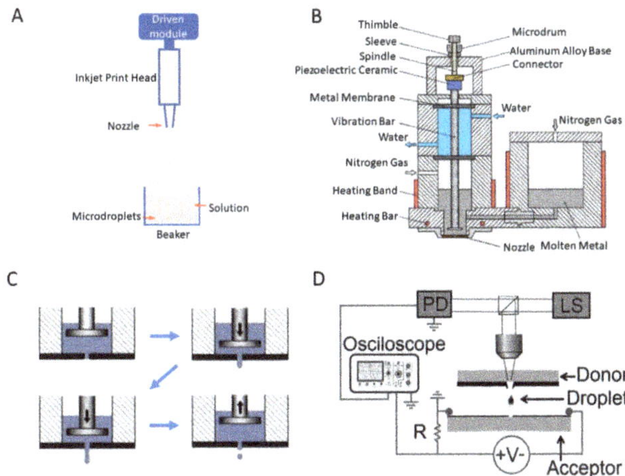

Figure 11. Illustrations of jetting mechanism and jetting systems. (**A**) Schematic of a typical jetting platform. (**B**) Schematic diagram of the piezoelectric membrane-piston-based jetting technology (PMPJT) system. (**C**) The principle of microdroplet formation based on the PMPJT. Reprinted with permission from ref [160]. (**D**) Schematic illustration of the aluminium (Al) microdroplets generation system combining supersonic laser-induced jetting and velocity measuring function. PD represents photodiode and LS is laser system. Reprinted with permission from ref [161].

4. Conclusions and Outlook

In this review, we have discussed the modular and integrated systems developed in recent years for producing a wide variety of NPs and MPs. For the synthesis of NPs, we have elucidated some representative systems based on top-down and bottom-up methods, which are summarised in the Tables 2 and 3, respectively. In terms of top-down approaches, the products can be synthesized rapidly by breaking the bulk materials into nanoscales. Sonication probe systems are suitable for producing LM NPs with small enough sizes, and the production performance can be optimised by incorporating appropriate control modules. However, it is still problematic for producing uniform-sized NPs as the disruptive force is uncontrollable. Laser ablation systems are ideal for the production of pure metal NPs like Au and Ag. Besides, these systems are capable of fabricating specific structures of NPs by combining external fields. Because industrial scale of production requires high power and multiple sets of expensive laser systems, the above-mentioned techniques are yet to be readily used in industry.

As for bottom-up approaches for producing NPs, microfluidics can precisely control the reaction time and volume to enable various NPs production with uniform and monodisperse sizes. In recent years, different groups had designed various structures of microfluidic chips and integrated multiple synthetic procedures into a single microfluidic device to produce NPs with desired functions and morphologies. Moreover, multichannel microreactor systems for providing continuous flow enable the scaling up of production, which would allow for expansion to achieve an industrial level of production. Nevertheless, the fabrication of microfluidic chips requires the photolithography techniques and specialised microfluidic facilities in cleanroom, which limits accessibility. In addition, the integrated systems usually need to be driven by external pumping systems—an inconvenience for settings other than laboratory. In addition to microfluidics, flame synthesis is a powerful bottom-up method. It does not need tedious steps and can easily form metastable compositions with a high purity in a short reaction time. However, such systems still face challenges. For example, it is difficult to produce more complicated NPs other than noble metals and carbon using flame synthesis platforms; also, the properties of substrate limit the productions and utilisation of NPs.

For the production of MPs (summarized in Table 4), microfluidic technology has many advantages, including the ability to control size, morphology, and composition of particles. In recent years, 3D printing technology further facilitates the fabrication of microfluidic devices without the need of photolithography. This technology lowers the entry requirements of using microfluidics and benefits the advancement of future integrated systems. In addition, other MPs generation systems have been developed using techniques including acoustic, centrifugal, spinning force, and jetting. By controlling parameters such as acoustic frequency, rotating speed, and driving force/frequency, MPs with various sizes, morphologies, and compositions, can be synthesized on demand. These integrated systems avoid complicated structures and reduce the difficulties of operation, which may lead to an expansion of the applications of MPs.

On the other hand, the development of modular and integrated systems for the versatile production of NPs/MPs await enhancement. For instance, some integrated systems require large space and hard to be decomposed into modules. In addition, in many cases, bulky peripherals are still needed to operate and control the platforms. Besides, the use of specialised and noncommercial components in many of the systems limits their accessibility. Regardless of these hardships, we still believe that further development of modular and integrated systems will bridge the gap between harnessing the vast potential of NPs/MPs research and the difficulty of synthesising NPs/MPs in simple, repeatable, and on-demand manners in different laboratories. We anticipate that researchers can readily select and establish synthesis systems for the production of customised NPs/MPs with a simplified process, thereby breaking the obstacle in exploring a wider range of applications of NPs/MPs. We envisage that some modular and integrated systems possess the potential to meet the industrial scale and quality, making it possible to further reduce the cost and complexity of synthesising NPs/MPs. As such, assisting the production of NPs and MPs using modular and integrated systems with the characteristics of automatic control, simple user interface, customisable functions, which only requires basic equipment which most of the laboratories can offer, can certainly facilitate interdisciplinary innovations for unleashing a wider range of research directions.

Funding: This research received no external funding.

Conflicts of Interest: The authors declare no conflict of interest.

References

1. Singh, A.K.; Xu, Q. Synergistic Catalysis over Bimetallic Alloy Nanoparticles. *Chemcatchem* **2013**, *5*, 652–676. [CrossRef]
2. Campelo, J.M.; Luna, D.; Luque, R.; Marinas, J.M.; Romero, A.A. Sustainable Preparation of Supported Metal Nanoparticles and Their Applications in Catalysis. *Chemsuschem* **2009**, *2*, 18–45. [CrossRef] [PubMed]
3. Vairavapandian, D.; Vichchulada, P.; Lay, M.D. Preparation and modification of carbon nanotubes: Review of recent advances and applications in catalysis and sensing. *Anal. Chim. Acta* **2008**, *626*, 119–129. [CrossRef]
4. Asadian, E.; Ghalkhani, M.; Shahrokhian, S. Electrochemical sensing based on carbon nanoparticles: A review. *Sens. Actuators B Chem.* **2019**, *293*, 183–209. [CrossRef]
5. Krishna, V.D.; Wu, K.; Su, D.; Cheeran, M.C.J.; Wang, J.P.; Perez, A. Nanotechnology: Review of concepts and potential application of sensing platforms in food safety. *Food Microbiol.* **2018**, *75*, 47–54. [CrossRef]
6. Moreira, A.F.; Rodrigues, C.F.; Reis, C.A.; Costa, E.C.; Correia, I.J. Gold-core silica shell nanoparticles application in imaging and therapy: A review. *Microporous Mesoporous Mater.* **2018**, *270*, 168–179. [CrossRef]
7. Padmanabhan, P.; Kumar, A.; Kumar, S.; Chaudhary, R.K.; Gulyás, B. Nanoparticles in practice for molecular-imaging applications: An overview. *Acta Biomater.* **2016**, *41*, 1–16. [CrossRef]
8. Daraee, H.; Eatemadi, A.; Abbasi, E.; Aval, S.F.; Kouhi, M.; Akbarzadeh, A. Application of gold nanoparticles in biomedical and drug delivery. *Artif. Cells Nanomed. Biotechnol.* **2016**, *44*, 410–422. [CrossRef]
9. Elahi, N.; Kamali, M.; Baghersad, M.H. Recent biomedical applications of gold nanoparticles: A review. *Talanta* **2018**, *184*, 537–556. [CrossRef]
10. Dickey, M.D. Emerging Applications of Liquid Metals Featuring Surface Oxides. *Acs Appl. Mater. Interfaces* **2014**, *6*, 18369–18379. [CrossRef]
11. Khoshmanesh, K.; Tang, S.-Y.; Zhu, J.Y.; Schaefer, S.; Mitchell, A.; Kalantar-Zadeh, K.; Dickey, M.D. Liquid metal enabled microfluidics. *Lab A Chip* **2017**, *17*, 974–993. [CrossRef] [PubMed]
12. Song, H.; Kim, T.; Kang, S.; Jin, H.; Lee, K.; Yoon, H.J. Ga-Based Liquid Metal Micro/Nanoparticles: Recent Advances and Applications. *Small* **2020**, *16*. [CrossRef] [PubMed]
13. Ali, A.; Ahmed, S. A review on chitosan and its nanocomposites in drug delivery. *Int. J. Biol. Macromol.* **2018**, *109*, 273–286. [CrossRef]
14. Yan, J.; Lu, Y.; Chen, G.; Yang, M.; Gu, Z. Advances in liquid metals for biomedical applications. *Chem. Soc. Rev.* **2018**, *47*, 2518–2533. [CrossRef]
15. Wang, Q.; Yu, Y.; Liu, J. Preparations, Characteristics and Applications of the Functional Liquid Metal Materials. *Adv. Eng. Mater.* **2018**, *20*, 1700781. [CrossRef]
16. Lin, Y.; Genzer, J.; Dickey, M.D. Attributes, Fabrication, and Applications of Gallium-Based Liquid Metal Particles. *Adv. Sci.* **2020**, *7*, 2000192. [CrossRef] [PubMed]
17. Lee, S.H.; Jun, B.-H. Silver Nanoparticles: Synthesis and Application for Nanomedicine. *Int. J. Mol. Sci.* **2019**, *20*, 865. [CrossRef]
18. Darabdhara, G.; Das, M.R.; Singh, S.P.; Rengan, A.K.; Szunerits, S.; Boukherroub, R. Ag and Au nanoparticles/reduced graphene oxide composite materials: Synthesis and application in diagnostics and therapeutics. *Adv. Colloid Interface Sci.* **2019**, *271*, 101991. [CrossRef]
19. Jeyaraj, M.; Gurunathan, S.; Qasim, M.; Kang, M.-H.; Kim, J.-H. A Comprehensive Review on the Synthesis, Characterization, and Biomedical Application of Platinum Nanoparticles. *Nanomaterials* **2019**, *9*, 1719. [CrossRef]
20. Azharuddin, M.; Zhu, G.H.; Das, D.; Ozgur, E.; Uzun, L.; Turner, A.P.; Patra, H.K. A repertoire of biomedical applications of noble metal nanoparticles. *Chem. Commun.* **2019**, *55*, 6964–6996. [CrossRef]
21. Rafique, M.; Shaikh, A.J.; Rasheed, R.; Tahir, M.B.; Bakhat, H.F.; Rafique, M.S.; Rabbani, F. A review on synthesis, characterization and applications of copper nanoparticles using green method. *Nano* **2017**, *12*, 1750043. [CrossRef]

22. Al-Hakkani, M.F. Biogenic copper nanoparticles and their applications: A review. *Sn Appl. Sci.* **2020**, *2*, 1–20. [CrossRef]

23. Din, M.I.; Rehan, R. Synthesis, characterization, and applications of copper nanoparticles. *Anal. Lett.* **2017**, *50*, 50–62. [CrossRef]

24. Cid, A.; Simal-Gandara, J. Synthesis, characterization, and potential applications of transition metal nanoparticles. *J. Inorg. Organomet. Polym. Mater.* **2020**, *30*, 1011–1032. [CrossRef]

25. Abo-Zeid, Y.; Williams, G.R. The potential anti-infective applications of metal oxide nanoparticles: A systematic review. *Wiley Interdiscip. Rev. Nanomed. Nanobiotechnol.* **2020**, *12*, e1592. [CrossRef]

26. George, J.M.; Antony, A.; Mathew, B. Metal oxide nanoparticles in electrochemical sensing and biosensing: A review. *Microchim. Acta* **2018**, *185*, 358. [CrossRef] [PubMed]

27. Raghunath, A.; Perumal, E. Metal oxide nanoparticles as antimicrobial agents: A promise for the future. *Int. J. Antimicrob. Agents* **2017**, *49*, 137–152. [CrossRef]

28. Wagner, A.M.; Knipe, J.M.; Orive, G.; Peppas, N.A. Quantum dots in biomedical applications. *Acta Biomater.* **2019**, *94*, 44–63. [CrossRef]

29. Wu, H.-L.; Li, X.-B.; Tung, C.-H.; Wu, L.-Z. Semiconductor Quantum Dots: An Emerging Candidate for CO2 Photoreduction. *Adv. Mater.* **2019**, *31*, 1900709. [CrossRef]

30. McHugh, K.J.; Jing, L.; Behrens, A.M.; Jayawardena, S.; Tang, W.; Gao, M.; Langer, R.; Jaklenec, A. Biocompatible Semiconductor Quantum Dots as Cancer Imaging Agents. *Adv. Mater.* **2018**, *30*, 1706356. [CrossRef]

31. Owen, J.; Brus, L. Chemical synthesis and luminescence applications of colloidal semiconductor quantum dots. *J. Am. Chem. Soc.* **2017**, *139*, 10939–10943. [CrossRef] [PubMed]

32. Boakye-Yiadom, K.O.; Kesse, S.; Opoku-Damoah, Y.; Filli, M.S.; Aquib, M.; Joelle, M.M.B.; Farooq, M.A.; Mavlyanova, R.; Raza, F.; Bavi, R. Carbon dots: Applications in bioimaging and theranostics. *Int. J. Pharm.* **2019**, *564*, 308–317. [CrossRef] [PubMed]

33. Alim, S.; Vejayan, J.; Yusoff, M.M.; Kafi, A.K.M. Recent uses of carbon nanotubes & gold nanoparticles in electrochemistry with application in biosensing: A review. *Biosens. Bioelectron.* **2018**, *121*, 125–136. [CrossRef] [PubMed]

34. Mazrad, Z.A.I.; Lee, K.; Chae, A.; In, I.; Lee, H.; Park, S.Y. Progress in internal/external stimuli responsive fluorescent carbon nanoparticles for theranostic and sensing applications. *J. Mater. Chem. B* **2018**, *6*, 1149–1178. [CrossRef]

35. Muhulet, A.; Miculescu, F.; Voicu, S.I.; Schütt, F.; Thakur, V.K.; Mishra, Y.K. Fundamentals and scopes of doped carbon nanotubes towards energy and biosensing applications. *Mater. Today Energy* **2018**, *9*, 154–186. [CrossRef]

36. Martins, C.; Sousa, F.; Araújo, F.; Sarmento, B. Functionalizing PLGA and PLGA derivatives for drug delivery and tissue regeneration applications. *Adv. Healthc. Mater.* **2018**, *7*, 1701035. [CrossRef]

37. Li, J.; Rao, J.; Pu, K. Recent progress on semiconducting polymer nanoparticles for molecular imaging and cancer phototherapy. *Biomaterials* **2018**, *155*, 217–235. [CrossRef]

38. Calzoni, E.; Cesaretti, A.; Polchi, A.; Di Michele, A.; Tancini, B.; Emiliani, C. Biocompatible polymer nanoparticles for drug delivery applications in cancer and neurodegenerative disorder therapies. *J. Funct. Biomater.* **2019**, *10*, 4. [CrossRef]

39. Ealias, A.M.; Saravanakumar, M.P. A review on the classification, characterisation, synthesis of nanoparticles and their application. *IOP Conf. Ser. Mater. Sci. Eng.* **2017**, *263*, 032019.

40. Xiao, J.; Liu, P.; Wang, C.X.; Yang, G.W. External field-assisted laser ablation in liquid: An efficient strategy for nanocrystal synthesis and nanostructure assembly. *Prog. Mater. Sci.* **2017**, *87*, 140–220. [CrossRef]

41. Chan, H.-K.; Kwok, P.C.L. Production methods for nanodrug particles using the bottom-up approach. *Adv. Drug Deliv. Rev.* **2011**, *63*, 406–416. [CrossRef]

42. Prow, T.W.; Grice, J.E.; Lin, L.L.; Faye, R.; Butler, M.; Becker, W.; Wurm, E.M.T.; Yoong, C.; Robertson, T.A.; Soyer, H.P.; et al. Nanoparticles and microparticles for skin drug delivery. *Adv. Drug Deliv. Rev.* **2011**, *63*, 470–491. [CrossRef]

43. Wang, J.; Li, Y.; Wang, X.; Wang, J.; Tian, H.; Zhao, P.; Tian, Y.; Gu, Y.; Wang, L.; Wang, C. Droplet microfluidics for the production of microparticles and nanoparticles. *Micromachines* **2017**, *8*, 22. [CrossRef]

44. Hao, N.; Nie, Y.; Zhang, J.X.J. Microfluidic synthesis of functional inorganic micro-/nanoparticles and applications in biomedical engineering. *Int. Mater. Rev.* **2018**, *63*, 461–487. [CrossRef]

45. Ma, J.; Lee, S.M.-Y.; Yi, C.; Li, C.-W. Controllable synthesis of functional nanoparticles by microfluidic platforms for biomedical applications—A review. *Lab A Chip* **2017**, *17*, 209–226. [CrossRef]

46. Zhang, H.; Tumarkin, E.; Sullan, R.M.A.; Walker, G.C.; Kumacheva, E. Exploring Microfluidic Routes to Microgels of Biological Polymers. *Macromol. Rapid Commun.* **2007**, *28*, 527–538. [CrossRef]

47. Biswas, S.; Miller, J.T.; Li, Y.; Nandakumar, K.; Kumar, C.S. Developing a millifluidic platform for the synthesis of ultrasmall nanoclusters: Ultrasmall copper nanoclusters as a case study. *Small* **2012**, *8*, 688–698. [CrossRef] [PubMed]

48. Tang, S.-Y.; Ayan, B.; Nama, N.; Bian, Y.; Lata, J.P.; Guo, X.; Huang, T.J. On-Chip Production of Size-Controllable Liquid Metal Microdroplets Using Acoustic Waves. *Small* **2016**, *12*, 3861–3869. [CrossRef]

49. Chen, Z.; Fu, Y.; Zhang, F.; Liu, L.; Zhang, N.; Zhou, D.; Yang, J.; Pang, Y.; Huang, Y. Spinning micropipette liquid emulsion generator for single cell whole genome amplification. *Lab A Chip* **2016**, *16*, 4512–4516. [CrossRef]

50. Choi, I.H.; Kim, J. A pneumatically driven inkjet printing system for highly viscous microdroplet formation. *Micro Nano Syst. Lett.* **2016**, *4*. [CrossRef]

51. Tang, S.Y.; Qiao, R.; Yan, S.; Yuan, D.; Zhao, Q.; Yun, G.; Davis, T.P.; Li, W. Microfluidic mass production of stabilized and stealthy liquid metal nanoparticles. *Small* **2018**, *14*, 1800118. [CrossRef] [PubMed]

52. Tang, S.-Y.; Joshipura, I.D.; Lin, Y.; Kalantar-Zadeh, K.; Mitchell, A.; Khoshmanesh, K.; Dickey, M.D. Liquid-Metal Microdroplets Formed Dynamically with Electrical Control of Size and Rate. *Adv. Mater.* **2016**, *28*, 604–609. [CrossRef] [PubMed]

53. Zhang, W.; Ou, J.Z.; Tang, S.Y.; Sivan, V.; Yao, D.D.; Latham, K.; Khoshmanesh, K.; Mitchell, A.; O'Mullane, A.P.; Kalantar-zadeh, K. Liquid metal/metal oxide frameworks. *Adv. Funct. Mater.* **2014**, *24*, 3799–3807. [CrossRef]

54. Lu, H.; Tang, S.-Y.; Dong, Z.; Liu, D.; Zhang, Y.; Zhang, C.; Yun, G.; Zhao, Q.; Kalantar-Zadeh, K.; Qiao, R. Dynamic Temperature Control System for the Optimized Production of Liquid Metal Nanoparticles. *Acs Appl. Nano Mater.* **2020**, *3*, 6905–6914. [CrossRef]

55. Tang, S.Y.; Qiao, R.; Lin, Y.; Li, Y.; Zhao, Q.; Yuan, D.; Yun, G.; Guo, J.; Dickey, M.D.; Huang, T.J. Functional Liquid Metal Nanoparticles Produced by Liquid-Based Nebulization. *Adv. Mater. Technol.* **2019**, *4*, 1800420. [CrossRef]

56. Sportelli, M.C.; Izzi, M.; Volpe, A.; Clemente, M.; Picca, R.A.; Ancona, A.; Lugarà, P.M.; Palazzo, G.; Cioffi, N. The Pros and Cons of the Use of Laser Ablation Synthesis for the Production of Silver Nano-Antimicrobials. *J. Antibiot.* **2018**, *7*, 67. [CrossRef]

57. Asahi, T.; Mafuné, F.; Rehbock, C.; Barcikowski, S. Strategies to harvest the unique properties of laser-generated nanomaterials in biomedical and energy applications. *Appl. Surf. Sci.* **2015**, *348*, 1–3. [CrossRef]

58. Wang, S.; Gao, L. Laser-driven nanomaterials and laser-enabled nanofabrication for industrial applications. In *Industrial Applications of Nanomaterials*; Elsevier: Amsterdam, The Netherlands, 2019; pp. 181–203.

59. Hu, X.; Takada, N.; Machmudah, S.; Wahyudiono; Kanda, H.; Goto, M. Ultrasonic-Enhanced Fabrication of Metal Nanoparticles by Laser Ablation in Liquid. *Ind. Eng. Chem. Res.* **2020**, *59*, 7512–7519. [CrossRef]

60. Liang, Y.; Liu, P.; Xiao, J.; Li, H.; Wang, C.; Yang, G. A general strategy for one-step fabrication of one-dimensional magnetic nanoparticle chains based on laser ablation in liquid. *Laser Phys. Lett.* **2014**, *11*, 056001. [CrossRef]

61. Streubel, R.; Barcikowski, S.; Gökce, B. Continuous multigram nanoparticle synthesis by high-power, high-repetition-rate ultrafast laser ablation in liquids. *Opt. Lett.* **2016**, *41*, 1486–1489. [CrossRef]

62. Yu, J.; Nan, J.; Zeng, H. Size control of nanoparticles by multiple-pulse laser ablation. *Appl. Surf. Sci.* **2017**, *402*, 330–335. [CrossRef]

63. Mahdieh, M.H.; Fattahi, B. Effects of water depth and laser pulse numbers on size properties of colloidal nanoparticles prepared by nanosecond pulsed laser ablation in liquid. *Opt. Laser Technol.* **2015**, *75*, 188–196. [CrossRef]

64. Kőrösi, L.; Rodio, M.; Dömötör, D.; Kovács, T.; Papp, S.; Diaspro, A.; Intartaglia, R.; Beke, S. Ultrasmall, ligand-free Ag nanoparticles with high antibacterial activity prepared by pulsed laser ablation in liquid. *J. Chem.* **2016**, *2016*, 4143560. [CrossRef]

65. Herbani, Y.; Nasution, R.; Mujtahid, F.; Masse, S. Pulse laser ablation of Au, Ag, and Cu metal targets in liquid for nanoparticle production. *J. Phys. Conf. Ser.* **2018**, *985*, 012005. [CrossRef]

66. Menazea, A. Femtosecond laser ablation-assisted synthesis of silver nanoparticles in organic and inorganic liquids medium and their antibacterial efficiency. *Radiat. Phys. Chem.* **2020**, *168*, 108616. [CrossRef]

67. Liu, P.; Wang, C.; Chen, X.; Yang, G. Controllable fabrication and cathodoluminescence performance of high-index facets GeO2 micro-and nanocubes and spindles upon electrical-field-assisted laser ablation in liquid. *J. Phys. Chem. C* **2008**, *112*, 13450–13456. [CrossRef]

68. Serkov, A.; Rakov, I.; Simakin, A.; Kuzmin, P.; Shafeev, G.; Mikhailova, G.; Antonova, L.K.; Troitskii, A.; Kuzmin, G. Influence of external magnetic field on laser-induced gold nanoparticles fragmentation. *Appl. Phys. Lett.* **2016**, *109*, 053107. [CrossRef]

69. Shafeev, G.; Rakov, I.; Ayyyzhy, K.; Mikhailova, G.; Troitskii, A.; Uvarov, O. Generation of Au nanorods by laser ablation in liquid and their further elongation in external magnetic field. *Appl. Surf. Sci.* **2019**, *466*, 477–482. [CrossRef]

70. Liu, P.; Liang, Y.; Lin, X.; Wang, C.; Yang, G. A general strategy to fabricate simple polyoxometalate nanostructures: Electrochemistry-assisted laser ablation in liquid. *Acs Nano* **2011**, *5*, 4748–4755. [CrossRef]

71. Liang, Y.; Liu, P.; Li, H.; Yang, G. ZnMoO4 micro-and nanostructures synthesized by electrochemistry-assisted laser ablation in liquids and their optical properties. *Cryst. Growth Des.* **2012**, *12*, 4487–4493. [CrossRef]

72. Liang, Y.; Liu, P.; Li, H.; Yang, G. Synthesis and characterization of copper vanadate nanostructures via electrochemistry assisted laser ablation in liquid and the optical multi-absorptions performance. *CrystEngComm* **2012**, *14*, 3291–3296. [CrossRef]

73. Streubel, R.; Bendt, G.; Gökce, B. Pilot-scale synthesis of metal nanoparticles by high-speed pulsed laser ablation in liquids. *Nanotechnology* **2016**, *27*, 205602. [CrossRef] [PubMed]

74. Guo, J.; Cheng, J.; Tan, H.; Sun, Q.; Yang, J.; Liu, W. Constructing a novel and high-performance liquid nanoparticle additive from a Ga-based liquid metal. *Nanoscale* **2020**, *12*, 9208–9218. [CrossRef] [PubMed]

75. Palazzo, G.; Valenza, G.; Dell'Aglio, M.; Giacomo, A.D. On the stability of gold nanoparticles synthesized by laser ablation in liquids. *J. Colloid Interface Sci.* **2017**, *489*, 47–56. [CrossRef] [PubMed]

76. Affandi, S.; Bidin, N. Pulse laser ablation in liquid induce gold nanoparticle production. *J. Teknol.* **2015**, *74*, 41–43. [CrossRef]

77. Valverde-Alva, M.A.; García-Fernández, T.; Villagrán-Muniz, M.; Sánchez-Aké, C.; Castañeda-Guzmán, R.; Esparza-Alegría, E.; Sánchez-Valdés, C.F.; Llamazares, J.L.S.; Herrera, C.E.M. Synthesis of silver nanoparticles by laser ablation in ethanol: A pulsed photoacoustic study. *Appl. Surf. Sci.* **2015**, *355*, 341–349. [CrossRef]

78. Tomko, J.; Naddeo, J.J.; Jimenez, R.; Tan, Y.; Steiner, M.; Fitz-Gerald, J.M.; Bubb, D.M.; O'Malley, S.M. Size and polydispersity trends found in gold nanoparticles synthesized by laser ablation in liquids. *Phys. Chem. Chem. Phys.* **2015**, *17*, 16327–16333. [CrossRef]

79. Bharati, M.S.S.; Chandu, B.; Rao, S.V. Explosives sensing using Ag–Cu alloy nanoparticles synthesized by femtosecond laser ablation and irradiation. *Rsc Adv.* **2019**, *9*, 1517–1525. [CrossRef]

80. Al-Antaki, A.H.M.; Luo, X.; Duan, X.; Lamb, R.N.; Hutchison, W.D.; Lawrance, W.; Raston, C.L. Continuous Flow Copper Laser Ablation Synthesis of Copper(I and II) Oxide Nanoparticles in Water. *ACS Omega* **2019**, *4*, 13577–13584. [CrossRef]

81. Khashan, K.S.; Sulaiman, G.M.; Abdulameer, F.A. Synthesis and Antibacterial Activity of CuO Nanoparticles Suspension Induced by Laser Ablation in Liquid. *Arab. J. Sci. Eng.* **2016**, *41*, 301–310. [CrossRef]

82. Mostafa, A.M.; Yousef, S.A.; Eisa, W.H.; Ewaida, M.A.; Al-Ashkar, E.A. Synthesis of cadmium oxide nanoparticles by pulsed laser ablation in liquid environment. *Optik* **2017**, *144*, 679–684. [CrossRef]

83. Mostafa, A.M.; Yousef, S.A.; Eisa, W.H.; Ewaida, M.A.; Al-Ashkar, E.A. Au@CdO core/shell nanoparticles synthesized by pulsed laser ablation in Au precursor solution. *Appl. Phys. A* **2017**, *123*. [CrossRef]

84. Hu, S.; Melton, C.; Mukherjee, D. A facile route for the synthesis of nanostructured oxides and hydroxides of cobalt using laser ablation synthesis in solution (LASIS). *Phys. Chem. Chem. Phys.* **2014**, *16*, 24034–24044. [CrossRef]

85. Ma, R.; Kim, Y.-J.; Reddy, D.A.; Kim, T.K. Synthesis of CeO2/Pd nanocomposites by pulsed laser ablation in liquids for the reduction of 4-nitrophenol to 4-aminophenol. *Ceram. Int.* **2015**, *41*, 12432–12438. [CrossRef]

86. Guo, Y.; Yang, X.; Li, G.; Dong, B.; Chen, L. Effect of ultrasonic intensification on synthesis of nano-sized particles with an impinging jet reactor. *Powder Technol.* **2019**, *354*, 218–230. [CrossRef]

87. Ismail, R.A.; Sulaiman, G.M.; Abdulrahman, S.A.; Marzoog, T.R. Antibacterial activity of magnetic iron oxide nanoparticles synthesized by laser ablation in liquid. *Mater. Sci. Eng. C* **2015**, *53*, 286–297. [CrossRef]

88. Pandey, B.K.; Shahi, A.K.; Shah, J.; Kotnala, R.K.; Gopal, R. Optical and magnetic properties of Fe2O3 nanoparticles synthesized by laser ablation/fragmentation technique in different liquid media. *Appl. Surf. Sci.* **2014**, *289*, 462–471. [CrossRef]

89. Chrzanowska, J.; Hoffman, J.; Małolepszy, A.; Mazurkiewicz, M.; Kowalewski, T.A.; Szymanski, Z.; Stobinski, L. Synthesis of carbon nanotubes by the laser ablation method: Effect of laser wavelength. *Phys. Status Solidi B Basic Solid State Phys.* **2015**, *252*, 1860–1867. [CrossRef]

90. Kazemizadeh, F.; Malekfar, R.; Parvin, P. Pulsed laser ablation synthesis of carbon nanoparticles in vacuum. *J. Phys. Chem. Solids* **2017**, *104*, 252–256. [CrossRef]

91. Nightingale, A.M.; Bannock, J.H.; Krishnadasan, S.H.; O'Mahony, F.T.; Haque, S.A.; Sloan, J.; Drury, C.; McIntyre, R.; deMello, J.C. Large-scale synthesis of nanocrystals in a multichannel droplet reactor. *J. Mater. Chem. A* **2013**, *1*, 4067–4076. [CrossRef]

92. Brobbey, K.J.; Haapanen, J.; Gunell, M.; Mäkelä, J.M.; Eerola, E.; Toivakka, M.; Saarinen, J.J. One-step flame synthesis of silver nanoparticles for roll-to-roll production of antibacterial paper. *Appl. Surf. Sci.* **2017**, *420*, 558–565. [CrossRef]

93. Mohapatra, D.; Badrayyana, S.; Parida, S. Facile wick-and-oil flame synthesis of high-quality hydrophilic onion-like carbon nanoparticles. *Mater. Chem. Phys.* **2016**, *174*, 112–119. [CrossRef]

94. Marre, S.; Park, J.; Rempel, J.; Guan, J.; Bawendi, M.G.; Jensen, K.F. Supercritical continuous-microflow synthesis of narrow size distribution quantum dots. *Adv. Mater.* **2008**, *20*, 4830–4834. [CrossRef]

95. Toyota, A.; Nakamura, H.; Ozono, H.; Yamashita, K.; Uehara, M.; Maeda, H. Combinatorial synthesis of CdSe nanoparticles using microreactors. *J. Phys. Chem. C* **2010**, *114*, 7527–7534. [CrossRef]

96. Vikram, A.; Kumar, V.; Ramesh, U.; Balakrishnan, K.; Oh, N.; Deshpande, K.; Ewers, T.; Trefonas, P.; Shim, M.; Kenis, P.J. A Millifluidic Reactor System for Multistep Continuous Synthesis of InP/ZnSeS Nanoparticles. *ChemNanoMat* **2018**, *4*, 943–953. [CrossRef]

97. Lignos, I.; Stavrakis, S.; Kilaj, A.; deMello, A.J. Millisecond-Timescale Monitoring of PbS Nanoparticle Nucleation and Growth Using Droplet-Based Microfluidics. *Small* **2015**, *11*, 4009–4017. [CrossRef]

98. Huang, X.; Neretina, S.; El-Sayed, M.A. Gold Nanorods: From Synthesis and Properties to Biological and Biomedical Applications. *Adv. Mater.* **2009**, *21*, 4880–4910. [CrossRef]

99. Duraiswamy, S.; Khan, S.A. Droplet-based microfluidic synthesis of anisotropic metal nanocrystals. *Small* **2009**, *5*, 2828–2834. [CrossRef]

100. Lohse, S.E.; Eller, J.R.; Sivapalan, S.T.; Plews, M.R.; Murphy, C.J. A simple millifluidic benchtop reactor system for the high-throughput synthesis and functionalization of gold nanoparticles with different sizes and shapes. *Acs Nano* **2013**, *7*, 4135–4150. [CrossRef]

101. Zhang, L.; Niu, G.; Lu, N.; Wang, J.; Tong, L.; Wang, L.; Kim, M.J.; Xia, Y. Continuous and scalable production of well-controlled noble-metal nanocrystals in milliliter-sized droplet reactors. *Nano Lett.* **2014**, *14*, 6626–6631. [CrossRef] [PubMed]

102. Cattaneo, S.; Althahban, S.; Freakley, S.J.; Sankar, M.; Davies, T.; He, Q.; Dimitratos, N.; Kiely, C.J.; Hutchings, G.J. Synthesis of highly uniform and composition-controlled gold–palladium supported nanoparticles in continuous flow. *Nanoscale* **2019**, *11*, 8247–8259. [CrossRef] [PubMed]

103. Thiele, M.; Knauer, A.; Csáki, A.; Mallsch, D.; Henkel, T.; Köhler, J.M.; Fritzsche, W. High-Throughput Synthesis of Uniform Silver Seed Particles by a Continuous Microfluidic Synthesis Platform. *Chem. Eng. Technol.* **2015**, *38*, 1131–1137. [CrossRef]

104. Kwak, C.H.; Kang, S.-M.; Jung, E.; Haldorai, Y.; Han, Y.-K.; Kim, W.-S.; Yu, T.; Huh, Y.S. Customized microfluidic reactor based on droplet formation for the synthesis of monodispersed silver nanoparticles. *J. Ind. Eng. Chem.* **2018**, *63*, 405–410. [CrossRef]

105. Niu, G.; Zhang, L.; Ruditskiy, A.; Wang, L.; Xia, Y. A droplet-reactor system capable of automation for the continuous and scalable production of noble-metal nanocrystals. *Nano Lett.* **2018**, *18*, 3879–3884. [CrossRef] [PubMed]

106. Sun, J.; Xianyu, Y.; Li, M.; Liu, W.; Zhang, L.; Liu, D.; Liu, C.; Hu, G.; Jiang, X. A microfluidic origami chip for synthesis of functionalized polymeric nanoparticles. *Nanoscale* **2013**, *5*, 5262–5265. [CrossRef]

107. Liu, D.; Cito, S.; Zhang, Y.; Wang, C.F.; Sikanen, T.M.; Santos, H.A. A versatile and robust microfluidic platform toward high throughput synthesis of homogeneous nanoparticles with tunable properties. *Adv. Mater.* **2015**, *27*, 2298–2304. [CrossRef]

108. Liu, D.; Zhang, H.; Cito, S.; Fan, J.; Mäkilä, E.; Salonen, J.; Hirvonen, J.; Sikanen, T.M.; Weitz, D.A.; Santos, H.l.A. Core/shell nanocomposites produced by superfast sequential microfluidic nanoprecipitation. *Nano Lett.* **2017**, *17*, 606–614. [CrossRef]

109. Kelesidis, G.A.; Goudeli, E.; Pratsinis, S.E. Flame synthesis of functional nanostructured materials and devices: Surface growth and aggregation. *Proc. Combust. Inst.* **2017**, *36*, 29–50. [CrossRef]

110. Mädler, L.; Kammler, H.; Mueller, R.; Pratsinis, S.E. Controlled synthesis of nanostructured particles by flame spray pyrolysis. *J. Aerosol Sci.* **2002**, *33*, 369–389. [CrossRef]

111. Chu, H.; Han, W.; Ren, F.; Xiang, L.; Wei, Y.; Zhang, C. Flame synthesis of carbon nanotubes on different substrates in methane diffusion flames. *ES Energy Environ.* **2018**, *2*, 73–81. [CrossRef]

112. Li, H.; Pokhrel, S.; Schowalter, M.; Rosenauer, A.; Kiefer, J.; Mädler, L. The gas-phase formation of tin dioxide nanoparticles in single droplet combustion and flame spray pyrolysis. *Combust. Flame* **2020**, *215*, 389–400. [CrossRef] [PubMed]

113. Esmeryan, K.D.; Castano, C.E.; Bressler, A.H.; Fergusson, C.P.; Mohammadi, R. Single-step flame synthesis of carbon nanoparticles with tunable structure and chemical reactivity. *RSC Adv.* **2016**, *6*, 61620–61629. [CrossRef]

114. Kathirvel, P.; Chandrasekaran, J.; Manoharan, D.; Kumar, S. Preparation and characterization of alpha alumina nanoparticles by in-flight oxidation of flame synthesis. *J. Alloy. Compd.* **2014**, *590*, 341–345. [CrossRef]

115. Hafshejani, L.D.; Tangsir, S.; Koponen, H.; Riikonen, J.; Karhunen, T.; Tapper, U.; Lehto, V.-P.; Moazed, H.; Naseri, A.A.; Hooshmand, A. Synthesis and characterization of Al2O3 nanoparticles by flame spray pyrolysis (FSP)—Role of Fe ions in the precursor. *Powder Technol.* **2016**, *298*, 42–49. [CrossRef]

116. Hirano, T.; Nakakura, S.; Rinaldi, F.G.; Tanabe, E.; Wang, W.-N.; Ogi, T. Synthesis of highly crystalline hexagonal cesium tungsten bronze nanoparticles by flame-assisted spray pyrolysis. *Adv. Powder Technol.* **2018**, *29*, 2512–2520. [CrossRef]

117. Thiele, M.; Knauer, A.; Malsch, D.; Csáki, A.; Henkel, T.; Köhler, J.M.; Fritzsche, W. Combination of microfluidic high-throughput production and parameter screening for efficient shaping of gold nanocubes using Dean-flow mixing. *Lab A Chip* **2017**, *17*, 1487–1495. [CrossRef]

118. Xu, L.; Peng, J.; Yan, M.; Zhang, D.; Shen, A.Q. Droplet synthesis of silver nanoparticles by a microfluidic device. *Chem. Eng. Process. Process Intensif.* **2016**, *102*, 186–193. [CrossRef]

119. Santana, J.S.; Gamler, J.T.L.; Skrabalak, S.E. Integration of Sequential Reactions in a Continuous Flow Droplet Reactor: A Route to Architecturally Defined Bimetallic Nanostructures. *Part. Part. Syst. Charact.* **2019**, *36*, 1900142. [CrossRef]

120. Du, L.; Li, Y.; Gao, R.; Yin, J.; Shen, C.; Wang, Y.; Luo, G. Controllability and flexibility in particle manufacturing of a segmented microfluidic device with passive picoinjection. *Aiche J.* **2018**, *64*, 3817–3825. [CrossRef]

121. Duong, A.D.; Ruan, G.; Mahajan, K.; Winter, J.O.; Wyslouzil, B.E. Scalable, semicontinuous production of micelles encapsulating nanoparticles via electrospray. *Langmuir* **2014**, *30*, 3939–3948. [CrossRef]

122. Xu, L.; Srinivasakannan, C.; Peng, J.; Zhang, D.; Chen, G. Synthesis of nickel nanoparticles by aqueous reduction in continuous flow microreactor. *Chem. Eng. Process.* **2015**, *93*, 44–49. [CrossRef]

123. Jiao, M.; Zeng, J.; Jing, L.; Liu, C.; Gao, M. Flow Synthesis of Biocompatible Fe$_3$O$_4$ Nanoparticles: Insight into the Effects of Residence Time, Fluid Velocity, and Tube Reactor Dimension on Particle Size Distribution. *Chem. Mater.* **2015**, *27*, 1299–1305. [CrossRef]

124. Paseta, L.; Seoane, B.; Julve, D.; Sebastián, V.; Téllez, C.; Coronas, J. Accelerating the Controlled Synthesis of Metal–Organic Frameworks by a Microfluidic Approach: A Nanoliter Continuous Reactor. *Acs Appl. Mater. Interfaces* **2013**, *5*, 9405–9410. [CrossRef] [PubMed]

125. Bomhard, S.V.; Schramm, J.; Bleul, R.; Thiermann, R.; Höbel, P.; Krtschil, U.; Löb, P.; Maskos, M. Modular Manufacturing Platform for Continuous Synthesis and Analysis of Versatile Nanomaterials. *Chem. Eng. Technol.* **2019**, *42*, 2085–2094. [CrossRef]

126. Gerdes, B.; Jehle, M.; Domke, M.; Zengerle, R.; Koltay, P.; Riegger, L. Drop-on-demand generation of aluminum alloy microdroplets at 950 °C using the StarJet technology. In Proceedings of the International Conference on Solid-State Sensors, Actuators and Microsystems, Kaohsiung, Taiwan, 1 June 2017; pp. 690–693.

127. Lee, S.W.; Choi, J.S.; Cho, K.Y.; Yim, J.-H. Facile fabrication of uniform-sized, magnetic, and electroconductive hybrid microspheres using a microfluidic droplet generator. *Eur. Polym. J.* **2016**, *80*, 40–47. [CrossRef]

128. Jeyhani, M.; Gnyawali, V.; Abbasi, N.; Hwang, D.K.; Tsai, S.S.H. Microneedle-assisted microfluidic flow focusing for versatile and high throughput water-in-water droplet generation. *J. Colloid Interface Sci.* **2019**, *553*, 382–389. [CrossRef] [PubMed]

129. Tang, S.-Y.; Wang, K.; Fan, K.; Feng, Z.; Zhang, Y.; Zhao, Q.; Yun, G.; Yuan, D.; Jiang, L.; Li, M.; et al. High-Throughput, Off-Chip Microdroplet Generator Enabled by a Spinning Conical Frustum. *Anal. Chem.* **2019**, *91*, 3725–3732. [CrossRef] [PubMed]

130. Kahkeshani, S.; Di Carlo, D. Drop formation using ferrofluids driven magnetically in a step emulsification device. *Lab A Chip* **2016**, *16*, 2474–2480. [CrossRef]

131. Lapierre, F.; Cameron, N.R.; Zhu, Y. Ready … set, flow: Simple fabrication of microdroplet generators and their use in the synthesis of PolyHIPE microspheres. *J. Micromech. Microeng.* **2015**, *25*, 035011. [CrossRef]

132. Varshney, R.; Sharma, S.; Prakash, B.; Laha, J.K.; Patra, D. One-Step Fabrication of Enzyme-Immobilized Reusable Polymerized Microcapsules from Microfluidic Droplets. *Acs Omega* **2019**, *4*, 13790–13794. [CrossRef]

133. Kim, C.M.; Kim, G.M. Fabrication of 512-Channel Geometrical Passive Breakup Device for High-Throughput Microdroplet Production. *Micromachines* **2019**, *10*, 709. [CrossRef] [PubMed]

134. Han, T.; Zhang, L.; Xu, H.; Xuan, J. Factory-on-chip: Modularised microfluidic reactors for continuous mass production of functional materials. *Chem. Eng. J.* **2017**, *326*, 765–773. [CrossRef]

135. Amstad, E.; Chemama, M.; Eggersdorfer, M.; Arriaga, L.R.; Brenner, M.P.; Weitz, D.A. Robust scalable high throughput production of monodisperse drops. *Lab A Chip* **2016**, *16*, 4163–4172. [CrossRef]

136. Kim, C.M.; Choi, H.J.; Kim, G.M. 512-Channel Geometric Droplet-Splitting Microfluidic Device by Injection of Premixed Emulsion for Microsphere Production. *Polymers* **2020**, *12*, 776. [CrossRef] [PubMed]

137. Ofner, A.; Moore, D.G.; Rühs, P.A.; Schwendimann, P.; Eggersdorfer, M.; Amstad, E.; Weitz, D.A.; Studart, A.R. High-Throughput Step Emulsification for the Production of Functional Materials Using a Glass Microfluidic Device. *Macromol. Chem. Phys.* **2017**, *218*. [CrossRef]

138. Mohamed, M.G.A.; Kheiri, S.; Islam, S.; Kumar, H.; Yang, A.; Kim, K. An integrated microfluidic flow-focusing platform for on-chip fabrication and filtration of cell-laden microgels. *Lab A Chip* **2019**, *19*, 1621–1632. [CrossRef] [PubMed]

139. Goff, G.C.L.; Lee, J.; Gupta, A.; Hill, W.A.; Doyle, P.S. High-Throughput Contact Flow Lithography. *Adv. Sci.* **2015**, *2*, 1500149. [CrossRef]

140. Wang, J.; Chao, P.H.; Slavik, R.; Dam, R.M.V. Multi-GBq production of the radiotracer [18F]fallypride in a droplet microreactor. *RSC Adv.* **2020**, *10*, 7828–7838. [CrossRef]

141. Zeng, W.; Jacobi, I.; Li, S.; Stone, H.A. Corrigendum: Variation in polydispersity in pump- and pressure-driven micro-droplet generators (Zeng et al. 2015 J. Micromech. Microeng. 25 115015). *J. Micromech. Microeng.* **2016**, *26*, 039501. [CrossRef]

142. Zeng, W.; Li, S.; Fu, H. Modeling of the pressure fluctuations induced by the process of droplet formation in a T-junction microdroplet generator. *Sens. Actuators A Phys.* **2018**, *272*, 11–17. [CrossRef]

143. Crawford, D.F.; Smith, C.A.; Whyte, G. Image-based closed-loop feedback for highly mono-dispersed microdroplet production. *Sci. Rep.* **2017**, *7*, 10545. [CrossRef] [PubMed]

144. Fu, H.; Zeng, W.; Li, S.; Yuan, S. Electrical-detection droplet microfluidic closed-loop control system for precise droplet production. *Sens. Actuators A Phys.* **2017**, *267*, 142–149. [CrossRef]

145. Mahdi, Y.; Daoud, K. Microdroplet size prediction in microfluidic systems via artificial neural network modeling for water-in-oil emulsion formulation. *J. Dispers. Sci. Technol.* **2017**, *38*, 1501–1508. [CrossRef]

146. Zhang, J.M.; Aguirre-Pablo, A.A.; Li, E.Q.; Buttner, U.; Thoroddsen, S.T. Droplet generation in cross-flow for cost-effective 3D-printed "plug-and-play" microfluidic devices. *RSC Adv.* **2016**, *6*, 81120–81129. [CrossRef]

147. Nguyen, H.V.; Nguyen, H.Q.; Nguyen, V.D.; Seo, T.S. A 3D printed screw-and-nut based droplet generator with facile and precise droplet size controllability. *Sens. Actuators B Chem.* **2019**, *296*. [CrossRef]

148. Hwang, Y.H.; Um, T.; Hong, J.; Ahn, G.N.; Qiao, J.; Kang, I.S.; Qi, L.; Lee, H.; Kim, D.P. Robust Production of Well-Controlled Microdroplets in a 3D-Printed Chimney-Shaped Milli-Fluidic Device. *Adv. Mater. Technol.* **2019**, *4*. [CrossRef]

149. He, X.; Wu, J.; Hu, T.; Xuan, S.; Gong, X. A 3D-printed coaxial microfluidic device approach for generating magnetic liquid metal droplets with large size controllability. *Microfluid. Nanofluidics* **2020**, *24*. [CrossRef]

150. Kishi, T.; Kiyama, Y.; Kanda, T.; Suzumori, K.; Seno, N. Microdroplet generation using an ultrasonic torsional transducer which has a micropore with a tapered nozzle. *Arch. Appl. Mech.* **2016**, *86*, 1751–1762. [CrossRef]

151. Fujimoro, N.; Yamada, T.; Murakami, T.; Kanda, T.; Mori, K.; Suzumori, K. Micro Droplets Generation in a Flowing Continuous Liquid Using an Ultrasonic Transducer. In Proceedings of the 2018 International Conference on Manipulation, Automation and Robotics at Small Scales (MARSS), Nagoya, Japan, 4–8 July 2018; p. 8481146. [CrossRef]

152. Maeda, K.; Onoe, H.; Takinoue, M.; Takeuchi, S. Controlled synthesis of 3D multi-compartmental particles with centrifuge-based microdroplet formation from a multi-barrelled capillary. *Adv. Mater.* **2012**, *24*, 1340–1346. [CrossRef]

153. Hayakawa, M.; Onoe, H.; Nagai, K.H.; Takinoue, M. Complex-shaped three-dimensional multi-compartmental microparticles generated by diffusional and Marangoni microflows in centrifugally discharged droplets. *Sci. Rep.* **2016**, *6*, 20793. [CrossRef]

154. Yamashita, H.; Morita, M.; Sugiura, H.; Fujiwara, K.; Onoe, H.; Takinoue, M. Generation of monodisperse cell-sized microdroplets using a centrifuge-based axisymmetric co-flowing microfluidic device. *J. Biosci. Bioeng.* **2015**, *119*, 492–495. [CrossRef]

155. Shin, D.-C.; Morimoto, Y.; Sawayama, J.; Miura, S.; Takeuchi, S. Centrifuge-based Step Emulsification Device for Simple and Fast Generation of Monodisperse Picoliter Droplets. *Sens. Actuators B Chem.* **2019**, *301*, 127164. [CrossRef]

156. Chen, Z.; Liao, P.; Zhang, F.; Jiang, M.; Zhu, Y.; Huang, Y. Centrifugal micro-channel array droplet generation for highly parallel digital PCR. *Lab Chip* **2017**, *17*, 235–240. [CrossRef] [PubMed]

157. Zhang, Y.; Tang, S.-Y.; Zhao, Q.; Yun, G.; Yuan, D.; Li, W. High-throughput production of uniformly sized liquid metal microdroplets using submerged electrodispersion. *Appl. Phys. Lett.* **2019**, *114*. [CrossRef]

158. Gao, Q.; He, Y.; Fu, J.-Z.; Qiu, J.-J.; Jin, Y.-A. Fabrication of shape controllable alginate microparticles based on drop-on-demand jetting. *J. Sol Gel Sci. Technol.* **2016**, *77*, 610–619. [CrossRef]

159. Zhao, L.; Yan, K.C.; Yao, R.; Lin, F.; Sun, W. Alternating Force Based Drop-on-Demand Microdroplet Formation and Three-Dimensional Deposition. *J. Manuf. Sci. Eng. Trans. Asme* **2015**, *137*. [CrossRef]

160. Ma, M.; Wei, X.; Shu, X.; Zhang, H. Producing solder droplets using piezoelectric membrane-piston-based jetting technology. *J. Mater. Process. Technol.* **2019**, *263*, 233–240. [CrossRef]

161. Zenou, M.; Sa'ar, A.; Kotler, Z. Supersonic laser-induced jetting of aluminum micro-droplets. *Appl. Phys. Lett.* **2015**, *106*. [CrossRef]

162. Liu, M.; Sun, X.-T.; Yang, C.-G.; Xu, Z.-R. On-chip preparation of calcium alginate particles based on droplet templates formed by using a centrifugal microfluidic technique. *J. Colloid Interface Sci.* **2016**, *466*, 20–27. [CrossRef]

163. Tsuchiya, M.; Kurashina, Y.; Heo, Y.J.; Onoe, H. One-Step Fabrication of Multi-Functional Core-Shell Janus Microparticles for Theranostics Application. In Proceedings of the International Conference on Micro Electro Mechanical Systems, Vancouver, BC, Canada, 18–22 January 2020.

164. Zou, W.; Yu, H.; Zhou, P.; Liu, L. Tip-assisted electrohydrodynamic jet printing for high-resolution microdroplet deposition. *Mater. Des.* **2019**, *166*. [CrossRef]

165. Zhong, S.-Y.; Qi, L.-H.; Xiong, W.; Luo, J.; Xu, Q.-X. Research on mechanism of generating aluminum droplets smaller than the nozzle diameter by pneumatic drop-on-demand technology. *Int. J. Adv. Manuf. Technol.* **2017**, *93*, 1771–1780. [CrossRef]

166. Wang, F.; Li, J.; Wang, Y.; Bao, W.; Chen, X.; Zhang, H.; Wang, Z. Feedback control of ejection state of a pneumatic valve-controlled micro-droplet generator through machine vision. In Proceedings of the International Conference on Machine Vision (ICMV 2018), Munich, Germany, 1–3 November 2018.

167. Sadeghian, H.; Hojjat, Y.; Ghodsi, M.; Sheykholeslami, M.R. An approach to design and fabrication of a piezo-actuated microdroplet generator. *Int. J. Adv. Manuf. Technol.* **2014**, *70*, 1091–1099. [CrossRef]

168. López-Iglesias, C.; Barros, J.; Ardao, I.; Gurikov, P.; Monteiro, F.J.; Smirnova, I.; Alvarez-Lorenzo, C.; García-González, C.A. Jet Cutting Technique for the Production of Chitosan Aerogel Microparticles Loaded with Vancomycin. *Polymers* **2020**, *12*, 273. [CrossRef]

169. Zhao, L.; Yan, K.C.; Yao, R.; Lin, F.; Sun, W. Modeling on Microdroplet Formation for Cell Printing Based on Alternating Viscous-Inertial Force Jetting. *J. Manuf. Sci. Eng. Trans. Asme* **2017**, *139*. [CrossRef]

Publisher's Note: MDPI stays neutral with regard to jurisdictional claims in published maps and institutional affiliations.

Perspective

Two-Phase Biocatalysis in Microfluidic Droplets

Lanting Xiang [1,2], Felix Kaspar [3,4], Anett Schallmey [2,3,5] and Iordania Constantinou [1,2,*]

[1] Institute for Microtechnology, Technische Universität Braunschweig, 38124 Braunschweig, Germany; l.xiang@tu-braunschweig.de
[2] Zentrum für Pharmaverfahrenstechnik (PVZ), Technische Universität Braunschweig, 38106 Braunschweig, Germany; a.schallmey@tu-braunschweig.de
[3] Institute for Biochemistry, Biotechnology and Bioinformatics, Technische Universität Braunschweig, 38106 Braunschweig, Germany; felix.kaspar@tu-braunschweig.de
[4] Chair of Bioprocess Engineering, Institute of Biotechnology, Faculty III Process Sciences, Technische Universität Berlin, 13355 Berlin, Germany
[5] Braunschweig Integrated Center of Systems Biology (BRICS), Technische Universität Braunschweig, 38106 Braunschweig, Germany
*   Correspondence: i.constantinou@tu-braunschweig.de; Tel.: +49-(0)-531-391-9769

Abstract: This Perspective discusses the literature related to two-phase biocatalysis in microfluidic droplets. Enzymes used as catalysts in biocatalysis are generally less stable in organic media than in their native aqueous environments; however, chemical and pharmaceutical compounds are often insoluble in water. The use of aqueous/organic two-phase media provides a solution to this problem and has therefore become standard practice for multiple biotransformations. In batch, two-phase biocatalysis is limited by mass transport, a limitation that can be overcome with the use of microfluidic systems. Although, two-phase biocatalysis in laminar flow systems has been extensively studied, microfluidic droplets have been primarily used for enzyme screening. In this Perspective, we summarize the limited published work on two-phase biocatalysis in microfluidic droplets and discuss the limitations, challenges, and future perspectives of this technology.

Keywords: microfluidics; two-phase biocatalysis; microfluidic droplets; enzyme

Citation: Xiang, L.; Kaspar, F.; Schallmey, A.; Constantinou, I. Two-Phase Biocatalysis in Microfluidic Droplets. *Biosensors* **2021**, *11*, 407. https://doi.org/10.3390/bios11110407

Received: 8 September 2021
Accepted: 18 October 2021
Published: 21 October 2021

Publisher's Note: MDPI stays neutral with regard to jurisdictional claims in published maps and institutional affiliations.

1. Introduction

Biocatalysis is an empowering technology in synthetic organic chemistry. The use of enzymes for chemical transformations often grants unparalleled chemo-, regio- and stereoselectivity, and even enables transformations that would be impossible to achieve with conventional chemical methods [1–4]. Owing to these advantages, recent years have witnessed a growing popularity of strategic enzymatic steps in the preparation of pharmaceuticals, natural products, and other fine chemicals [5–10]. By reducing the length of synthetic routes and the demand for laborious purification processes, biocatalytic transformations typically exhibit more favorable green chemistry metrics than their corresponding chemical counterparts [11,12]. As such, biocatalysis continues to evolve into the intuitive choice for the development of environmentally friendly chemical processes, both in academic as well as industrial settings [13]. However, the efficient application of enzymes is often hampered by the low water solubility of desired starting materials. As many compounds of interest in organic chemistry exhibit only limited hydrophilicity, substrate loadings in biocatalytic transformations are often restricted to the low millimolar range, far below the desired titers for industrial settings [11]. To increase the substrate concentrations and thereby decrease solvent use, several strategies and process designs have been established. Among these, the use of two immiscible liquid phases, an aqueous and a non-aqueous phase, the latter usually comprising an organic solvent [14–16] or ionic liquid [17,18], has emerged as an elegant and remarkably versatile approach. The enzymatic reaction typically takes place in the aqueous phase under substrate concentrations near the solubility limit, while the non-aqueous (commonly organic) phase acts as a substrate reservoir, continuously delivering the substrate to the aqueous phase. At the same time, the non-aqueous phase serves as a product sink, removing the reaction product from the aqueous phase. Beyond enabling exceptionally high net substrate concentrations, two-phase biocatalysis has several key advantages, including the prevention of inhibition effects or shift of reaction equilibria via continuous product removal, as well as straightforward product isolation via phase separation [19,20]. Despite these advantages, biphasic processes are yet to become a true standard technique in biocatalysis, as the design of efficient biphasic enzymatic transformations is challenging and requires non-trivial engineering. Another major challenge is the mixing of the two liquid phases which maximizes mass transfer but can also compromise the stability and hence the activity of the enzyme catalyst [21]. Moreover, stable emulsions can form when performing liquid–liquid two-phase reactions in batch under vigorous stirring, complicating phase separation and thus product isolation during subsequent down-stream processing [22]. The aforementioned challenges can be overcome with the use of microfluidic systems.

2. Biocatalysis at the Microscale

Flow chemistry has been performed in organic chemistry laboratories in academia and industry for the development, optimization, and intensification of chemical processes for several decades [23]. More recently, the first biocatalytic reactions performed in microfluidics were reported [24,25]; however, flow biocatalysis today remains in its infancy compared to flow chemistry. This is surprising considering the many advantages microfluidics offer, such as the possibility of continuously operated reactions, much higher surface-to-volume ratios and hence short diffusion paths, and tremendously increased mass and heat transfer rates [26,27]. These advantages are especially promising for liquid–liquid two-phase biocatalytic reactions that are often limited by mass transfer in batch [28]. As different physical effects influence fluid flow at the microscale compared to larger scales [29], fluid flow regimes unachievable on the macroscale (such as laminar flow) can be achieved in microchannels. Moreover, process development significantly benefits from microscale technology due to reduced consumption of reagents and time, speeding up process development and reducing development costs [26]. In the biocatalysis field, this has led to the development of microfluidics-based screening methods for activity-based screening of metagenomic libraries [30,31] or mutant libraries derived from enzyme evolution campaigns [32,33]. Additionally, enzyme characterization and process parameter estimation and optimization are also facilitated by the use of microfluidic devices (often microfluidic droplets) due to ultra-high throughput and low reagent consumption [23,34–36]. In addition to high-throughput screening and characterization possibilities, miniaturization provides a tool that allows researchers to gain an understanding of fundamental process principles and reaction characteristics. This is made possible by the fact that experimental conditions can be precisely and independently controlled in microfluidic systems. Although some processes optimized at the microscale cannot be directly scaled up, fundamental process understanding gained at the microscale can still facilitate process optimization in the macroscale. Finally, smaller volumes and better spatial and temporal reaction control in microsystems increase safety in biocatalytic reactions that involve, e.g., toxic reagents [37].

3. Droplet Generation in Microfluidic Systems

For two-phase biocatalysis in microfluidics, the two liquid phases need to be in contact to ensure fast mass transport across the interface. With the absence of turbulence at the microscale, molecular diffusion becomes the dominant mechanism of mass transfer. A liquid–liquid, two-phase flow in microchannels can be achieved through laminar flow or using microfluidic droplets.

Laminar flow is dominant when Reynold numbers are low and is characterized by fluid streamlines parallel to the channel walls. The two phases flow next to each other in the form of two layers forming one interface (Figure 1a) or three layers forming two interfaces (Figure 1b). Two-layer flow profiles are commonly produced using T-junctions and three-layer laminar flow profiles using Ψ-junctions. The width of each phase, i.e., the volume each phase occupies in the channel, can be adjusted by changing the flow rate ratio between the two liquids. Such microsystems are often used for two-phase biocatalysis using immiscible hydrophobic substrates, as they provide advantages like short diffusion paths across the aqueous/non-aqueous interface and easy phase separation at the outlets [38,39].

Figure 1. Laminar flow in a microfluidic channel: (**a**) two liquid phases flowing next to each other and forming two layers and one interface; (**b**) two liquid phases flowing next to each other and forming three layers and two interfaces. The microfluidic device architectures used to achieve these flow patterns are shown as inserts.

Similar to laminar flows, microfluidic droplets have also been employed as a tool for biphasic enzymatic transformations, however, to a much lesser extent. Microfluidic droplet formation requires two immiscible fluids (gas–liquid or liquid–liquid), referred to as the continuous phase (medium in which droplets flow) and the dispersed phase (droplets). In the context of two-phase biocatalysis, liquid–liquid droplets are formed using an organic and an aqueous phase. Each microdroplet can be regarded as an independent microreactor, with fast mass transfer taking place across the droplet boundary. In addition to efficient mass transfer, reactions in microfluidic droplets can greatly reduce the consumption of expensive reagents due to their tiny volumes (commonly pL to nL) and at the same time limit cross-contamination. There are two common droplet geometries; ellipsoidal (Figure 2a) and spherical (Figure 2b). Ellipsoidal droplets can be produced using T-, Y- and Ψ-junctions. They are larger compared to spherical microdroplets and their width is typically determined by the width of the microfluidic channel. The flow of large ellipsoidal droplets (slugs) in a microchannel is commonly referred to as slug flow. Spherical microdroplets are produced using flow focusing or co-flowing microfluidic structures and often have diameters much smaller than the width of the microchannel. Device geometries and droplet generation methods are discussed in detail below.

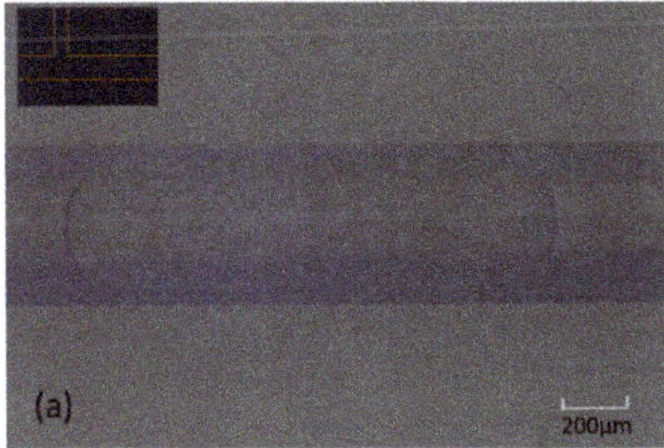
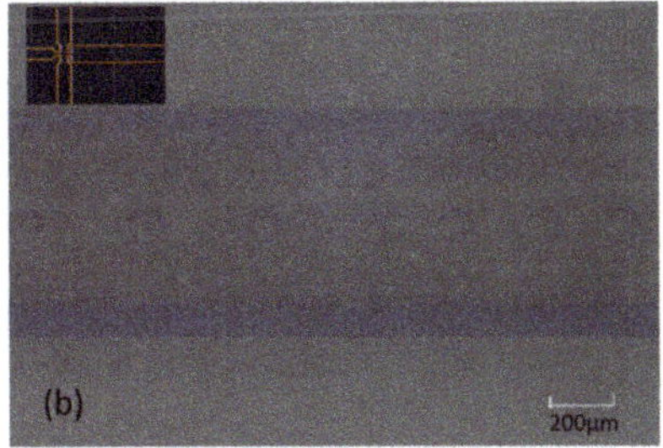

Figure 2. (**a**) Ellipsoidal microdroplet in a microfluidic channel/slug flow. (**b**) Spherical microdroplets in a microfluidic channel/droplet flow. The microfluidic device architectures used to achieve these flow patterns are shown in the inserts and discussed in detail below.

3.1. Microfluidic Droplet Production and Their Applications in Two-Phase Biocatalysis

Various applications of microfluidic droplets employed in the field of biocatalysis have been reported in the literature. These can be divided into three main categories: (1) microfluidic droplets for ultrahigh-throughput library screening, (2) microfluidic droplets for enzyme characterization and parameter estimation, and (3) intensification of two-phase biocatalytic reactions using microfluidic droplets or droplet-like fluid flows. The first two categories have already been reviewed comprehensively in recent literature and will not be further discussed in this perspective [23,30,33,34]. Herein, we focus on synthetically useful biocatalytic reactions employing two liquid phases, in an effort to highlight the benefits and limitations associated with droplet microfluidics when used to perform these reactions. Before we discuss specific examples of two-phase biocatalysis in droplets, we offer a relevant introduction into the production technology of droplet microfluidics.

3.2. Production Technology of Droplet Microfluidics

Droplet formation methods can be classified as passive or active. Passive methods primarily use microchannel architecture and flow rates to control the formation of droplets, while active methods employ external force fields (e.g., light, electricity). In droplet-based biocatalysis, active droplet formation methods are generally avoided as the used external force fields could harm the enzymes or cells used to facilitate the reactions. The most common passive hydrodynamic methods used to produce microdroplets can be categorized into three groups according to the microchannel architecture: T-junction, flow-focusing, and co-flowing devices. The geometry of the intersection where the two flows meet, the flow rate of each phase and the properties of the fluid (surface tension, viscosity, etc.) determine the local stresses that deform the liquid interface and cause droplet formation.

T-junction: T-junctions were first proposed by Thorsen et al. in 2001 [40]. Droplets in this device architecture are formed when two mutually immiscible fluids meet at the T-shaped microchannel intersection. Under the action of pressure and shear, the continuous phase cuts off the dispersed phase to form droplets. As seen in Figure 3, T-junctions have a simple architecture and are easy to fabricate, which makes them a popular choice for applications in biocatalysis. However, the droplet width is determined by the width of the microfluidic channel, and therefore, spherical droplets with smaller diameters cannot be produced. This can be a limiting factor when using these devices for two-phase biocatalysis, as the available droplet surface area can be quite small compared to that of smaller droplets that float in the middle of the channel. This can partly inhibit mass transfer across the droplet interface (e.g., diffusion of substrate from the organic to the aqueous phase), a critical step during biocatalysis.

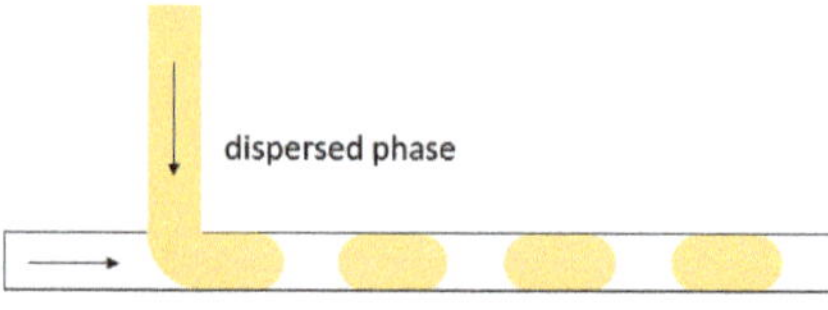

Figure 3. Droplet generation in a T-junction. Large droplets are formed in the channel, resulting in slug flow.

T-junctions, and similarly Y-junctions, have been the most commonly used microfluidic architectures to perform biphasic biocatalysis experiments. Some examples include work from Pohar et al., who used microfluidic devices of this architecture for the synthesis of isoamyl acetate [41], and Tušek et al., in which Y-junction microfluidic devices were used to study hexanol oxidation [42]. In the work of Tušek, two-phase biocatalysis was performed in a droplet microreactor and a cuvette under the same conditions. A kinetic model and a reactor model were established based on the results of the biotransformation. Comparing the kinetic parameters estimated by the two measurements, the maximum reaction rate in the microreactor was estimated to be approximately 33 times higher than in the cuvette. Additionally, in the experiments performed in microfluidic droplets, almost no reaction inhibition or reverse reaction were observed.

Flow focusing: The first micromachined flow-focusing system was introduced in 2001 by Ganan-Calvo et al. [43]. In 2003, Anna et al. used the flow-focusing geometry to form microfluidic water droplets in oil (water-in-oil droplets) and oil droplets in water (oil-in-water droplets) [44]. To generate microfluidic droplets, the continuous phase is injected into the microfluidic device through two side channels and is used to squeeze the dispersed phase, flowing through the central channel (Figure 4). The formation of microfluidic droplets is also facilitated by geometric constraints, such as the introduction of an orifice (a neck). The flow-focusing method can produce spherical microfluidic droplets with a diameter much smaller than the channel width.

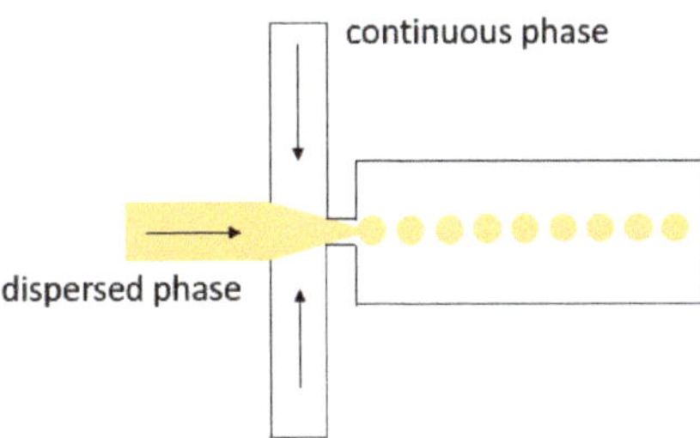

Figure 4. Droplet generation in a flow-focusing device. Small spherical droplets are formed in the channel resulting in droplet flow.

The use of flow-focusing microfluidic devices for two-phase biocatalysis has been very limited despite the many advantages this method provides, such as controlled generation of spherical microdroplets (as opposed to ellipsoidal, see Figure 2), that float in the center of the channel, away from the walls, enabling mass transfer across the droplet interface. The only example of biocatalysis in flow focusing systems was published by Novak et al., who used this method for the synthesis of isoamyl acetate enzymes in microdroplets and the recovery of ionic liquids [45]. The use of microfluidic droplets enabled a significantly higher productivity of isoamyl acetate compared to the respective batch process. More details about this work can be found in Section 4.

Co-flowing: Co-flowing is a method in which a capillary is embedded in a microchannel, and the dispersed and continuous phases flow through the capillary and the microchannel, respectively [46]. Unlike flow-focusing systems, co-flowing channels do not have a "neck-like" structure. The continuous phase does not squeeze the dispersed phase, but it is itself squeezed around the dispersed phase (Figure 5). This causes a destabilization at the dispersed phase front, which causes it to break into microdroplets. The size of droplets generated in co-flowing devices is controlled by the channel widths, the diameter of the capillary, and the flow rates used in experiments. In general, the co-flowing method produces well-controlled droplets with a diameter smaller than the width of the microfluidic channel. To date, no co-flowing system has been used in two-phase biocatalysis, presumably due to the complicated droplet generation principle and device fabrication process.

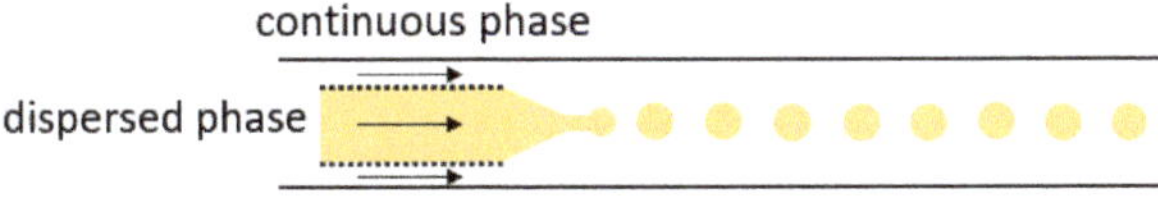

Figure 5. Droplet generation in a co-flowing microfluidic system. Small spherical droplets are formed in the channel resulting in droplet flow.

3.3. Droplet Generation and the Use of Surfactants

Droplet generation in microfluidic systems generally requires the use of surfactants, especially when droplets are generated using the flow-focusing or co-flowing methods. Surfactants are amphiphilic molecules commonly added to the continuous phase at small concentrations to promote the formation of new interfaces [47]. These molecules are adsorbed at the liquid–liquid interface to promote droplet stabilization and prevent coalescence [48]. Therefore, the addition of surfactants is important for stable droplet generation and manipulation and can also affect droplet properties such as their size.

However, as these amphiphilic molecules line the interface between the aqueous and organic phase, i.e., the droplet perimeter, it is expected that they also play a key role in the control of molecular exchange across the interface. Indeed, the rate of solute transport across a droplet interface through a surfactant monolayer has been shown to depend on many parameters such as the solute and surfactant size and structure, the density of the monolayer, and the curvature of the interface [49]. As mass transport across the droplet interface is a critical step during biocatalysis, the choice of suitable surfactants is also critical for high conversion rates. Additionally, surfactants have been shown to affect the catalytic activity of enzymes [50], and in turn, enzymes themselves can potentially serve as "surfactants" that stabilize droplet formation, as reported in the examples of lipase-catalyzed isoamyl acetate synthesis discussed in more detail below [41,45].

Most examples on biphasic biocatalysis in microreactors have relied on the use of polysorbate 20 (also known commercially as Tween 20) [28,50] or simply on the amphiphilic properties of the enzyme catalyst itself [41,51]. Previous work on enzyme evolution in microdroplets has primarily relied on fluorinated polyethers as surfactants [52,53]. Apart from available synthetic surfactants, a broad range of biological surface-active agents (biosurfactants) are also known today, ranging from various glycolipids or lipopeptides to surface-active proteins (e.g., hydrophobins) [54–56]. Biosurfactants have found various applications in industry due to their high interface activity [57]; their use in microfluidic droplet stabilization remains however limited.

3.4. Microfluidic Device Materials

When microfluidics are intended to host biotransformations, the choice of device material is critical. Aspects such as surface tolerance to solvents, channel hydrophobicity and gas permeability need to be carefully considered. The materials most commonly used in the fabrication of microreactors are glass, polycarbonate, and polytetrafluoroethylene (PTFE). Glass has the advantage of good transparency, chemical stability, and good biocompatibility. However, the fabrication of microchannels in glass requires expertise in microtechnology, as advanced processing using photolithography or laser microstructuring are necessary. Low entry manufacturing technologies such as 3D printing have also been investigated as they can be easily adopted by non-experts and used to create inexpensive devices [58]. However, the resolution that can be achieved using 3D printing methods such as stereolithography is limited (feature size in the order of a few hundred micrometers), and therefore, larger devices are produced, which requires a higher consumption of sample and reagents. Additionally, synthetic resins often display low tolerance against organic solvents. Polycarbonate offers the advantages of low cost, low water absorption, and good processing performance. It is the material of choice for microfluidics used in biomedical research and biological analysis, but its tendency to absorb some organic solvents makes it difficult to use for two-phase biocatalysis applications. However, surface coatings offer a solution to this problem. Mohr et al. used polycarbonate-based microfluidics coated with a 25-µm film of fluorinated ethylene propylene to create a recirculating, two-phase flow microbioreactor. The fluoropolymer coating was used to increase the chemical resistance and hydrophobicity of the microchannels [59]. Finally, PTFE offers inertness to chemicals and solvents, antifouling properties, moderate gas permeability, and low non-specific protein adsorption compared to polycarbonate. It is therefore widely used in the field of biphasic biocatalysis, especially used in microreactors made from microcapillary tubing [60].

4. Examples of Two-Phase Biocatalysis in Microdroplets

Biocatalytic esterifications, redox reactions, and lyase-catalyzed conversions making use of two liquid phases of different polarity have been performed in microdroplets (see Figure 6). These biphasic biotransformations have been used to increase substrate concentration in the reaction, efficiently extract product into the non-aqueous phase for simplified downstream processing, or reduce substrate and/or product inhibition of the enzyme. In the following paragraphs, notable examples of two-phase biocatalysis in microdroplets are discussed. These examples include work performed in both microfabricated reactors, as well as droplets generated in similar device architectures using microcapillaries (capillary-based microreactors).

In 2009, Znidarsic–Plazl and coworkers reported the CalB-catalyzed synthesis of isoamyl acetate, a flavor, and fragrance compound used in food, cosmetics, and pharmaceutical industry, from isoamyl alcohol and acetic anhydride in a continuously-operated Ψ-shaped droplet microreactor made out of glass [41]. The employed two-phase system consisted of a water-miscible ionic liquid and n-heptane. While all reagents and the aqueous enzyme solution were suspended in the ionic liquid phase, the product (isoamyl acetate) was efficiently extracted into the hydrophobic organic phase. Using a 3:1 ratio of alcohol to acetic anhydride and a microfluidic flow pattern that resulted in a combination of long slugs (ellipsoidal droplets) and circulating fine droplets of n-heptane in the continuous ionic liquid phase, enabled an almost 3-fold productivity increase compared to the respective batch process. This significant improvement was mainly attributed to more efficient mixing and hence a much higher interfacial area between liquids within the microchannel. When a flow-focusing microfluidic chip was used for the same enzymatic reaction, uniform n-heptane droplets were formed in the ionic liquid phase [45]. This resulted in an even higher interfacial area, across which the reaction and product extraction takes place, with the amphiphilic lipase positioned at the ionic liquid-n-heptane interface. Moreover, the integration of a membrane-based phase separator allowed the reuse of the enzyme-containing ionic liquid over several cycles.

In another approach, the performance of capillary microreactors with various inner diameters was compared, when *Rhizomucor miehei* lipase was used for the esterification of oleic acid with 1-butanol for biodiesel production [60]. The aqueous phase containing the dissolved enzyme and the organic phase comprising fatty acids and butanol in n-heptane were introduced through a Y-junction resulting in a slug-flow profile. Two microreactor materials were used: hydrophobic PTFE or hydrophilic stainless steel. In PTFE channels, aqueous droplets were formed within the organic slug. In stainless steel channels, droplets of the organic phase were formed within the continuous aqueous phase. An increase in the volumetric organic-to-aqueous phase flow ratio significantly improved enzyme performance in the stainless steel microreactor in contrast to the PTFE-based one. This difference was explained by a larger increase in the interfacial area when the organic phase is moving through the continuous aqueous phase in the form of long droplets, as compared to the short aqueous droplets in a continuous organic phase observed in the PTFE-based microreactor [60].

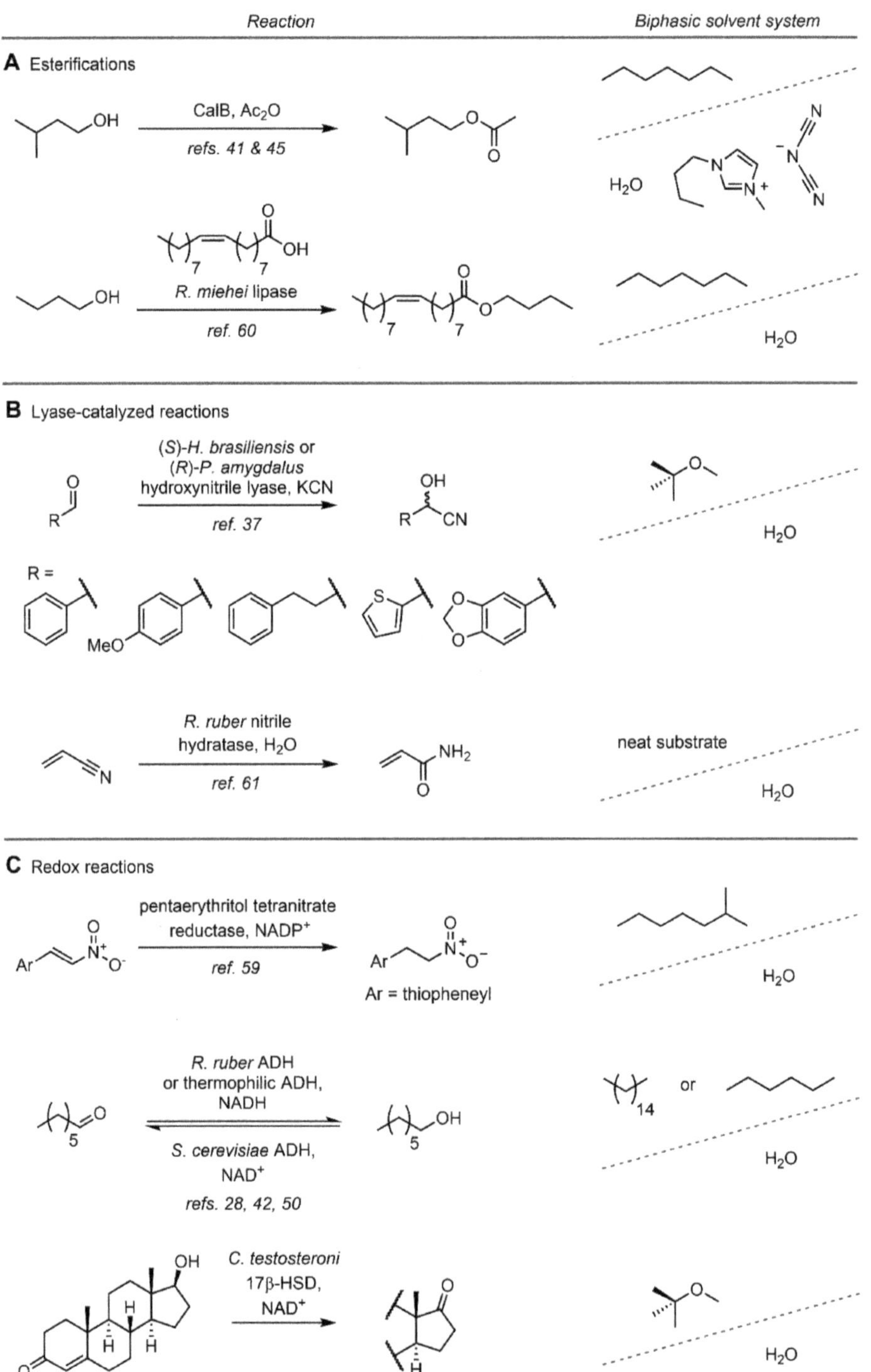

Figure 6. Biocatalytic two-phase reaction examples performed in microfluidic droplets.

Next to those biocatalytic esterifications, Koch et al. reported the successful formation of C-C bonds using crude lysates of two different hydroxynitrile lyases in specially designed microreactors [37]. Enzyme-catalyzed addition of HCN to different aldehydes using a two-liquid phase system under undefined slug flow in Y-shape microfluidics yielded α-cyanohydrins with excellent enantioselectivity. The organic phase made of MTBE contained the respective aldehyde, while the cyanide (in form of KCN) was provided via the enzyme-containing aqueous phase. The authors further demonstrated that this setup could be applied successfully for the screening of different reaction parameters using only minute amounts of reagents and a significantly shorter time compared to batch experiments. As another example, the industrially relevant hydration of acrylonitrile by nitrile hydratase to form acrylamide was reported using a membrane dispersion, stainless steel microreactor [61]. In membrane dispersion microreactors, the two phases are separated by a membrane. Droplets are formed when one phase is pushed into the other through the membrane. In the work discussed here, acrylonitrile was added as a substrate to the continuous aqueous phase, which contained whole *Rhodococcus ruber* TH3 cells as the catalyst. Membrane dispersion was used to produce acrylonitrile droplets with 25–35 μm diameter. The large surface area of these microdroplets significantly enhanced mass transfer and led to high acrylamide concentration (45.8 wt%) within 35 min of reaction.

Selected biocatalytic redox reactions in microfluidic droplets have also been reported, which utilize key advantages of microdroplets and have served as model systems for process optimization. Mohr and colleagues examined the performance of a promiscuous pentaerythritol tetranitrate reductase under anaerobic conditions in a recirculating continuous-flow microbioreactor [59]. Coating the polycarbonate bioreactor walls with thick hydrophobic fluoropolymer enabled the formation of stable aqueous droplets in isooctane using a T-junction. Installation of spectroscopic cells in the recirculating organic and aqueous channels allowed UV-spectroscopic reaction monitoring based on the absorbance of the substrate/product and the NADPH cofactor. By implementing a NADPH recycling system in the aqueous phase, the authors also investigated the performance of the bioreactor when it comes to the reduction of selected olefins. They found an improved performance compared to analogous reactions in batch, presumably due to a combination of increased mass transfer granted by a higher surface-to-volume ratio, as well as efficient product removal from the aqueous phase. To further study mass transfer limitations in biphasic redox reactions, Buehler and coworkers examined the effect of various process variables on the reduction of 1-heptaldehyde by a thermophilic alcohol dehydrogenase [28]. An aqueous buffer containing the enzyme, the cofactor, and recycling system based on formate dehydrogenation was mixed with a hexadecane phase at a T-junction and formed slugs in the organic phase. The authors probed the effects of flow rate, capillary diameter, phase ratio, as well as enzyme and substrate concentrations under segmented flow conditions. By optimizing these parameters, the system could be tuned for efficient enzyme usage, while maintaining high productivity. Remarkably, despite the enzyme's inhibition at >1 mM substrate in the aqueous phase, the mass transfer optimization performed in this capillary system enabled the efficient feeding of substrate and concurrent product removal, achieving product titers of almost 40 mM. In an additional study, the same authors utilized slugs formed in a capillary T-junction to address the low stability of the dehydrogenase under operational conditions [50]. Inactivation of the enzyme at the liquid interface was successfully eliminated through the application of low concentrations of Tween 20 as a surfactant, most likely via stabilization of the phase barrier. Finally, Zelic and colleagues used the reverse reaction as a model system for the evaluation of kinetic parameters in a continuously operated Y-shaped tubular microreactor [42]. *Using Saccharomyces cerevisiae* alcohol dehydrogenase, the authors examined the oxidation of hexanol in a biphasic system composed of buffer and hexane. Compared to batch experiments, the authors found a drastic increase in the reaction rate as well as a complete elimination of product inhibition.

A further example comes from our own recent work, where we used flow-focusing microfluidic devices to perform testosterone dehydrogenation catalyzed by the

17β-hydroxysteroid dehydrogenase (17β-HSD) from *Comamonas testosteroni* [62] in microfluidic droplets. Microfluidic devices were fabricated in glass using femtosecond laser ablation. The buffer solution containing the enzyme was used as the aqueous carrier phase and methyl tert-butyl ether (MTBE) containing testosterone was used as the organic phase confined in the microdroplets. For reactions performed in microfluidics with 10 mM steroid concentration, a 91% conversion of testosterone to androstenedione was obtained in only 40 s of reaction time, resulting in a space-time yield (STY) of 234 g L^{-1} h^{-1}. In comparison, respective batch reactions yielded 85% conversion after 10 min reaction time, resulting in a STY of only 14.6 g L^{-1} h^{-1}.

The works discussed above underscore the value of microdroplets in the testing and optimization of biphasic biocatalytic processes that typically suffer from product inhibition and low mass transfer rates. At the same time, many questions arise about the utility of droplet microreactors and the root of the underutilization of such systems in biocatalytic process design.

5. Conclusions and Future Perspective

As presented in detail above, microfluidic systems offer a unique set of advantages that make them particularly suitable as reaction vessels for two-phase biocatalysis. These advantages include continuously operated reactions, short diffusion paths, and tremendously increased mass transfer rates. Biphasic biocatalytic processes have been reported in microfluidic devices, primarily ones utilizing laminar flow configurations [38,39]. Despite the additional advantages offered by droplet-based microfluidics, such as dramatically increased surface-to-volume ratios, their use in two-phase biocatalysis remains limited. This is mainly due to the complexity associated with microfluidic device fabrication and the establishment of a stable droplet flow. Commercial droplet generation and control systems (such as the Elveflow Microfluidic Droplet Pack) offer an out-of-the-box solution to this problem as their use requires no expertise in microtechnology. The few examples of two-phase biocatalysis in microdroplets discussed above primarily utilize simple T-, Y-, or Ψ-junctions to establish slugs [37,41,50,59,60]. Slug flow can be realized at lower flow rates compared to droplet flow and does not typically require the use of surfactants or channel surface treatment. Additionally, slug flow can be established in microcapillary tubes, making this method more accessible to researchers with no access to microfabrication facilities or specialized equipment. However, slugs usually have large, variable diameters that are difficult to control. This means lower surface-area-to-volume ratio compared to microfluidic droplets generated in flow-focusing systems and less controlled experiments. Flow-focusing devices have a relatively complicated architecture and droplet generation often requires device and process optimization. However, the droplets produced in these systems have a uniform, defined spherical shape and a diameter that is smaller than the channel size. This ensures a large interfacial area between the two phases in biphasic biocatalysis, which in turn should lead to enhanced conversion and reaction rates because of increased mass transport across the interface. Stable microdroplets that do not coalesce usually require the use of a surfactant. Surfactants line the interface between the aqueous and organic phase, and it is thus expected that they would also control molecular exchange across the interface. On the other hand, as enzymes can be inactivated when in contact with organic solvents, the addition of surfactants could have a positive effect, by eliminating enzyme inactivation at the liquid-liquid interface as shown in one literature example [50]. Similarly, a large interfacial area between the aqueous and organic phases in microdroplet systems might introduce an additional risk for enzyme inactivation, as enzyme inactivation in liquid−liquid bubble column systems was found to be proportional to the interfacial area [63]. A large interfacial area is, however, a requirement for enhanced mass transfer. The tradeoff between maximized mass transfer and high enzyme inactivation rates in small droplets with large surface-to-volume ratios remains to be investigated, as it was never explicitly studied in droplet microfluidics. It should be noted here, however, that such enzyme inactivation will largely depend on the solvent system used as well as the employed enzyme and it hence needs to be addressed on a case-by-case basis.

To directly compare and evaluate the effectiveness of each method when employed in two-phase biocatalysis is a difficult task. Although individual studies often compare results obtained in batch process to results obtained in droplet-based processes, a comparison between two-phase biocatalysis in batch, laminar flow, slug flow, and droplet flow has not been performed. Nevertheless, comparisons between batch processes and droplet processes consistently show higher conversion in droplets, explained by enhanced mass transport, product removal (lower product inhibition), and reaction rates.

As evident by the literature review and discussion in this manuscript, microfluidic droplets have the potential to improve biphasic biocatalysis but have so far been grossly underutilized. Biocatalytic reactions that are product inhibited could particularly benefit from fast mass transfer that allows in situ product extraction into microdroplets. Similarly, reactions that are substrate inhibited benefit from chemical compartmentalization and short diffusion paths. Although slow biocatalytic reactions might not be compatible with conventional continuous-flow microfluidics due to the short liquid residence time in these devices (seconds to few minutes), longer channels, stationary reaction chambers, or the use of long capillaries offer alternatives that can make long reactions in microdroplets feasible.

In summary, several questions related to the application of this technology in biocatalysis remain unanswered, including:

- Which method of droplet generation produces the best results and why?
- Which type of flow (laminar, slug, droplet) can deliver the best results and why?
- What are the effects of surfactants on biotransformation?
- Is there a tradeoff between enzyme inactivation and enhanced mass transfer when the interfacial area increases?

Answers to these questions would provide researchers with the tool kits necessary to make informed decisions about the most suitable technology to use for their reactions. This would lower the barrier for the adaptation of droplet microfluidics in biocatalysis laboratories and would fuel development in this direction.

Author Contributions: Conceptualization, I.C. and L.X.; writing—original draft preparation, L.X., F.K., A.S. and I.C.; writing—review and editing, L.X., F.K., A.S. and I.C.; funding acquisition, I.C. All authors have read and agreed to the published version of the manuscript.

Funding: This work has been carried out within the framework of the SMART BIOTECS alliance between the Technische Universität Braunschweig and the Leibniz Universität Hannover. This initiative is supported by the Ministry of Science and Culture (MWK) of Lower Saxony, Germany. L.X. was funded through the 12plus6 initiative at the Faculty of Mechanical Engineering at Technische Universität Braunschweig.

Institutional Review Board Statement: Not applicable.

Informed Consent Statement: Not applicable.

Acknowledgments: We acknowledge support by the Open Access Publication Funds of Technische Universität Braunschweig.

Conflicts of Interest: The authors declare no conflict of interest.

References

1. Arnold, F.H. Directed Evolution: Bringing New Chemistry to Life. *Angew. Chem. Int. Ed.* **2018**, *57*, 4143–4148. [CrossRef]
2. Bell, E.L.; Finnigan, W.; France, S.P.; Green, A.P.; Hayes, M.A.; Hepworth, L.J.; Lovelock, S.L.; Niikura, H.; Osuna, S.; Romero, E.; et al. Biocatalysis. *Nat. Rev. Methods Primers* **2021**, *1*, 46. [CrossRef]
3. Sandoval, B.A.; Hyster, T.K. Emerging strategies for expanding the toolbox of enzymes in biocatalysis. *Curr. Opin. Chem. Biol.* **2020**, *55*, 45–51. [CrossRef]
4. Winkler, C.K.; Schrittwieser, J.H.; Kroutil, W. Power of Biocatalysis for Organic Synthesis. *ACS Cent. Sci.* **2021**, *7*, 55–71. [CrossRef]
5. Chakrabarty, S.; Romero, E.O.; Pyser, J.B.; Yazarians, J.A.; Narayan, A.R.H. Chemoenzymatic Total Synthesis of Natural Products. *Acc. Chem. Res.* **2021**, *54*, 1374–1384. [CrossRef]

6. Devine, P.N.; Howard, R.M.; Kumar, R.; Thompson, M.P.; Truppo, M.D.; Turner, N.J. Extending the application of biocatalysis to meet the challenges of drug development. *Nat. Rev. Chem.* **2018**, *2*, 409–421. [CrossRef]

7. Stout, C.N.; Renata, H. Reinvigorating the Chiral Pool: Chemoenzymatic Approaches to Complex Peptides and Terpenoids. *Acc. Chem. Res.* **2021**, *54*, 1143–1156. [CrossRef]

8. Fryszkowska, A.; Devine, P.N. Biocatalysis in drug discovery and development. *Curr. Opin. Chem. Biol.* **2020**, *55*, 151–160. [CrossRef]

9. Hughes, G.; Lewis, J.C. Introduction: Biocatalysis in Industry. *Chem. Rev.* **2018**, *118*, 1–3. [CrossRef]

10. Wu, S.; Snajdrova, R.; Moore, J.C.; Baldenius, K.; Bornscheuer, U.T. Biocatalysis: Enzymatic Synthesis for Industrial Applications. *Angew. Chem. Int. Ed.* **2021**, *60*, 88–119. [CrossRef]

11. Ni, Y.; Holtmann, D.; Hollmann, F. How Green is Biocatalysis? To Calculate is To Know. *ChemCatChem* **2014**, *6*, 930–943. [CrossRef]

12. Kaspar, F.; Stone, M.R.L.; Neubauer, P.; Kurreck, A. Route efficiency assessment and review of the synthesis of β-nucleosides via N-glycosylation of nucleobases. *Green Chem.* **2021**, *23*, 37–50. [CrossRef]

13. de María, P.; Hollmann, F. On the (Un)greenness of Biocatalysis: Some Challenging Figures and Some Promising Options. *Front. Microbiol.* **2015**, *6*, 1257.

14. Carrea, G. Biocatalysis in water-organic solvent two-phase systems. *Trends Biotechnol.* **1984**, *2*, 102–106. [CrossRef]

15. Bühler, B.; Bollhalder, I.; Hauer, B.; Witholt, B.; Schmid, A. Use of the two-liquid phase concept to exploit kinetically controlled multistep biocatalysis. *Biotechnol. Bioeng.* **2003**, *81*, 683–694. [CrossRef]

16. Antonini, E.; Carrea, G.; Cremonesi, P. Enzyme catalysed reactions in water—Organic solvent two-phase systems. *Enzym. Microb. Technol.* **1981**, *3*, 291–296. [CrossRef]

17. Oppermann, S.; Stein, F.; Kragl, U. Ionic liquids for two-phase systems and their application for purification, extraction and biocatalysis. *Appl. Microbiol. Biotechnol.* **2011**, *89*, 493–499. [CrossRef]

18. Sheldon, R.A. Biocatalysis in Ionic Liquids. In *Catalysis in Ionic Liquids: From Catalyst Synthesis to Application*; The Royal Society of Chemistry: London, UK, 2014; pp. 20–43. ISBN 978-1-84973-603-9.

19. Brink, L.E.S.; Tramper, J.; Luyben, K.C.A.M.; Van 't Riet, K. Biocatalysis in organic media. *Enzym. Microb. Technol.* **1988**, *10*, 736–743. [CrossRef]

20. Grundtvig, I.P.R.; Heintz, S.; Krühne, U.; Gernaey, K.V.; Adlercreutz, P.; Hayler, J.D.; Wells, A.S.; Woodley, J.M. Screening of organic solvents for bioprocesses using aqueous-organic two-phase systems. *Biotechnol. Adv.* **2018**, *36*, 1801–1814. [CrossRef]

21. Ghatorae, A.S.; Bell, G.; Halling, P.J. Inactivation of enzymes by organic solvents: New technique with well-defined interfacial area. *Biotechnol. Bioeng.* **1994**, *43*, 331–336. [CrossRef]

22. Mathys, R.G.; Schmid, A.; Witholt, B. Integrated two-liquid phase bioconversion and product-recovery processes for the oxidation of alkanes: Process design and economic evaluation. *Biotechnol. Bioeng.* **1999**, *64*, 459–477. [CrossRef]

23. Žnidaršič-Plazl, P. Let the Biocatalyst Flow. *Acta Chim. Slov.* **2021**, *68*, 1–16. [CrossRef]

24. Bolivar, J.M.; Wiesbauer, J.; Nidetzky, B. Biotransformations in microstructured reactors: More than flowing with the stream? *Trends Biotechnol.* **2011**, *29*, 333–342. [CrossRef]

25. Burgahn, T.; Pietrek, P.; Dittmeyer, R.; Rabe, K.S.; Niemeyer, C.M. Evaluation of a Microreactor for Flow Biocatalysis by Combined Theory and Experiment. *ChemCatChem* **2020**, *12*, 2452–2460. [CrossRef]

26. Wohlgemuth, R.; Plazl, I.; Žnidaršič-Plazl, P.; Gernaey, K.V.; Woodley, J.M. Microscale technology and biocatalytic processes: Opportunities and challenges for synthesis. *Trends Biotechnol.* **2015**, *33*, 302–314. [CrossRef]

27. Rabe, K.S.; Müller, J.; Skoupi, M.; Niemeyer, C.M. Cascades in Compartments: En Route to Machine-Assisted Biotechnology. *Angew. Chem. Int. Ed.* **2017**, *56*, 13574–13589. [CrossRef]

28. Karande, R.; Schmid, A.; Buehler, K. Miniaturizing Biocatalysis: Enzyme-Catalyzed Reactions in an Aqueous/Organic Segmented Flow Capillary Microreactor. *Adv. Synth. Catal.* **2011**, *353*, 2511–2521. [CrossRef]

29. Krühne, U.; Heintz, S.; Ringborg, R.; Rosinha, I.P.; Tufvesson, P.; Gernaey, K.V.; Woodley, J.M. Biocatalytic process development using microfluidic miniaturized systems. *Green Process. Synth.* **2014**, *3*, 23–31. [CrossRef]

30. Mair, P.; Gielen, F.; Hollfelder, F. Exploring sequence space in search of functional enzymes using microfluidic droplets. *Curr. Opin. Chem. Biol.* **2017**, *37*, 137–144. [CrossRef]

31. Colin, P.-Y.; Kintses, B.; Gielen, F.; Miton, C.M.; Fischer, G.; Mohamed, M.F.; Hyvönen, M.; Morgavi, D.P.; Janssen, D.B.; Hollfelder, F. Ultrahigh-throughput discovery of promiscuous enzymes by picodroplet functional metagenomics. *Nat. Commun.* **2015**, *6*, 10008. [CrossRef]

32. Kintses, B.; Hein, C.; Mohamed, M.F.; Fischlechner, M.; Courtois, F.; Lainé, C.; Hollfelder, F. Picoliter Cell Lysate Assays in Microfluidic Droplet Compartments for Directed Enzyme Evolution. *Chem. Biol.* **2012**, *19*, 1001–1009. [CrossRef] [PubMed]

33. Stucki, A.; Vallapurackal, J.; Ward, T.R.; Dittrich, P.S. Droplet Microfluidics and Directed Evolution of Enzymes: An Intertwined Journey. *Angew. Chemie Int. Ed.* **2021**, 1433–7851. [CrossRef]

34. Žnidaršič-Plazl, P. The Promises and the Challenges of Biotransformations in Microflow. *Biotechnol. J.* **2019**, *14*, 1800580. [CrossRef]

35. Badenhorst, C.P.S.; Bornscheuer, U.T. Droplet microfluidics: From simple activity screening to sophisticated kinetics. *Chem* **2021**, *7*, 835–838. [CrossRef]

36. Holland-Moritz, D.A.; Wismer, M.K.; Mann, B.F.; Farasat, I.; Devine, P.; Guetschow, E.D.; Mangion, I.; Welch, C.J.; Moore, J.C.; Sun, S.; et al. Mass Activated Droplet Sorting (MADS) Enables High-Throughput Screening of Enzymatic Reactions at Nanoliter Scale. *Angew. Chemie Int. Ed.* **2020**, *59*, 4470–4477. [CrossRef]

37. Koch, K.; van den Berg, R.J.F.; Nieuwland, P.J.; Wijtmans, R.; Schoemaker, H.E.; van Hest, J.C.M.; Rutjes, F.P.J.T. Enzymatic enantioselective C–C-bond formation in microreactors. *Biotechnol. Bioeng.* **2008**, *99*, 1028–1033. [CrossRef]
38. Meng, S.-X.; Xue, L.-H.; Xie, C.-Y.; Bai, R.-X.; Yang, X.; Qiu, Z.-P.; Guo, T.; Wang, Y.-L.; Meng, T. Enhanced enzymatic reaction by aqueous two-phase systems using parallel-laminar flow in a double Y-branched microfluidic device. *Chem. Eng. J.* **2018**, *335*, 392–400. [CrossRef]
39. Bolivar, J.M.; Nidetzky, B. Multiphase biotransformations in microstructured reactors: Opportunities for biocatalytic process intensification and smart flow processing. *Green Process. Synth.* **2013**, *2*, 541–559. [CrossRef]
40. Thorsen, T.; Roberts, R.W.; Arnold, F.H.; Quake, S.R. Dynamic pattern formation in a vesicle-generating microfluidic device. *Phys. Rev. Lett.* **2001**, *86*, 4163–4166. [CrossRef] [PubMed]
41. Pohar, A.; Plazl, I.; Žnidaršič–Plazl, P. Lipase-catalyzed synthesis of isoamyl acetate in an ionic liquid/n–heptane two-phase system at the microreactor scale. *Lab Chip* **2009**, *9*, 3385–3390. [CrossRef] [PubMed]
42. Tušek, A.; Šalić, A.; Kurtanjek, Ž.; Zelić, B. Modeling and kinetic parameter estimation of alcohol dehydrogenase-catalyzed hexanol oxidation in a microreactor. *Eng. Life Sci.* **2012**, *12*, 49–56. [CrossRef]
43. Gañán-Calvo, A.M.; Gordillo, J.M. Perfectly monodisperse microbubbling by capillary flow focusing. *Phys. Rev. Lett.* **2001**, *87*, 274501. [CrossRef] [PubMed]
44. Anna, S.; Bontoux, N.; Stone, H. Formation of dispersions using "flow focusing" in microchannels. *Appl. Phys. Lett.* **2003**, *82*, 364–366. [CrossRef]
45. Novak, U.; Žnidaršič-Plazl, P. Integrated lipase-catalyzed isoamyl acetate synthesis in a miniaturized system with enzyme and ionic liquid recycle. *Green Process. Synth.* **2013**, *2*, 561–568. [CrossRef]
46. Cramer, C.; Fischer, P.; Windhab, E.J. Drop formation in a co-flowing ambient fluid. *Chem. Eng. Sci.* **2004**, *59*, 3045–3058. [CrossRef]
47. Seemann, R.; Brinkmann, M.; Pfohl, T.; Herminghaus, S. Droplet based microfluidics. *Rep. Prog. Phys.* **2012**, *75*, 16601. [CrossRef]
48. Baret, J.-C. Surfactants in droplet-based microfluidics. *Lab Chip* **2012**, *12*, 422–433. [CrossRef]
49. Ahn, Y.N.; Gupta, A.; Chauhan, A.; Kopelevich, D.I. Molecular Transport through Surfactant-Covered Oil–Water Interfaces: Role of Physical Properties of Solutes and Surfactants in Creating Energy Barriers for Transport. *Langmuir* **2011**, *27*, 2420–2436. [CrossRef]
50. Karande, R.; Schmid, A.; Buehler, K. Enzyme Catalysis in an Aqueous/Organic Segment Flow Microreactor: Ways to Stabilize Enzyme Activity. *Langmuir* **2010**, *26*, 9152–9159. [CrossRef]
51. Novak, U.; Lavric, D.; Žnidaršič-Plazl, P. Continuous lipase B-catalyzed isoamyl acetate synthesis in a two-liquid phase system using corning® AFRTM module coupled with a membrane separator enabling biocatalyst recycle. *J. Flow Chem.* **2016**, *6*, 33–38. [CrossRef]
52. Fallah-Araghi, A.; Baret, J.-C.; Ryckelynck, M.; Griffiths, A.D. A completely in vitro ultrahigh-throughput droplet-based microfluidic screening system for protein engineering and directed evolution. *Lab Chip* **2012**, *12*, 882–891. [CrossRef] [PubMed]
53. Obexer, R.; Godina, A.; Garrabou, X.; Mittl, P.R.E.; Baker, D.; Griffiths, A.D.; Hilvert, D. Emergence of a catalytic tetrad during evolution of a highly active artificial aldolase. *Nat. Chem.* **2017**, *9*, 50–56. [CrossRef] [PubMed]
54. Ron, E.Z.; Rosenberg, E. Natural roles of biosurfactants. *Environ. Microbiol.* **2001**, *3*, 229–236. [CrossRef] [PubMed]
55. Sunde, M.; Pham, C.L.L.; Kwan, A.H. Molecular Characteristics and Biological Functions of Surface-Active and Surfactant Proteins. *Annu. Rev. Biochem.* **2017**, *86*, 585–608. [CrossRef] [PubMed]
56. Linder, M.B.; Szilvay, G.R.; Nakari-Setälä, T.; Penttilä, M.E. Hydrophobins: The protein-amphiphiles of filamentous fungi. *FEMS Microbiol. Rev.* **2005**, *29*, 877–896. [CrossRef]
57. Santos, D.K.F.; Rufino, R.D.; Luna, J.M.; Santos, V.A.; Sarubbo, L.A. Biosurfactants: Multifunctional Biomolecules of the 21st Century. *Int. J. Mol. Sci.* **2016**, *17*, 401. [CrossRef]
58. Grösche, M.; Zoheir, A.E.; Stegmaier, J.; Mikut, R.; Mager, D.; Korvink, J.G.; Rabe, K.S.; Niemeyer, C.M. Microfluidic Chips for Life Sciences-A Comparison of Low Entry Manufacturing Technologies. *Small* **2019**, *15*, e1901956. [CrossRef]
59. Mohr, S.; Fisher, K.; Scrutton, N.S.; Goddard, N.J.; Fielden, P.R. Continuous two-phase flow miniaturised bioreactor for monitoring anaerobic biocatalysis by pentaerythritol tetranitrate reductase. *Lab Chip* **2010**, *10*, 1929–1936. [CrossRef]
60. Hommes, A.; de Wit, T.; Euverink, G.J.W.; Yue, J. Enzymatic Biodiesel Synthesis by the Biphasic Esterification of Oleic Acid and 1-Butanol in Microreactors. *Ind. Eng. Chem. Res.* **2019**, *58*, 15432–15444. [CrossRef]
61. Li, J.; Chen, J.; Wang, Y.; Luo, G.; Yu, H. Hydration of acrylonitrile to produce acrylamide using biocatalyst in a membrane dispersion microreactor. *Bioresour. Technol.* **2014**, *169*, 416–420. [CrossRef]
62. Benach, J.; Filling, C.; Oppermann, U.C.T.; Roversi, P.; Bricogne, G.; Berndt, K.D.; Jörnvall, H.; Ladenstein, R. Structure of Bacterial 3β/17β-Hydroxysteroid Dehydrogenase at 1.2 Å Resolution: A Model for Multiple Steroid Recognition. *Biochemistry* **2002**, *41*, 14659–14668. [CrossRef] [PubMed]
63. Ghatorae, A.S.; Guerra, M.J.; Bell, G.; Halling, P.J. Immiscible organic solvent inactivation of urease, chymotrypsin, lipase, and ribonuclease: Separation of dissolved solvent and interfacial effects. *Biotechnol. Bioeng.* **1994**, *44*, 1355–1361. [CrossRef] [PubMed]

MDPI

St. Alban-Anlage 66

4052 Basel

Switzerland

Tel. +41 61 683 77 34

Fax +41 61 302 89 18

www.mdpi.com

Biosensors Editorial Office

E-mail: biosensors@mdpi.com

www.mdpi.com/journal/biosensors